Inverse Imaging with Poisson Data

From cells to galaxies

Inverse Imaging with Poisson Data

From cells to galaxies

Mario Bertero

Patrizia Boccacci

DIBRIS, Università di Genova, Italia

Valeria Ruggiero

Dipartimento di Matematica e Informatica, Università di Ferrara, Italia

IOP Publishing, Bristol, UK

Multimedia content is available for this book from http://iopscience.iop.org/book/978-0-7503-1437-4.

ISBN 978-0-7503-1437-4 (ebook)
ISBN 978-0-7503-1438-1 (print)
ISBN 978-0-7503-1439-8 (mobi)

DOI 10.1088/2053-2563/aae109

Version: 20181201

IOP Expanding Physics
ISSN 2053-2563 (online)
ISSN 2054-7315 (print)

British Library Cataloguing-in-Publication Data: A catalogue record for this book is available from the British Library.

Published by IOP Publishing, wholly owned by The Institute of Physics, London

IOP Publishing, Temple Circus, Temple Way, Bristol, BS1 6HG, UK

US Office: IOP Publishing, Inc., 190 North Independence Mall West, Suite 601, Philadelphia, PA 19106, USA

To our families.

Contents

Preface

In 2009 the journal *Inverse problems* published a special issue of reviews dedicated to the 25th anniversary of the journal. As stated by the Editor-in-Chief W W Symes '*Inverse Problems* finishes its 25th year of publication with this special issue, composed of topical reviews and one paper invited by the Editorial Board. Our instructions to the writers of topical review articles advise that comprehensive coverage is not mandatory, so this issue is not comprehensive. Instead, it presents in-depth surveys of a representative selection of central topics, from pure and applied mathematics and from several of the many areas of science and technology that give rise to inverse problems'. In this issue we published a paper entitled *Image deblurring with Poisson data: from cells to galaxies*, focusing on deconvolution problems appearing in microscopy and astronomy. The paper was successful and, for this reason, in 2016 Daniel Jopling, at that time Publisher of *Inverse Problems*, sent a message exploring the possibility of expanding the paper into an ebook within the IOP ebook program. Our reaction was positive for two reasons. The first is the considerable advances, in the years from 2009 to now, both in the numerical and theoretical treatment of the topic. The second is that a book on Poisson data inversion should contain at least an introduction to the applications in medical imaging, even if, as a consequence of the enormous literature on the subject, a comprehensive treatment of this application is not possible. The present book is the result of our efforts and we hope that it will be useful as an introduction and reference book to young researchers.

Acknowledgements

The authors would like to thank several colleagues and friends for their continuous support during the preparation of the manuscript and for their different and useful contributions: material for numerical experiments, references, critical reading of chapters, etc.

We mention them in alphabetic order: Carmelo Arcidiacono, Alessandro Benfenati, Silvia Bonettini, Al Conrad, Christine De Mol, Imke de Pater, Andrea La Camera, Marco Prato, Giorgio Talenti, Giuseppe Vicidomini, Gaetano Zanghirati, Luca Zanni.

Author biographies

Mario Bertero

Mario Bertero, born in 1938, received the advanced degree in Physics from the University of Genova in 1960 and the *libera docenza* in Theoretical Physics in 1968. He held research and teaching appointments at Institute für Theoretische Kernphysic (University of Bonn), École Polytechnique (Paris), Université Libre de Bruxelles and University of Genova. In 1981 he became full professor in Mathematics and from 1997 full professor in Computer Science, University of Genova. He retired in 2010 from teaching (but not from research).

From 1990 to 1994 he was Editor-in-Chief of the journal *Inverse Problems*, published by the Institute of Physics, and is currently a member of the International Advisory Panel of the same journal.

He organized a NATO Advanced Research Workshop on *Inverse Problems in Scattering and Imaging*, held in Cape Cod, USA, 14–19 April 1991, and an International Conference on Inverse Problems, sponsored by the Italian Research Council (CNR), held in Vietri, Italy, 28 September–2 October 1998. Moreover, he was a promoter and chairman of the first conference *Applied Inverse Problems* held in Montecatini (Italy), the first of a series organized every two years in Europe and North America (http://aip.disi.unige.it).

He is author and co-author of about 150 articles, co-editor, with E R Pike, of the book *Inverse Problems in Scattering and Imaging* and co-author with P Boccacci of the book *Introduction to Inverse Problems in Imaging*.

His first research activity concerned potential scattering in quantum theory. Around 1975 his interest was addressed to the theory and the applications of inverse and ill-posed problems. He developed methods based on the singular value decomposition of linear operators for the solution of inverse problems in various domains of applied science. In particular, the finite Laplace transform for application to polydispersity analysis by means of photon correlation spectroscopy and integral operators modeling confocal scanning microscopy. The study suggested a new version of the confocal scanning microscope; a prototype was realized by E R Pike at King's College in London. Other applications were to particle sizing, microwave thermography, seismic tomography, chirp-pulse microwave tomography and astronomical imaging. He is a member of the Institute of Physics and of Accademia Ligure di Scienze ed Arti.

Patrizia Boccacci

Patrizia Boccacci, born in 1955, received the advanced degree in Physics from the University of Genova, Italy, in 1980. From 1981 to 1983 she worked on a project of multi-directional holographic interferometry, funded by ESA (European Space Agency). From 1985 to 2000 she was technical research manager with the Department of Physics (DIFI) of the University of Genova, Italy. From May 2004 she has been associate professor at the Department of Informatics, Bioengineering, Robotics and System Engineering (DIBRIS) of the University of Genova. She is co-author of several articles and co-author of one book (http://scholar.google.com/citations?hl=en&user=M-vBV5kAAAAJ). Her main research interest is in numerical methods for the solution of inverse ill-posed problem and their applications, such as optical tomography for the investigation of crystal growth in microgravity, confocal microscopy, particle sizing, seismic tomography and astrophysics. More recently her main research interest is in image restoration problems related to the LBT (large binocular telescope) project and in medical imaging.

Valeria Ruggiero

Valeria Ruggiero, born in 1956, received the advanced degree in Mathematics from the University of Ferrara, Italy, in 1978. From 1992 to 1998 she was associate professor in Numerical Analysis at the University of Modena and from 1998 to 2000 at the University of Ferrara. From 2000, she has been full professor in Numerical Analysis at the University of Ferrara. Presently she is Director of the National Group for Scientific Computation of the Istituto Nazionale di Alta Matematica (INdAM).

She was national coordinator of research projects (PRIN 97 'Numerical Analysis: Methods and Mathematical Software' dmi.unife.it/it/ricerca-dmi/gruppi-di-ricerca-1/annum97, FIRB 2001 Parallel algorithms and Nonlinear Numerical Optimization 'dm.unife.it/pn2o/', PRIN 2008 'Optimization methods and software for inverse problems' http://www.unife.it/prisma) funded by MIUR (Ministero Italiano per l'Università e la Ricerca). Her research activity concerns the development and the analysis of numerical methods for large scale systems, parallel computing, nonlinear optimization and related applications. In particular, theoretical and computational results have been obtained about the inexact Newton interior point method for nonlinear programming problems and nonlinear systems, including the analysis of different iterative solvers for inner linear symmetric indefinite systems. Non-monotone strategies are analyzed also for the semi-smooth case, with application to optimal control problems and variational inequalities.

Her more recent research interests are in variational methods for inverse problems, with particular attention to variable-metric forward–backward and primal–dual first order algorithms for image restoration. The results of her research activity are described in more than 50 publications in international scientific journals, in proceedings and books, in several software packages and in a number of communications to conferences. She is member of the Editorial Board of *Computational Optimization and Applications* and referee for scientific journals in the areas of numerical analysis, optimization and scientific computing (https://sites.google.com/a/unife.it/valeria-ruggiero).

Acronyms

ADMM	alternating direction method of multipliers
AEM	alternating extragradient method
AO	adaptive optics
BB	Barzilai and Borwein
BCD	block coordinate descent
CP	Chambolle and Pock
CT	computed tomography
CCD	charged couple device
CdTe	cadmium telluride
CLSM	confocal laser scanning microscope
COSTAR	corrective optics space telescope axial replacement
CZT	cadmium zinc telluride
DOE	diffractive optical elements
EM	expectation maximization
EM–RL	expectation maximization–Richardson–Lucy
FB	forward–backward
FBP	filtered back-projection
FFT	fast Fourier transform
FISTA	fast iterative shrinkage-thresholding algorithm
FOM	figure of merit
FoV	field of view
FT	Fourier transform
FWHM	full width half maximum
HR	high-resolution
HS	hypersurface regularization
HST	Hubble space telescope
ISM	image scanning microscopy
KKT	Karush–Khun–Tucker
KL	Kullback–Leibler
LBT	large binocular telescope
LBTI	large binocular telescope interferometer
LBTO	large binocular telescope observatory
LOR	line of response
l.s.c.	lower semi-continuous
LR	low-resolution
MAP	maximum *a posteriori*
MIS	MISTRAL
ML	maximum likelihood
ML–EM	maximum likelihood–expectation maximization
MM	majorize–minorize
MRF	Markov random field
MRI	magnetic resonance imaging
NA	numerical aperture of a lens
OSEM	ordered subset expectation maximization
OSL	one-step late
OTF	optical transfer function
PDHG	primal–dual hybrid gradient

PET	positron emission tomography
PHA	pulse height analyzer
PMT	photo multiplier tube
PSF	point spread function
PSWF	prolate spheroidal wave functions
RL	Richardson–Lucy
rms	root mean square
ROI	region of interest
RON	read out noise
ROR	radius of rotation
r.v.	random variable
SFISTA	scaled fast iterative shrinkage-thresholding algorithm
SF-PSF	space variant-point spread function
SGM	split-gradient method
SGP	scaled gradient projection
SIM	structured illumination microscopy
SNR	signal to noise ratio
SPDHG	scaled primal–dual hybrid gradient
SPECT	single photon emission computerized tomography
SPIRAL	sparse Poisson intensity reconstruction algorithm
SR	Strehl ratio
SSIM	ztructural similarity
STED	stimulated emission depletion microscopy
SVD	singular value decomposition
T0	Tikhonov regularization of order 0
T1	Tikhonov regularization of order 1
T2	Tikhonov regularization of order 2
TOF	time of flight
TV	total variation regularization
VMILA	variable-metric inexact line-search algorithms
WFM	wide-field microscope

Symbols

H	imaging matrix
h	point spread function
b	background emission
$\bar{x}$	true source object
x	generic source object
$\bar{y}$	noise-free image
y	detected image or blurred and noisy image
y_s	background subtracted image
$f_0(x; y)$	data-fidelity function
$f_1(x)$	regularization function
β	regularization parameter
$f_\beta(x; y)$	regularized objective function
x^*	minimizer of $f_0(x; y)$
x_β^*	minimizer of $f_\beta(x; y)$
#	cardinality

Chapter 1

Introduction

The title of the book contains two expressions which deserve a few comments. The first one is *inverse imaging*. This expression defines methods of image processing which are very frequently named in other ways, such as image deblurring, image deconvolution, image reconstruction, image restoration, image reconstruction from projections etc. All these methods imply the solution of an inverse problem. Therefore, the goal of inverse imaging is to provide an image of an unknown object by mathematical inversion of observed data.

The second one is *Poisson data*. This expression indicates data obtained by photon counting. Indeed, the number of photons arriving on the detector from a source, in a given time interval, is a random variable: if we observe the number of photons arriving in different time intervals of the same length, we obtain different values. If suitable conditions are satisfied, their probability distribution is a Poisson distribution. In conclusion the book deals with inverse problems in imaging with the additional assumption that data have specific statistical properties.

We must point out that the assumption of Poisson statistics is an approximation. Indeed, a photon can be detected and counted if it interacts with the detector, an example of the interaction of a quantum entity (the photon) with a macroscopic object (the detector). The effect of this interaction must be registered and interaction plus registration can modify the statistics, sometimes in a way which can be hardly modeled. However, the assumption of Poisson statistics provides a well-defined mathematical model that can be sufficiently accurate in several instances.

It is well-known that inverse problems are ill-posed and that specific techniques must be devised for their treatment. A classic approach is provided by Tikhonov regularization theory [2, 4, 11]: a considerable collection of methods, mathematical results and practical applications is now available in this framework. However, it was early recognized (see, for instance [6, 12]) that Tikhonov regularization is based on the assumption of data perturbed by additive Gaussian noise. As a consequence, Tikhonov theory cannot take into account the statistical properties of photon counting. To our knowledge, the paper of Shepp and Vardi [9] is the first where an attempt to go beyond Tikhonov theory is considered. Indeed, the paper deals with emission tomography where data are acquired by photon counting; therefore the assumption of data satisfying Poisson statistics is introduced. The approach of maximum likelihood is proposed for implementing this assumption in the image reconstruction problem. The method of expectation–maximization [3] is applied for solving the problem. Nowadays, the resulting iterative method is a classic one and is used also in class exercises, since it is easily implemented.

At first glance, the approach of Shepp and Vardi does not provide the basis of a new regularization theory. However, the successive approach of Bayesian estimation introduced by Geman and Geman [7], combined with manipulations transforming the approaches into the solution of minimization problems, indicate the way to a new treatment of inverse problems.

1.1 Scope of the book and topic selection

A previous book by Bertero and Boccacci [2] is an introduction to linear inverse problems, to their ill-posedness and to Tikhonov regularization theory, with a focus on imaging applications. In that book only a couple of chapters is devoted to statistical methods and in particular to the maximum-likelihood approach of Shepp and Vardi. The present book is completely dedicated to the research line initiated by that paper. The final goal of this research should be to produce an appropriate regularization theory which should replace Tikhonov theory in the case of data acquired by photon counting. In our opinion this goal is not yet completely reached. The approaches and methods proposed for going beyond the original paper of Shepp and Vardi are a mixture of probabilistic and analytic techniques.

Anyway, a considerable collection of methods, results and applications is already available and the main scope of the book is to provide an introduction to the topic and an account of the state-of-the art of the research. This research is still very active and therefore it is necessary to clarify what is already a classic and well-established result and what is still research in action. Sometimes the development of the research in different applied domains has followed different paths.

A couple of examples. The approach of Shepp and Vardi leads to a data-fidelity function (or data misfit) given in terms of a generalized Kullback–Leibler divergence [8] which replaces the standard least-squares function of Tikhonov regularization theory. However, since in the case of a sufficiently large number of photons Poisson distribution can be approximated by a Gaussian distribution, several proposed methods are based on a weighted least-squares approach derived from this approximation. We do not consider these methods and therefore, we automatically exclude a large part of methods developed in important application areas. Regarding medical imaging we mention, for instance, a paper of Fessler [5] where weighted least-squares methods are theorized. Another approach leading to least-squares methods is proposed in astronomy [10] and is based on the so-called Anscombe transform [1], a variance stabilizing transform. Indeed, in the case of photon counting, the variance of the pixel values is signal dependent, hence is pixel dependent. The effect of Anscombe transform is that, after its applications, the pixel values have zero mean and variance 1.

Our choice is not dictated by a belief of superiority of the methods derived from the approach of Shepp and Vardi but is the result of a mathematical choice. Nor is a purpose of our book to compare least-squares methods with methods based on the assumption of Poisson statistics. The ultimate aim is to delineate a well-defined research topic and to provide an account of both well-established results and recent research activity. Therefore, the book concerns the line of thought generated by the paper of Shepp and Vardi and its achievements both theoretical and practical. We must also point out that we restrict the analysis to linear problems of inverse imaging that are the most important ones in applied science.

1.2 Structure of the book

We briefly describe the contents of the book together with a concise reading guide.

The reader interested in the main applications of the methods described in the book can have a look at chapter 2. Here they will find the description of the main imaging systems based on photon counting, ranging from microscopy to emission tomography and astronomy: an account of the very wide field of practical and scientific applications of the methods that are described in the book. Here they will also find the origin of the sub-title 'From cells to galaxies'.

Chapters 3 and 4, contain the basic results which should be familiar to everyone who intends to work in this research area. Chapter 3 concerns the mathematical modeling of the imaging process, frequently called *forward model* or *forward problem* in inverse problems theory. The reader mainly interested in the discrete models can skip or only look at sections from 3.1–3.3 (and ignore section 3.6) that concern models in infinite dimension (i.e. in terms of functions rather than vectors); on the other hand he/she must concentrate on sections 3.4 and 3.5 where noise properties and discrete models are discussed.

Chapter 4 is fundamental since it concerns the maximum-likelihood approach of Shepp and Vardi and its reformulation as a minimization problem. In such a way an analogy is established with least-squares problems, the least-squares function being replaced by a generalized Kullback–Leibler divergence [8]. A definition of data-fidelity function (or data misfit) is possible in this way. The ill-conditioning of the problem is discussed as well as the Bayes approach which allows the introduction of prior information on the solution. This approach leads to a penalization of the data-fidelity function thus suggesting a further analogy with Tikhonov regularization theory. The approach introduces a parameter which plays a role similar to the well-known regularization parameter of Tikhonov theory and, for this reason, it is named in the same way. Section 4.6 contains results that are not essential for the understanding of the content of the chapter but may be useful to the reader interested in understanding the proof of some basic results.

Chapter 5 is also fundamental because the iterative method proposed by Shepp and Vardi for the solution of the maximum-likelihood approach is given as well as its generalization which is used in microscopy and astronomy. Several properties of this algorithm (that will be called EM–RL for reasons explained in section 4.1) are discussed as well as its acceleration denoted OSEM. Finally, iterative methods for the minimization of the penalized functionals, obtained with simple modifications of the EM–RL algorithm, are introduced and discussed. Sections 5.5.1 and 5.5.2 contain convergence proofs and therefore can be ignored by the reader not interested in these mathematical questions.

The subsequent chapters are more specialized and contain recent results. Therefore, they may suffer from a bias due to the personal experience of the authors.

Chapter 6 is devoted to optimization methods. Indeed the iterative methods described in chapter 5, that we call *classic methods*, can provide accurate results but are not very efficient: sometimes an extremely large number of iterations is required. The increase of the size of the images to be processed demands more efficiency. In addition, the classic methods can be applied only to differentiable regularization and implement only the constraint of non-negativity while other constraints may be useful in the applications. Since the basic problems in the maximum-likelihood/Bayes approach are formulated as minimization problems, it is quite natural to look at numerical optimization theory for finding answers to the need for efficiency. A lesson learned from the classic methods is that they are scaled gradient methods; this property suggests that an appropriate scaling of the gradient may be relevant for increasing the efficiency of gradient methods. The chapter attempts to organize the several proposed methods into a few large classes, discussing first the methods for differentiable regularization and subsequently those, more complex, for non-differentiable regularization. Only methods tested on numerical examples are considered. No convergence proof is given; the relevant literature is provided to the interested reader. It is obvious that this chapter is mainly recommended to the reader interested to the application of the optimization methods to the specific problems considered in this book. In any case he should have a good background in optimization theory.

Chapter 7 is strictly related to the previous one; it is intended to show in action some of the methods discussed in chapter 6 and to provide a few comparisons between them. No systematic comparison is attempted. Another scope is to provide a few examples of the quality of the reconstructions of both simulated and real images.

In chapter 8 specific problems in image deconvolution are considered. Also in this chapter a bias due to the specific activity of the authors can be remarked. Miscellaneous topics are considered such as super-resolution, correction of boundary effects, blind deconvolution, deconvolution of objects consisting of bright point-sources superimposed to a smooth and unknown background and finally space-varying blur. It is obvious that this chapter may be interesting to the reader who is mainly concerned with deconvolution problems.

Finally, the last chapter is the shortest one even though it concerns presumably the most difficult problem in the case of inverse problems with Poisson data: we mean the formulation of a regularization theory in infinite dimension that should replace Tikhonov regularization theory applicable to the case of additive Gaussian noise. We subdivide the approaches into two classes: deterministic and statistical. We discuss the basic ideas and the main results without giving mathematical details because the mathematical background required for understanding the methods used by the authors is beyond that required to the reader of this book. We provide the basic references to the interested reader.

References

[1] Anscombe F J 1948 The transformation of Poisson, binomial and negative-binomial data *Biometrika* **35** 246–54

[2] Bertero M and Boccacci P 1998 *Introduction to Inverse Problems in Imaging* (Bristol: Institute of Physics)

[3] Dempster A P, Laird N M and Rubin D B 1977 Maximum likelihood from incomplete data via the EM algorithm *J. R. Stat. Soc.* B **39** 1–38

[4] Engl H W, Hanke M and Neubauer A 1996 *Regularization of Inverse Problems* (Dordrecht: Kluwer)

[5] Fessler J A 1994 Penalized weighted least-squares image reconstruction for positron emission tomography *IEEE Trans. Med. Imaging* **13** 290–300

[6] Franklin J N 1970 Well-posed stochastic extensions of ill-posed linear problems *J. Math. Anal. Appl.* **31** 682–716

[7] Geman S and Geman D 1984 Stochastic relaxation, Gibbs distributions, and the Bayesian restoration of images *IEEE Trans. Pattern Anal. Mach. Intell.* **6** 721–41

[8] Kullback S and Leibler R A 1951 On information and sufficiency *Ann. Math. Stat.* **1** 79–86

[9] Shepp L A and Vardi Y 1982 Maximum likelihood reconstruction for emission tomography *IEEE Trans. Med. Imaging* **1** 113–22

[10] Stark J L, Murtagh F and Bijaoui A 1998 *Image Processing and Data Analysis: The Multiscale Approach* (Cambridge: Cambridge University Press)

[11] Tikhonov A N and Arsenin V Y 1977 *Solution of Ill-Posed Problems* (New York: Wiley) (Russian edition in 1974)

[12] Turchin V F, Kozlov V P and Malkevich M S 1971 The use of mathematical-statistics methods in the solution of incorrectly posed problems *Sov. Phys. Uspekhi* **13** 681–703

IOP Publishing

Inverse Imaging with Poisson Data

From cells to galaxies

Mario Bertero, Patrizia Boccacci and Valeria Ruggiero

Chapter 2

Examples of applications

In this chapter we describe some of the most important imaging techniques based on the acquisition of data by photon counting. The order of presentation follows the increase in the 'size' of the physical objects imaged by these techniques. We start with microscopy (the 'cells'), focusing on fluorescence microscopy and in particular on the new powerful techniques such as confocal laser scanning microscopy (CLSM) and stimulated emission depletion (STED) microscopy. We continue with medical imaging (the 'human body'), introducing x-ray computed tomography (CT) which stimulated the evolution of scintigraphy into emission tomography, namely single photon emission computed tomography (SPECT) and positron emission tomography (PET), the basic imaging modalities in nuclear medicine. We conclude with astronomy (the 'galaxies'), focusing on observations in the visible and near infrared domains and explaining the importance of observing the same target at different wavelengths.

The image formation process in a wide range of applications can be described by a three-step chain:

1. the radiation source that can be the object to be imaged (fluorescence microscopy, emission tomography and astronomy) or a residue radiation due to an absorbing object (x-ray tomography, tomosynthesis, digital radiography);
2. the imaging system that can be described by medium, lenses, mirrors, collimators, etc;
3. the detecting system that acquires the data: CCD (charged coupled device) cameras, CMOS (complementary metal-oxide–semiconductor) sensors, photo-tubes, scintillators, etc.

2.1 Fluorescence microscopy

The optical microscope is a type of microscope which uses visible light and a system of lenses to magnify images of small samples. Fluorescence is one of the most

doi:10.1088/2053-2563/aae109ch2

commonly used physical phenomena in biological microscopy, mainly because of its high sensitivity and high specificity. Fluorescence is a form of luminescence. The absorption and subsequent re-radiation of light by organic and inorganic specimens is typically the result of physical phenomena described as being either fluorescence or phosphorescence.

In fluorescence microscopy the specimen (cells or living tissues) is stained by fluorophores (or fluorocromes), fluorescent chemical compounds that can re-emit light upon light excitation. The specimen is illuminated by light with a specific wavelength; the light absorption by the fluorophores causes the nearly simultaneous emission of light with a longer wavelength (Stokes shift).

The illumination light is separated from the weaker emitted fluorescence by means of a suitable spectral filter. In fluorescence microscopy, the specimen under investigation is itself the light source and the object imaged by the microscope is essentially the distribution of the fluorophores within the specimen.

2.1.1 Wide-field microscope

Wide-field fluorescence microscopy is a very largely used technique to obtain both structural and dynamic information. It is based on the illumination of the whole sample. A simple scheme of this type of microscope is shown in figure 2.1.

The illumination system is based on a mercury or xenon lamp and a specific wavelength is selected by means of a filter (the 'excitation filter' in the scheme). Excitation light is directed to the sample via a 'dichroic mirror' (i.e. a mirror that reflects some wavelengths but is transparent to others) and the 'objective lens' focuses the excitation light on the specimen which emits fluorescent light. Since this has a different wavelength, it can be separated from the excitation light by a filter

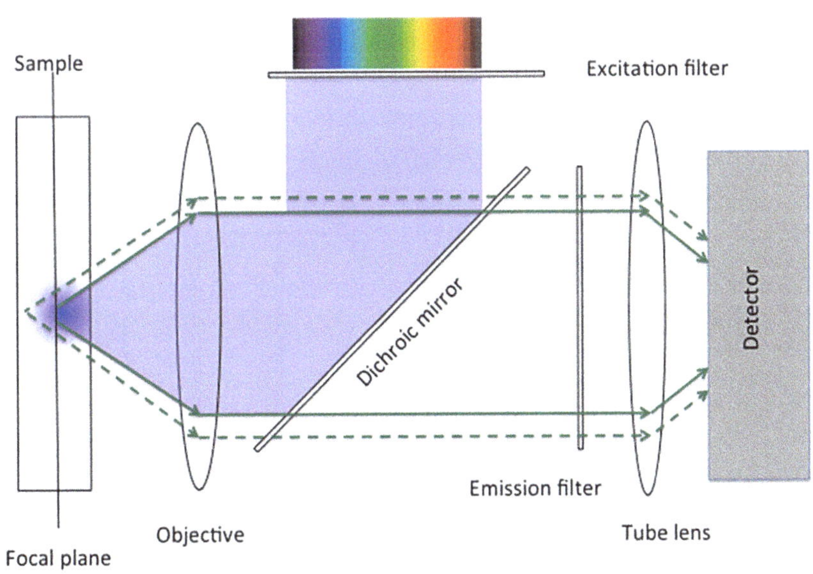

Figure 2.1. Optical scheme of a wide-field microscope.

(indicated as the 'emission filter' in the figure) and is projected by the ocular (the 'tube lens' in the scheme) onto a plane. It can be viewed directly by the eye of the user or detected by a camera (usually a CCD camera).

When fluorescent specimens are imaged using such a microscope, out-of focus fluorescence (indicated by a dotted green line in the figure) emitted by the specimen away from the region of interest often degrades the resolution of those features that are in focus. This situation is especially problematic for specimens having a thickness greater than about 2 μm. Thus, such a microscope is not an intrinsic 3D microscope.

2.1.2 Confocal microscope

Confocal fluorescence microscopy is a microscopic technique developed to overcome the drawbacks of the wide-field microscopy and to provide true three-dimensional (3D) optical resolution. In this technique, often named confocal laser scanning microscopy (CLSM), 3D resolution is accomplished by actively suppressing any signal coming from out-of-focus planes. This is achieved by means of a pinhole in front of the detector as schematically shown in figure 2.2. Light originating from an in-focus plane of the specimen is imaged by the microscope objective, whereas light coming from out-of-focus planes is blocked by the pinhole.

In a confocal fluorescence microscope, as shown in figure 2.2, the specimen is generally illuminated, or, more precisely, excited, by a laser. The light coming from the laser (typically a Gaussian laser beam) is expanded by a lens system, is reflected by a dichroic mirror, and focused by a microscope objective to produce a small spot in the specimen.

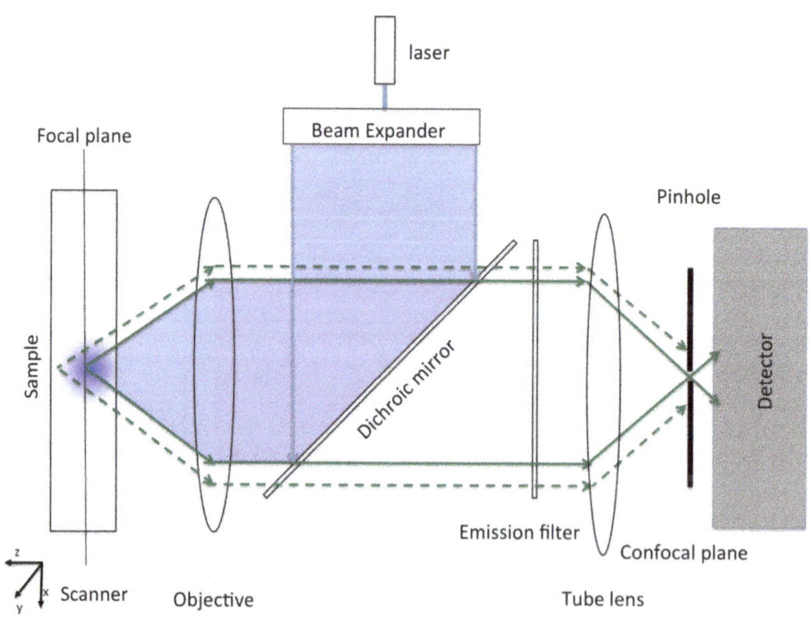

Figure 2.2. Optical scheme of a confocal microscope.

The emitted light comes back with a longer wavelength through the objective, is transmitted by the dichroic mirror and is focused by a lens (called 'tube lens') on its focal plane where a pinhole is located. The focal plane of the tube lens is conjugate to the focal plane of the objective, i.e. a point on the first plane is imaged into a point of the second one.

In this configuration, light from under-focal-plane (dashed green lines) will be focused at a plane behind the pinhole and therefore is blocked by the pinhole plate. Only the light from the focal plane is just focused at the pinhole and can reach the image detector placed behind the pinhole. This process simulates what one can do with a micro-tome for cutting some unwanted tissue away, but you do it here optically; indeed this process is called *optical sectioning*. The size of the pinhole determines how thick an optical slice will be. The smaller the pinhole, the thinner the slice.

In wide-field microscopy the image formation is instantaneous. The photographic film, the CCD camera or the retina in the human eye, detects the whole image from the exit of the microscope immediately. In confocal microscopy only a dot-portion of the sample is imaged. Therefore, a scanning procedure in the focal plane is required for illuminating the specimen point by point; the 3D image is obtained by moving the focal plane and repeating the scanning. The light emitted from each dot passes the pinhole and reaches a photomultiplier tube (PMT) or any other single-point detector. In conclusion, both object illumination and image detection depend on the scanning process, i.e. the image is generated through scanning.

The PMT has a large active area compared to the small cell of a CCD chip, thus it can accommodate more photo-electrons and has a wide dynamic range (range between maximum and minimum detectable intensity). It also has a high photon sensitivity suitable for detecting both strong and weak signals, providing enough SNR (signal-to-noise ratio). Since there is no charge accumulation as in a CCD chip, PMT has a very high read-out bandwidth (up to GHz); furthermore, it can work in the so called single photon mode which consists in recording the arrival time of a photon with a precision of a few tens or hundreds of picoseconds.

We conclude by remarking that confocal microscopy does not only provide optical sectioning but also an improvement of resolution, as we will discuss in the subsequent subsection.

2.1.2.1 Confocal resolution and PSF

In optics, the numerical aperture (NA) of an optical system is a dimensionless number that characterizes the range of angles over which the system can accept or emit light. In most areas of optics, and especially in microscopy, the numerical aperture of an optical system such as an objective lens is defined by

$$\mathrm{NA} = n \sin \alpha \tag{2.1}$$

where n is the index of refraction of the medium in which the lens is immersed (1.00 for air, 1.33 for pure water, and typically 1.52 for immersion oil), and α is the maximal half-angle of the cone of light that can enter or exit the lens. Therefore, in microscopy it is critical to use high quality immersion lenses, with high NA and high correction of aberrations. This point is basic for improving the 'resolution' of a microscope as we discuss in a moment. Indeed, light microscopes, including wide-field and confocal microscopes, are fundamentally limited in the resolution they can achieve by diffraction phenomenon.

Diffraction of light occurs when a wave passes through a hole which, in the case of a lens, is just the circular aperture defined by its size. Since the size of this disc is large with respect to the visible wavelengths, this effect is small but it is sufficient to produce a significant effect in the imaging of specimens which are also very small.

In the case of a monochromatic plane wave (corresponding to a point source at infinity) incident on a circular aperture, the intensity pattern created by diffraction is known as an Airy pattern. It is given by

$$I(\theta) = I_0\left(\frac{2J_1(ka\sin\theta)}{ka\sin\theta}\right)^2, \tag{2.2}$$

where I_0 is the maximum intensity at the center of the pattern, J_1 is the Bessel function of first kind of order 1, $k = 2\pi/\lambda$ is the wave number, λ is the wavelength, a is the radius of the circular aperture, and θ is the angle of observation, i.e. the angle between the axis of the circular aperture and the line between the aperture center and the observation point. The Airy pattern is shown in figure 2.3; the Airy disc is the circular region defined by the first circle where the light pattern is zero (the first black circle in the figure).

If a perfect converging lens is placed in front of the circular aperture, the diffraction pattern in the focal plane of the lens is given by equation (2.2) with $\sin\theta \simeq \tan\theta = \frac{r}{f}$,

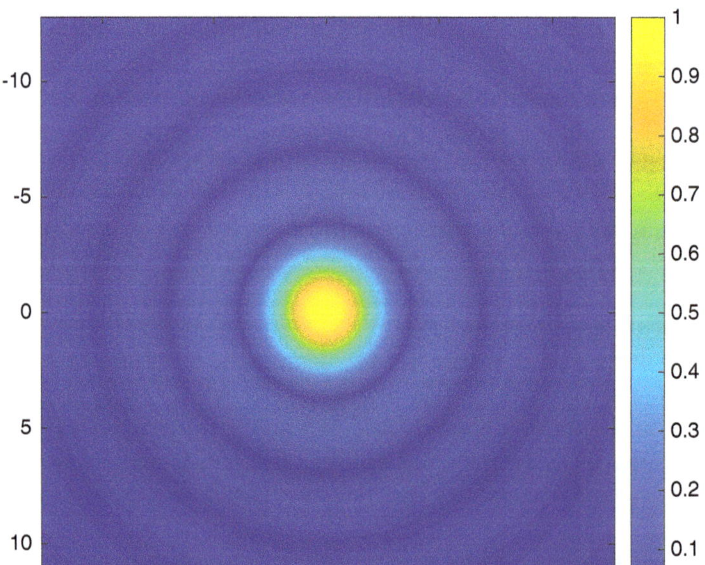

Figure 2.3. A computer generated Airy pattern. In order to enhance the brightness of the rings, the intensities are represented using a square-root gray scale. The image domain is sampled using arbitrary units.

where r is the distance of a point in the focal plane from the optical axis (see figure 2.4). The approximation holds true if the size of the diffraction pattern is small with respect to the focal length f. Since the first zero of $J_1(x)$ is $x_1 \simeq 3.831\,7$, the radius of the first circle, where the diffraction pattern is zero, is given by

$$r_1 \simeq \frac{x_1}{2\pi}\lambda\frac{f}{a} \simeq 0.61\frac{\lambda}{\mathrm{NA}}; \tag{2.3}$$

indeed, in the approximation of small α, if $n = 1$ we have $\mathrm{NA} = \sin\alpha \simeq \frac{a}{f}$. It is remarkable that the previous formula holds true also in the case of immersion lenses, i.e. when $\mathrm{NA} = n \sin\alpha$, since the *Abbe diffraction limit*, discovered by Ernst Abbe in 1873, is given by

$$r_1 = 0.5\frac{\lambda}{n \sin\alpha}. \tag{2.4}$$

The accepted criterion for determining the diffraction limit to resolution was developed by Lord Rayleigh in the 19th century. The Rayleigh criterion, illustrated by the video in figure 2.5, states that the images of two objects are resolvable when the center of the diffraction pattern of one is directly over the first minimum of the diffraction pattern of the other. Therefore, the resolution distance is just given by equation (2.3) and it turns out that an increase of NA provides an improvement of resolution. For example, for a point source radiating light at a wavelength of 510 nm and a microscope objective with a numerical aperture of 1.4, the value of r_1 will be

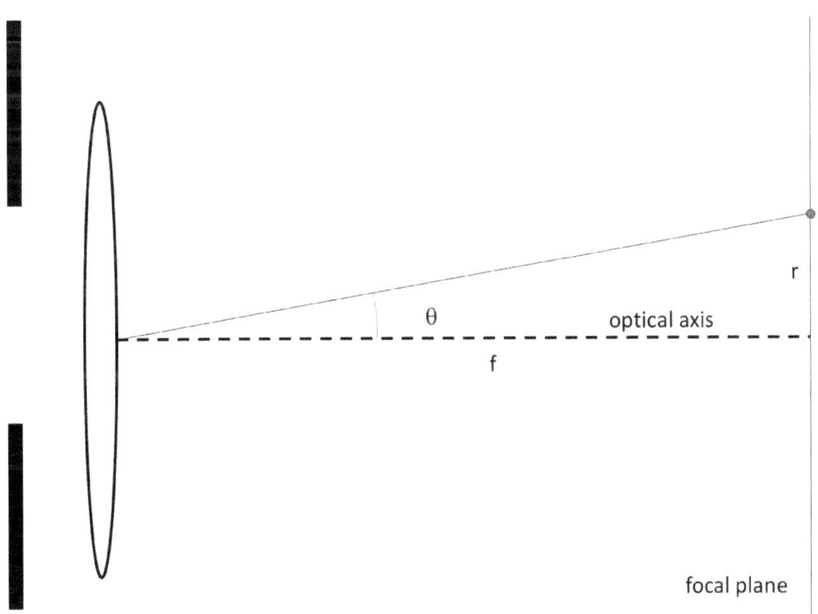

Figure 2.4. A perfect converging lens, with focal length f, placed in front of a circular aperture.

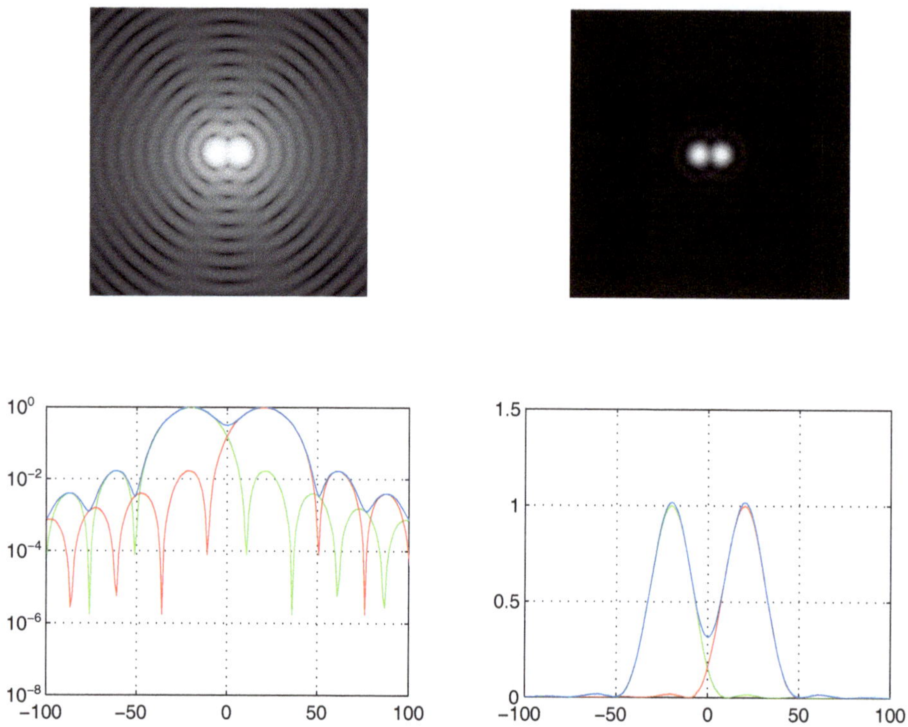

Figure 2.5. A frame of a video clip illustrating the diffraction limit of resolution. In the two top panels the image of two point sources acquired by a microscope are shown. In the left side the image is represented using a logarithmic scale, while to the right it is represented using a square-root scale. In the two bottom panels the profile of the intensity through the center of the image is displayed with a log (left) and a linear scale (right). Video available at http://iopscience.iop.org/book/978-0-7503-1437-4.

222 nm. This value is small compared to the size of biological cells (about 1 μm) but large compared to the size of most sub-cellular structures (tens of nanometers) and the size of a single protein (a few nanometers).

In order to quantitatively describe the performance of a microscope, an important tool is the *point spread function* (PSF). The amplitude of the PSF is the amplitude of the image of a point-like source in the image plane; in the scalar approximation of light propagation [9] the square modulus of the light amplitude is the intensity of the electromagnetic field. It is difficult to observe directly the PSF amplitude but it is possible to observe its square modulus, namely the intensity PSF. In confocal microscopy, which allows the detection of 3D images, it is possible to define a 3D PSF and its shape, in the lateral and axial directions with respect to the optical axis, providing information about lateral and axial resolution. The PSF can be measured, by recording the image of an unresolved bead, or can be calculated by an optical model. To this purpose the following assumptions are made:

1. the optical system is linear;
2. the optical system is space-invariant, i.e. the response of the system is independent of the position of the source.

As treated in detail in [4], the amplitude PSF of a well-corrected lens in its focal region is given by

$$h_{\mathrm{lens}}(u, v) = -i\frac{2\pi\, na \sin^2\alpha}{\lambda} e^{iu/\sin^2\alpha} \int_0^1 J_0(v\rho) e^{-iu\rho^2/2} \rho d\rho, \tag{2.5}$$

where, λ is the wavelength, n is the refractive index of the medium in which the source is embedded, $n \sin\alpha$ is the numerical aperture (NA) of the objective lens that collects light, a is the radius of the detection aperture and J_0 is the Bessel function of first kind of zero order; moreover, if z is the coordinate along the optical axis and $r = \sqrt{x^2 + y^2}$ is the lateral distance from the optical axis in a coordinate system with the origin in the focus of the lens, u and v are the dimensionless variables

$$v = 2\pi\, n \sin\alpha \frac{r}{\lambda}, \quad u = 2\pi\, n \sin^2\alpha \frac{z}{\lambda}. \tag{2.6}$$

Then, the intensity PSF $|h_{\mathrm{lens}}(u, v)|^2$ is given by

$$|h_{\mathrm{lens}}(u, v)|^2 = h_{\mathrm{lens}}(u, v) h^*_{\mathrm{lens}}(u, v). \tag{2.7}$$

A confocal fluorescence microscope uses both point-wise illumination and point-wise detection and the objective lens forms an image in the common conjugate plane. Only the fluorophores that are in the volume shared by the illuminating and detection PSFs are detected. Therefore, the PSF of a confocal microscope can be modeled by the product of illuminating intensity PSF and detection intensity PSF and is given by

$$|h_{\mathrm{confocal}}(u, v)|^2 = |h_{\mathrm{ill}}(u, v)|^2 |h_{\mathrm{det}}(u, v)|^2. \tag{2.8}$$

The first factor, $|h_{\mathrm{ill}}|^2$ depends on the wavelength of the illuminating beam while the second factor, $|h_{\mathrm{det}}|^2$ depends on the wavelength of the radiation emitted by the fluorophores. The full width half maximum (FWHM) of a confocal microscope PSF in the lateral direction is reduced by a factor of $1/\sqrt{2}$ with respect to that of the wide-field microscope and also the side lobes are reduced. Moreover, the axial resolution distance is wider than the lateral one by about a factor of 3. In figure 2.6 we show the intensity PSF calculated for a confocal microscope, assuming that the difference between the illumination and the detection wavelengths is negligible [3].

2.1.2.2 Confocal SNR

As discussed in [21], in many biological applications of confocal microscopy the signal level is low. Indeed, as described above, the size of the pinhole should be exceedingly small in order to select a very thin section of a thick specimen; but, if the size of the pinhole is too small, then the signal level is too low. On the other hand a very large pinhole cancels the optical sectioning and a confocal microscope becomes a wide-field microscope.

An increase of the signal level by increasing the exposure time is not possible due to the photo-bleaching effect that reduces the light emitted by fluorophores during the illumination. Photo-bleaching is the photo-chemical alteration of a fluorophore

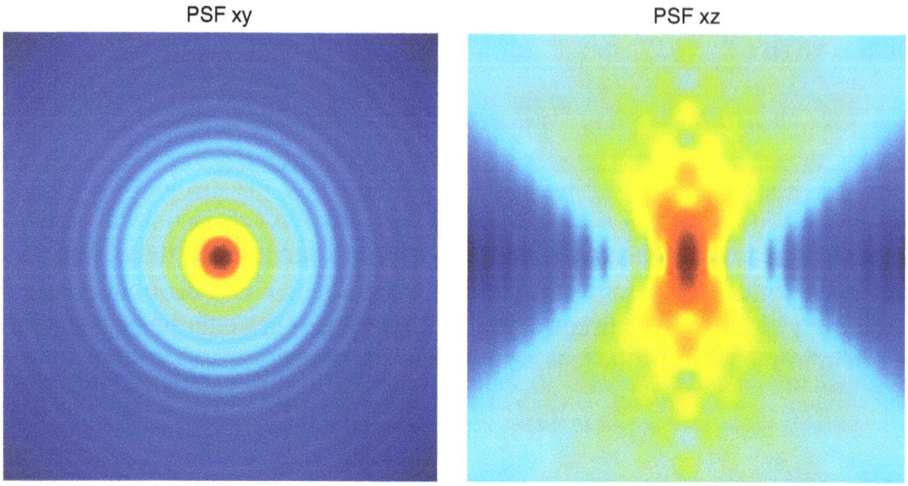

Figure 2.6. Intensity PSF of a confocal microscope (log-scale): (a) x–y section in the focal plane, orthogonal to the optical axis; (b) x–z section in a plane through the optical axis (corresponding to the vertical direction).

molecule such that it becomes permanently unable to fluoresce. It is caused by the interaction with the surrounding molecules or by an excess in the light exposure used to stimulate the fluorophores. Obviously it may alter observations in fluorescence microscopy.

An increase of the signal level by increasing the incident laser power can produce both a saturation of fluorophores and a nonlinear relationship between fluorophore concentration and detected signal. In all cases, an estimate of the SNR is relevant. It can be obtained as follows.

As stated by the corpuscular theory of light, a beam of power P, with a wavelength λ (and frequency $\nu = c/\lambda$), incident for a *dwell time* τ, i.e. the time the laser dwells on each position, contains n_p photons with energy $E = h\nu$, given by

$$n_p = \frac{P\tau}{h\nu} = \frac{P\tau\lambda}{hc} \tag{2.9}$$

h being the Planck constant ($6.626\,07 \times 10^{-34}$ m^2 kg s^{-1}) and c the velocity of light (2.998×10^8 m s^{-1} in vacuum). Due to the relatively small number of photons, the photon-noise (or shot noise or photon counting noise) is the most important source of noise in CLSM. It results from the statistical fluctuations in the number of detected photons and it is modeled by a Poisson distribution, as discussed in section 3.4. This distribution has the property that its *mean* is equal to its *variance* so that we can approximately estimate the SNR as follows:

$$S/N = \frac{n_p}{\sqrt{n_p}} = \sqrt{n_p}\,. \tag{2.10}$$

If the light is detected by a PMT (or another type of detector) with quantum efficiency Q_e, i.e. the fraction of incident photons which are detected by the detector, the SNR becomes

$$S/N = \frac{Q_e n_p}{\sqrt{Q_e n_p + n_s^2}} \tag{2.11}$$

where n_s^2 is the detector noise (due, mainly, to the dark current noise or to the read-out noise).

2.1.3 STED microscope

STED microscopy is a method able to resolve structures well beyond the diffraction limit, possibly until nano-scale; therefore it is a super-resolution technique (see also section 8.1). STED is based on a modification of the confocal microscope and uses two different diffraction patterns for illuminating the sample, one exciting and the other de-exciting fluorocromes in a selected region.

It was developed by Stefan W Hell and Jan Wichmann in 1994 [11], and was first experimentally demonstrated by Hell and Thomas Klar in 1999 [15]. Thanks to this basic invention and its development, Hell was awarded the Nobel Prize in Chemistry in 2014. V A Okhonin (Institute of Biophysics, USSR Academy of Sciences, Siberian Branch, Krasnoyarsk) patented the STED idea in 1986 but presumably this patent was unknown to Hell and Wichmann in 1994. A wide review of this technique is contained in a recent paper of G Vicidomini, P Bianchini and A Diaspro [23].

In figure 2.7 the optical scheme of a STED microscope is shown. The key point is the downsizing of the fluorescent spot used to scan the sample in confocal microscopy. The most typical STED microscope uses a pair of synchronized laser pulses.

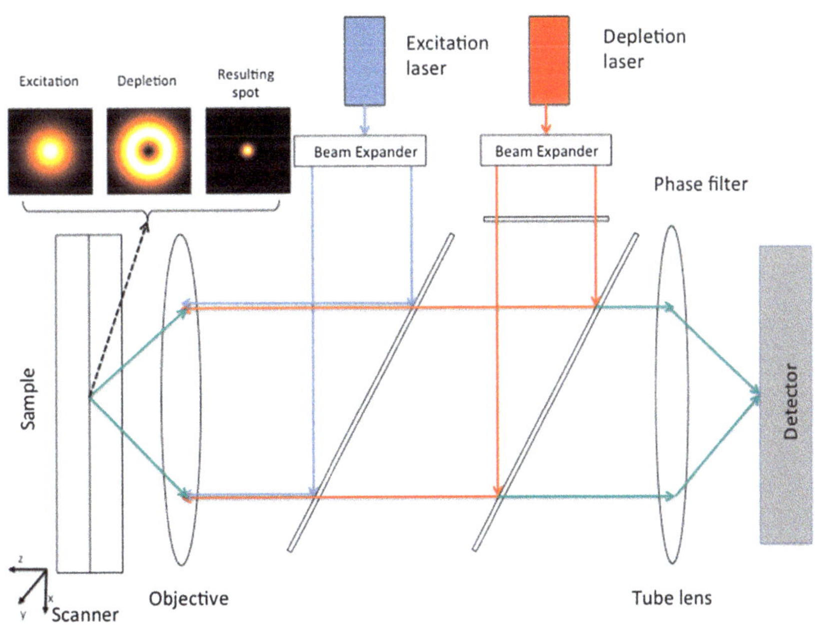

Figure 2.7. Optical scheme of the STED microscope. In the upper-left panel: excitation spot (2D, left), doughnut-shaped de-excitation spot (center) and remaining area allowing fluorescence (right).

The first, generated, for instance, by a picosecond pulsed diode laser, is used to excite the fluorophores of the sample as in a conventional confocal microscope and produces an ordinary diffraction limited focal point. The excitation pulse is followed by a depletion pulse of a red-shifted laser. The spatial extension of this laser beam is determined by a pair of perpendicular polarized beams (created by the phase filter) which induces a depletion beam with a doughnut-like shape by stimulating the emission in this region. In this way the fluorescence is inhibited only in the outer regions of the illuminating spot. The result is a small, tightly focused, sub-diffraction sized spot.

We outline a simplified presentation of the STED principle using a very simplified Jablonski diagram shown in figure 2.8.

- **Depletion of an outer region of the illuminating spot**

  We schematize the fluorescence as an excitation of the fluorophore molecule from the electronic ground state S_0 to the excited singlet state S_1, induced by the excitation beam (the blue one in figure 2.7). After initial vibrational relaxation, the electron could decay from S_1 to an excited vibrational sub-state of S_0 (see the left panel of figure 2.8), so that the fluorophore can emit light as fluorescence. However, spontaneous emission of fluorescence is not the only deactivation process of the S_1 state. It can be induced by an incident photon. Therefore, in a suitable selected spatial region, the spontaneous decay of the electron can be anticipated by a stimulated decay induced by the second laser (the red one in figure 2.7), forcing the electron to decay into a higher vibrational sub-state of the ground state S_0 (indicated as S_0^*, right panel of figure 2.8); in such a way a stimulated photon is emitted with an energy smaller, and therefore with a wavelength

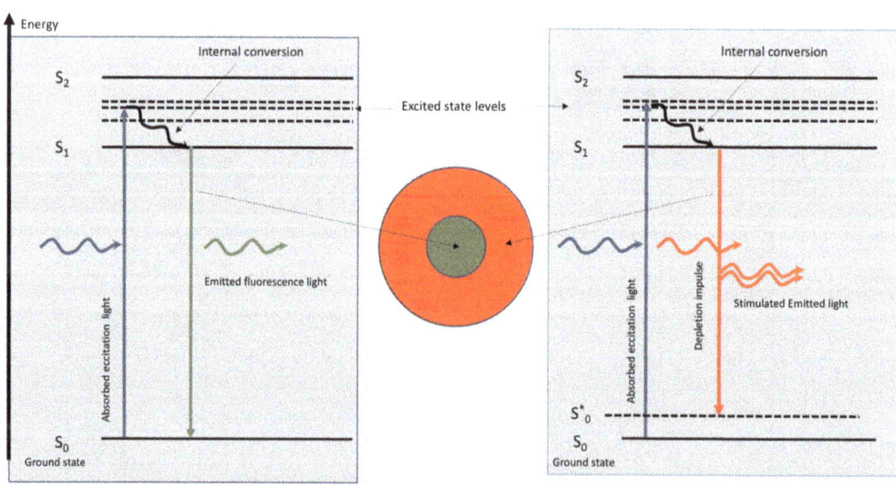

Figure 2.8. Very simplified Jablonski schemes. Left panel: spontaneous emission of a fluorescent photon referring to the central spot of the annulus (center of the figure). Right panel: stimulated emission of a photon in the outer region of the annulus. Notice the energy difference between the left and right emission, allowing the suppression of the stimulated photons by filtering.

longer than that of a fluorescent photon (left panel of figure 2.8). The stimulated photon is red-shifted with respect to the previous one and can be removed by suitable filtering. This is the stimulated emission step removing electrons from S_1. Complete depletion of the excited state in the selected spatial region can occur if a sufficient number of stimulating photons reach the spot generated by the first laser. If this event occurs then this region does not contribute to the imaging of the sample since all the photons originating from this region are removed by the filtering. To achieve the large number of incident photons needed to suppress fluorescence, the laser used to generate the photons must be of high intensity. Unfortunately, this high intensity can lead to photo-bleaching of the fluorophores. The strategies for solving this problem are beyond the scope of this elementary presentation. A concise description can be found for instance in [16].

- **Diffraction optical elements**

  STED works by depleting fluorescence in specific regions of the sample while leaving a central focal spot active to emit fluorescence. This focal area can be modeled by altering the properties of the pupil plane of the objective lens using phase-filters. It is possible to design a set of diffractive optical elements (DOEs) for a variety of depletion patterns. The most common example of these diffractive optical elements, or DOEs, is a torus shape. This DOE is generated by a circular polarization of the depletion laser, combined with a helical phase ramp. In other words, fluorescence is removed in an annulus with its center in the central point of the exciting beam. The radius of the emitting region can be reduced by increasing the intensity of the laser inducing depletion (even if the issue of photo-bleaching must be considered again). In conclusion, the result is the reduction of the size of the central spot. Its shape, inserted in equation (2.8) as the new $|h_{\text{ill}}(u, v)|^2$, provides the reduction of the size of the PSF of a STED microscope.

Super-resolution requires small pixels, i.e. a larger number of pixels for a given field-of-view, hence a longer acquisition time than standard CLSM. However, the size of the focal spot depends on the intensity of the laser used for depletion. As a result, it can be tuned by changing its size and the imaging speed. A compromise can be reached between these two factors for each specific imaging task. Rates of 80 frames per second have been recorded, with focal spots around 60 nm. Up to 200 frames per second can be reached for small fields-of-view.

In figure 2.9 a confocal and a STED image of the same sample are shown, in order to compare the resolution achieved in both cases.

We conclude by recalling that several improvements in microscopy techniques have been proposed in the last years: two-photon excitation microscopy (TPEM) [7] and selective plane illumination microscopy (SPIM) [13]. All these methods remove out-of-focus light by rejecting it before reaching the detector or by precluding its generation. Further hybrid techniques, which remove out-of-focus light by

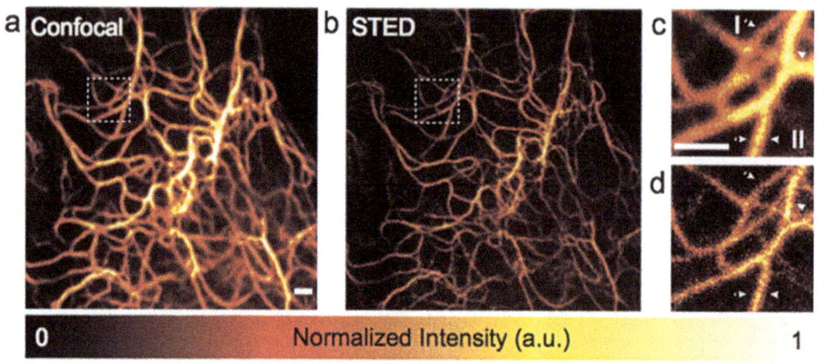

Figure 2.9. Bidimensional CLSM (a) and STED microscopy (b) images of the microtubulin network of PtK2 cell (512 × 512 pixel, dwell time 10 μs, pixel size 32 nm (courtesy of G Vicidomini).

combining optical and computational methods are 4Pi microscopy [10] and structured illumination microscopy (SIM) [17].

2.2 Medical imaging (tomography)

The term *tomography*, derived from the Ancient Greek τόμοζ (slice, section) and γραφό (write), refers to any non-destructive method producing the three-dimensional image of the internal structure of an object from many measurements taken from different angles, allowing the interior to be imaged without cutting the object. This method of sectioning, using different kinds of penetrating probes, is used in medical applications, archeology, geophysics and in several other applications.

In this section we deal only with medical imaging distinguishing transmission from emission tomography. Indeed the first is fundamental for understanding the basic ideas of medical tomography while the second is the main objective of the image reconstruction methods proposed for dealing with Poisson data.

A first difference between the two methodologies is that, in transmission tomography, the radiation used for inspecting the human body is an external x-ray beam and the reconstructed volume represents a map of the density variations inside the human body; in the second case, instead, a radioactive tracer is injected into the patient and the reconstructed volume represents a map of the tracer density. As transmission tomography was a revolution in radiology, similarly emission tomography was a revolution in nuclear medicine.

The two different modalities have different and complementary clinical applications. While transmission tomography can be applied, in principle, to non-living objects and, in the case of medical imaging, provides information on the morphology of internal organs, emission tomography assumes an active metabolism of the subject under examination so that it provides fundamental metabolic information by recovering the map of the radioactive tracer.

2.2.1 Radiation–matter interaction

By *radiation* here we mean an electromagnetic wave that propagates in space, with frequencies higher than those of the visible or ultraviolet spectrum; typically from x-rays and beyond. Ultraviolet, visible, infrared, microwave, and radio frequencies are excluded. Therefore, for our purposes, radiation can be described by a flow of quanta, the photons, with energy $E = h\nu$ and momentum $p = h\nu/c$, h being the Planck constant, ν the radiation frequency and c the light velocity. The advantages of x- or γ-ray applications are based on the following properties:

- small wavelength provides a resolution comparable with the characteristic size of the electron density distribution in atoms;
- x- or γ-rays weakly interact with the matter providing non-destructive characterization of the objects;
- radiation propagation takes place along straight lines.

We recall that photons are individual units of energy. As an x- or γ-ray passes through an object, three possible fates await each single photon:

- it can penetrate the section of matter without interacting;
- it can interact with the matter and be completely absorbed by transferring its energy to the body;
- it can interact and be deflected from its original direction, by transferring part of its energy.

As a photon makes its way through the matter, there is no way to predict precisely how far it will travel before interacting. In clinical applications we are generally not concerned with the fate of an individual photon but rather with the collective interaction of a large number of photons. In most instances we are interested in the overall rate at which photons interact as they make their way through a specific material.

Let us observe what happens when a group of monochromatic photons (i.e. with a fixed energy) encounters a slice of material (with a uniform density) with a Δx thickness, as illustrated in figure 2.10. Some of the photons interact with the material, and some pass on through. The interactions, either due to photoelectric or Compton scattering, remove some of the photons from the beam in a process known as *attenuation*. The *linear attenuation coefficient* μ is the actual fraction of photons interacting with a thick unit of material. For example, if the fraction that interacts in the 1 cm thickness is 0.1, or 10%, the value of the linear attenuation coefficient is 0.1 cm^{-1}.

In the case of a constant attenuation coefficient it is easy to obtain the number of photons as a function of the depth x

$$N(x) = N_0 e^{-\mu x}, \tag{2.12}$$

where N_0 is the number of photons entering at $x = 0$. If the attenuation coefficient depends on the position x, equation (2.12) becomes

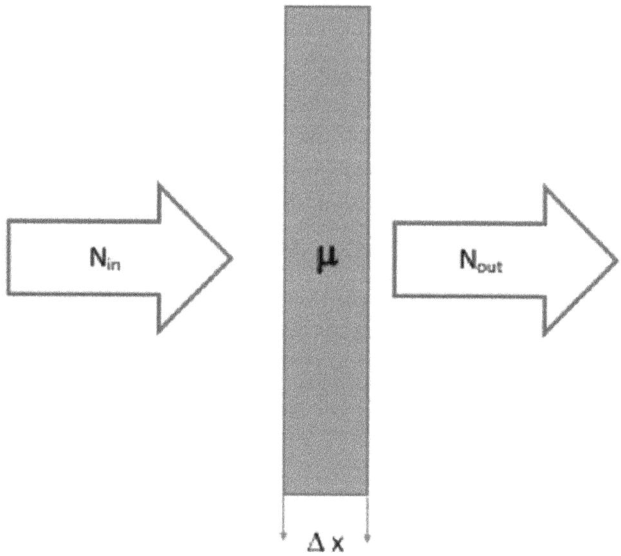

Figure 2.10. Linear attenuation coefficient and its effect on a photon flux with fixed energy.

$$N(x) = N_0 e^{-\int_L \mu(x')dx'} \tag{2.13}$$

where L is the path of the ray from 0 to x inside the attenuating body.

2.2.2 X-ray transmission tomography

X-ray computed tomography (CT) is a diagnostic technique that uses ionizing radiation (x-rays) to produce cross-sectional images (or slices) of specific areas of a scanned object, i.e. the human body. In 1972 diagnostic imaging was revolutionized by the introduction of tomography in clinical practice. In fact it overcomes the limits of traditional radiology, i.e. the representation of a projection of the human body on a single plane with overlap between organs and lesions.

CT was conceived and realized by a British engineer Godfrey Hounsfield and theoretically investigated by a South African physicist Allan Cormack; both received the Nobel Prize in Physiology or Medicine in 1979. The first CT-scanner was realized in 1971 and it was dedicated to brain imaging. In 1974 CT scanners were developed also for the thorax and abdomen.

To explain the working principle of tomography we refer to the inspection and consequent reconstruction of an *axial (transverse)* slice of the human body, i.e. a slice orthogonal to a hypothetical axis along the patient body (head–feet), which also appears to be the rotation axis of the tomograph. By moving and rotating the source–detector system, the tomograph acquires, for different angles, the amount of residual radiation due to the body absorption. Therefore, for each direction the number of photons, as given by equation (2.13), is detected. To be more precise, this equation gives the mean value of the number of photons which satisfies Poisson statistics. It is clear that the ratio N/N_0 provides information about the integral of the linear attenuation coefficient along the line joining the detector and the receiver. The relationship is nonlinear but we obtain a linear relationship by simply taking the negative logarithm of N/N_0. This is what is usually done in practice even if, in this way, the Poisson statistics of the data is lost. Anyway, it is obvious that the data must be processed by a computer using some suitable method for estimating a map of the linear attenuation coefficient μ and, for this reason the methodology is called *computed tomography*.

Since the reconstruction of μ can be performed off-line, the reduction of the time required for data acquisition is an important issue for reducing patient inconvenience. This issue generated an evolution of the acquisition schemes which are shown in figure 2.11 and can be described as follows.

- First generation: translate–rotate (parallel beam) with single detector (A).
- Second generation: translate–rotate with row of detectors (B).
- Third generation: rotate–rotate with continuous rotation of a row of detectors (C).
- Fourth generation: rotate-fixed with a complete ring of fixed detectors (D).

The four generations, with different scanning geometries, are characterized by an increase in the number of detectors and consequent decrease in the scanning time. The evolution of the technology, allowing the reduction of the acquisition time, has also reduced the dose of radiation absorbed by the patient as well as the artifacts due to their respiration and movements. While, in the first three generations of scanners, the acquisition was performed slice by slice, in the fourth generation the rotation around the patient is combined with a slow translation along the axis. In this case, the detector array is made up of several layers and the mode is named *spiral tomography*.

X-rays are not directly detectable. A solid state x-ray detector consists of a scintillator that converts x-rays into visible light and a photo-diode that converts light into electrical signals. For transmission imaging systems that use solid state detectors, such as current clinical x-ray CT systems, the data statistics is considerably more complicated than that of the ideal photon counting case. There are numerous sources of variability affecting data statistics. In the case of a monochromatic source they can be summarized as follows [18].

- Usually the current of the x-ray tube fluctuates slightly (but noticeably) around its mean value.
- Each injected photon may be absorbed or scattered, remaining within the object with a different energy; both are random processes.
- For a given x-ray tube current, the number of x-ray photons transmitted towards a given detector element is a random variable, typically modeled by a Poisson distribution around some mean, as we already remarked.
- X-ray photons that reach a given detector element may interact with it, or may pass through the detector element without interacting.

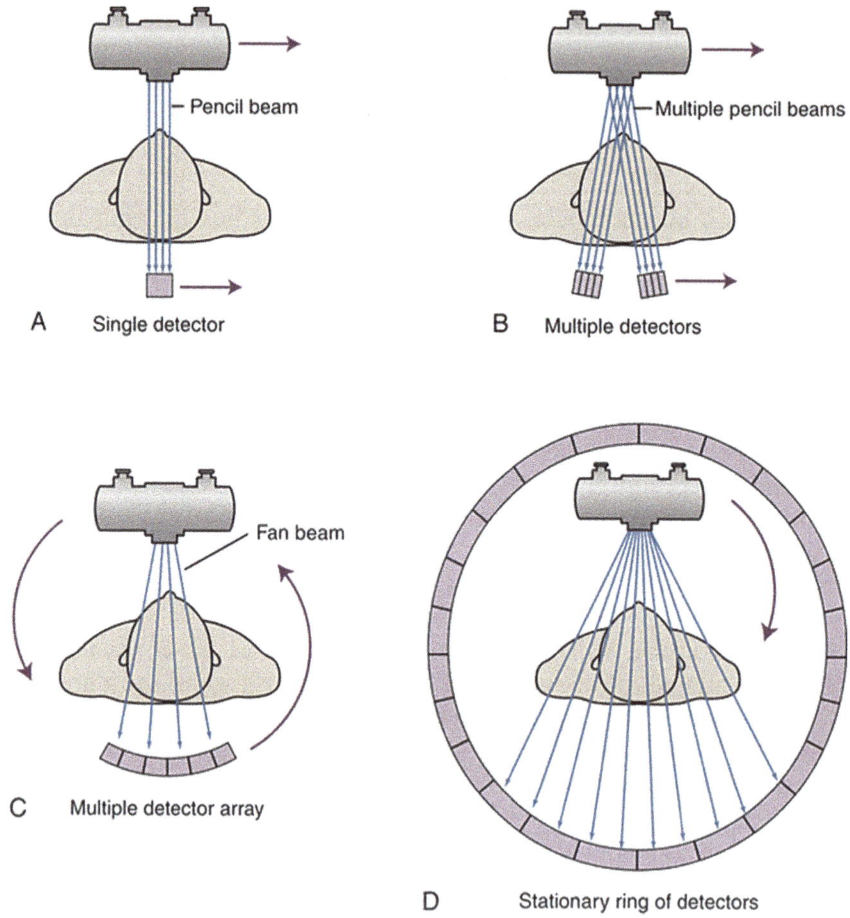

Figure 2.11. The four generations of x-ray computed tomography. (A) First generation: an x-ray tube and a detector scan the human body section first translating and then rotating. (B) Second generation: the x-ray tube generates a little fan and the data are collected by an array of detectors; the system rotates around the body. (C) Third generation: the x-ray tube generates a fan beam and rotates continuously; the detector array has a curved geometry and rotates with the x-ray beam. (D) Fourth generation: the detector array becomes a fixed ring and only the x-ray tube rotates.

- An x-ray photon that interacts with the detector can do it via Compton scattering and/or photoelectric absorption.
- The amount of the x-ray photons energy that is transferred to the electrons in the scintillator is a random variable because the x-ray photon may scatter within the scintillator and then exit having deposited only part of its energy.
- The conversion of light photons into photo-electrons involves random processes.
- Electronic noise in the data acquisition system, including quantization in the analog-to-digital converter that yield the final (raw) measured values, adds further random variability to the measurements.

2.2.3 Emission tomography

In x-ray tomography, the source of radiation is external to the body under examination and the attenuation of the radiation measured by the detectors allows the reconstruction of the density of the tissues that produced the attenuation. On the other hand, in emission tomography, the source is internal, i.e. γ-radiation is produced by the decay of a radionuclide (tracer) injected to the patient. In this case, the measurement of the emitted radiation allows the reconstruction of the concentration of the tracer, related to tissue metabolic activity.

There are two kinds of emission tomography using different kinds of radionuclide and the main difference is related to the emitted radiation and its detection. Single photon emission computed tomography (SPECT) uses a radionuclide which emits single photons, while positron emission tomography (PET) uses tracers which emit positrons. Both techniques are used for cancer and/or metabolic studies on living people or animals.

One of the most important clinical uses of SPECT and PET is oncology. SPECT and PET, according to the tracer used, allow, in a non-invasive way, numerous patho-physiological parameters to be evaluated and quantified. They provide information on the 'metabolic-molecular' state of the lesion, allow a more accurate definition of the real extent of the neoplasm, and provide accurate information on the biologically more active parts of the tumor. All this information can be used to guide the stereo-tactic biopsy or to address therapeutic treatments.

Other clinical applications of these imaging techniques are: cerebral or myocardial perfusion studies, epilepsy, Alzheimer, Parkinson, dementia and brain trauma.

2.2.3.1 SPECT imaging

Figure 2.12 and 2.13 show the main components and imaging devices of a SPECT system. A radioactive-labeled pharmaceutical (radio-pharmaceutical) is administered to a patient. Depending on the properties of the radio-pharmaceutical, it is taken up by different organs and/or tissue types. Radio-pharmaceuticals, used in SPECT, are labeled with a radionuclide that emits a γ-photon. Typically, a scintillation camera system is used as the imaging device.

The scintillation camera (gamma camera) consists of: (i) a collimator, a slab made of materials with a high density and a high atomic number, such as lead or tungsten, with holes that allow only those photons traveling along a given direction to pass through; (ii) a large-area scintillator (commonly NaI(Tl) crystal) that converts the energy of a γ-photon to lower-energy photons which are in turn converted into electric signals by photomultiplier tubes (PMTs). The signals from an array of PMTs are processed by electronic circuitry to provide information about the position at which a photon interacts with the crystal. The scintillation camera provides a 2D image of the 3D radioactivity distribution or radio-pharmaceutical uptake within the patient [14]. This image is basically the image produced by scintigraphy in a process similar to the capture of x-ray images. The gamma camera is rotated around the body to collect data along different directions able to produce a 3D image after suitable computational treatment.

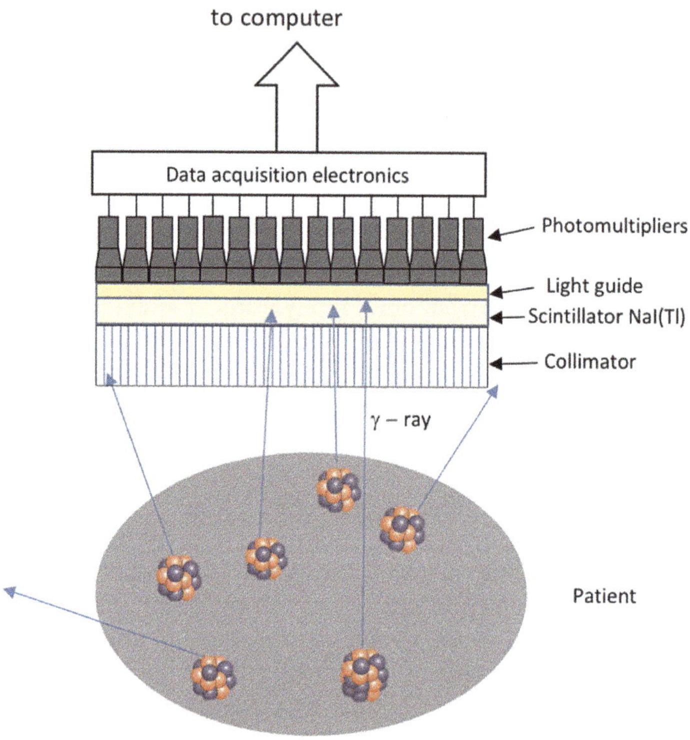

Figure 2.12. View of a slice of the components of a SPECT system. A γ-ray emitted by the radionuclide and traveling in a fixed direction, passes through a collimator hole and reaches the scintillator that converts it into ultraviolet-light. Then the low energy photons are converted in electric signals by the array of PMTs.

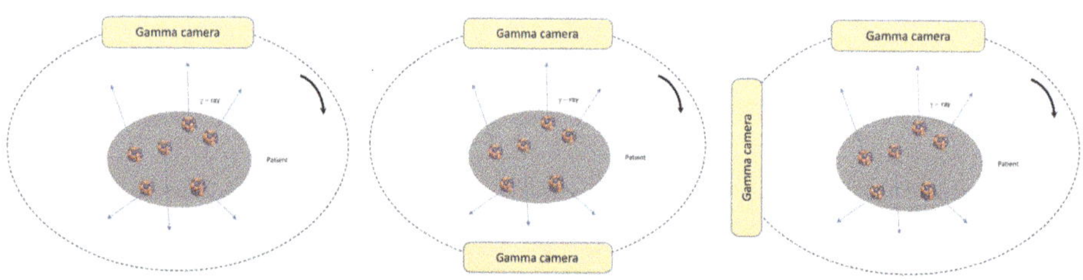

Figure 2.13. Three different schemes of the geometry of SPECT acquisition. From left to right: a scheme with only one gamma camera (one head), a scheme with two parallel gamma-cameras (two heads—H geometry) and a scheme with two orthogonal gamma-cameras (two heads—L geometry).

In recent years, alternatives to the gamma camera have been introduced into clinical practice. Semiconductor materials, like cadmium telluride (CdTe) and cadmium zinc telluride (CZT), convert the gamma photons directly into electrical signals and offer better performance. For example, for cardiac examination, these detectors allow shorter acquisition times and considerably diminished patient dose, improving signal-to-noise ratio and image quality. Unfortunately, the CZT technique is expensive compared to NaI and photo-multipliers, and this restricts its usage to smaller systems that are designed for special applications like cardiac or preclinical imaging. Moreover, these detectors make it possible to use SPECT simultaneously with magnetic resonance imaging (MRI), being compatible with strong magnetic fields.

The goal of SPECT is to determine accurately the 3D radioactivity distribution resulting from the radio-pharmaceutical uptake inside the patient (instead of the attenuation coefficient distribution from different tissues as obtained from x-ray CT). Two main factors influence the quality of SPECT images: the collimator and the attenuation of the radiation inside the patient body.

The amount of radio-pharmaceutical is limited by the allowable dose of radiation that can be administered to the patient; therefore a very limited number of photons can be used for imaging. Indeed, the direction of arrival of the γ-rays must be known for imaging purposes and this is possible only if the γ-ray passes through one of the collimator holes. Therefore, the acceptance angle or geometric response of the collimator further limits the fraction of photons that are acceptable for imaging. In conclusion, only a small fraction (typically 10^{-4}–10^{-2}) of emitted photons pass through the holes of the collimator and can be detected, thus seriously limiting the sensitivity of the images. The collimator can be designed to allow detection of more photons, but increased detection efficiency usually can be achieved only with a concurrent loss of spatial resolution.

There are different types of collimators, differentiated by the thickness of the septa, holes diameter and orientation of the septa (with parallel flat holes, converging and diverging collimators, pinhole collimators, fan beam collimators). The simplest is the collimator with parallel holes, shown in figure 2.14. In this simple

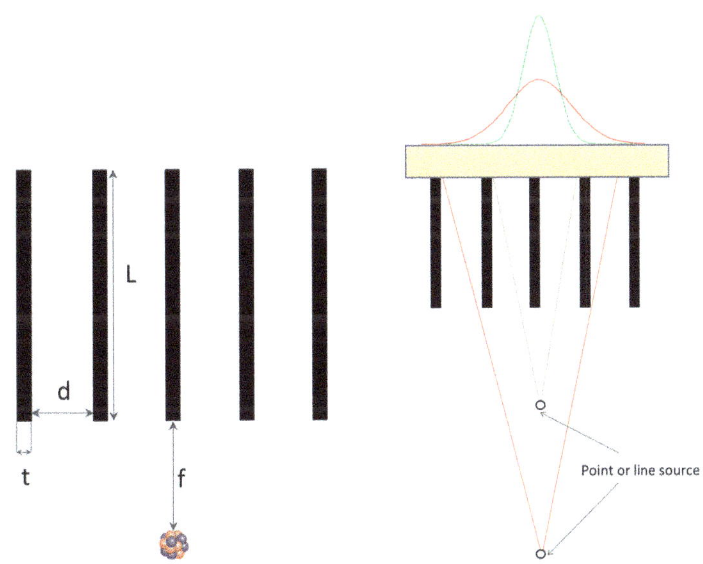

Figure 2.14. Left panel: collimator geometry (t = septa thickness, d = septa distance, L = collimator depth, f = event distance). Right panel: system response as a function of source–collimator distance.

case we illustrate the collimator characteristics considering only its geometry represented in figure 2.14.

- **Geometric efficiency**: this can be defined as the ratio between the number of photons passing through the collimator and the total number of photons arriving on the collimator and is given by [1, 22]

$$\epsilon = \left(K\frac{d}{L}\right)^2 \left(\frac{d}{d+t}\right)^2, \tag{2.14}$$

  where K is a constant which depends on the shape of the holes and their distribution pattern. It has been determined numerically and confirmed by experiments (i.e. $K = 0.24$ for round hole in hexagonal array, and 0.26 for hexagonal hole in hexagonal array). Larger holes increase the sensitivity but degrade the resolution. This effect is often called the resolution-sensitivity trade-off and depends, in a complicated manner, on many parameters, such as the size of the region of interest (ROI) to be imaged, the type of collimator (pinhole, parallel-hole, fan beam, etc), the energy of the photons to be detected, the detector intrinsic spatial resolution, the size of the detector, and the radius of rotation (ROR).
- **Resolution**: the geometric resolution is given by [1]

$$R = d\left(1 + \frac{2f}{L}\right) \tag{2.15}$$

  and this quantity, which influences the overall resolution, is related to the FWHM of the acquired signal and depends on the distance of the source from the collimator (or from the gamma camera).

The system response of two sources, one near and one far from the collimator, is illustrated in figure 2.14; a simple 2D Gaussian model of the response is considered in [8] and successfully compared with data from experimental acquisitions.

The second major factor that affects the quantitative accuracy and quality of SPECT images is the attenuation due to possible photoelectric absorption and Compton scattering of the γ-rays emitted by the radionuclides before reaching the γ-camera. The degree of attenuation is determined by:

- the path length between the source and the edge of the attenuating material;
- the linear attenuation coefficient, which is a function of photon energy and the amount and types of materials contained in the attenuating medium.

For example, the attenuation coefficient for the 140 keV photons (from the commonly used isotope Tc-99m) in water or soft tissue is 0.15 cm^{-1}. So the thickness of material that attenuates half of the incident photons is 4.5 cm, in water.

The attenuation effect is further complicated by the fact that different regions of the body have different attenuation coefficients. In particular, the non-uniform attenuation distribution in the thorax is a major problem in cardiac SPECT image reconstructions.

One approach for reducing both the space-variant response of the system and the effect of attenuation by the body tissues is the *'conjugate counting'*. Conjugate counting consists in acquiring data (or image profiles) from directly opposing views and then combining these data into a single data set or line of response. A source that is located relatively close to the detector in one view will be relatively far away in the opposing view, as illustrated in figure 2.14. Hence, the response profile will be narrower and attenuation by overlying tissues will be smaller in the first view and larger in the second, with partially offsetting effects. Conjugate counting requires the views be obtained over a full 360° angular range around the object. Then data from opposing views are combined to yield the equivalent of a single 180° scan. Conjugate views generally are combined with pixel-by-pixel geometric mean, since geometric mean seems to provide better results than arithmetic mean [5].

Photons that have been scattered before reaching the detector give erroneous spatial information about the origin of the radioactive source, and a significant fraction of the photons detected in SPECT have been scattered. Typical ratios of scattered to un-scattered photons are about 20%–30% in brain studies and about 30%–40% in cardiac and body studies.

Overall resolution of SPECT systems is about 4–5 mm and the resulting images are significantly blurred. This relatively low resolution affects image quality as well as quantitative accuracy.

For this reason SPECT reconstructions are often used in multi-modal observations. SPECT and CT (or MRI) images are merged by means of a co-registration technique, in order to obtain both anatomic and functional information of the studied region. An example is shown in figure 2.15. Moreover, if SPECT/CT images are acquired, the tissue attenuation in SPECT may be corrected by considering the density information derived from CT reconstruction.

2.2.3.2 PET imaging

PET uses radioactive tracers obtained by marking molecules normally present in biological tissues (such as sugars, amino acids, water) with radionuclides which decay β^+, i.e. emit a positron, the antiparticle of the electron with the same mass but

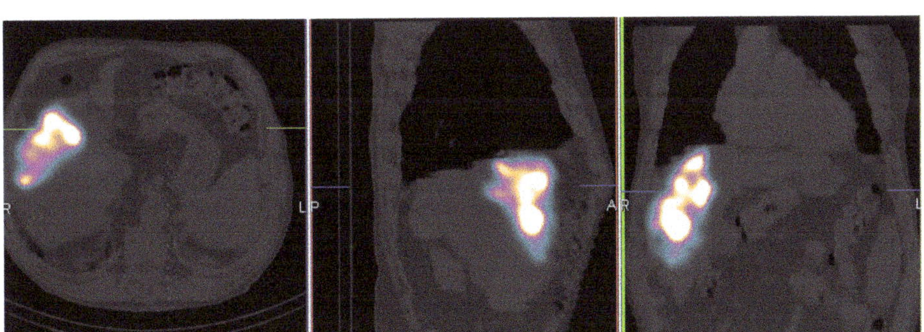

Figure 2.15. Axial (transverse), sagittal and coronal slices of a SPECT/CT image which highlights a tumor in the liver parenchyma.

a positive electric charge. The radionuclides used in PET have two important characteristics: they have a short half-life (of the order of minutes or a few hours) and are isotopes of main elements constituting the biological matter (oxygen, fluorine, carbon, etc). The short half-life results in the need for a cyclotron, the accelerator of particles used for the production of PET radionuclides, in close proximity of the diagnostic center. For instance, in oncology applications, the routinely used tracer is fluorine-18 with a half-life of 199.77 min.

The procedure begins with the injection of a radio-pharmaceutical consisting of a radio-isotope tracer with a short half-life, chemically bound to a molecule active at the metabolic level (vector), for example fluorodeoxyglucose (^{18}F-FDG). After a waiting time, during which the metabolically active molecule (often a sugar) reaches a certain concentration within the organic tissues to be analyzed, the subject is placed in the scanner. The isotope decays, emitting a positron. After a path that can reach a maximum of a few millimeters, the positron is annihilated by an electron, producing a pair of gamma photons, both with an energy of 511 keV, corresponding to the mass of the electron, emitted in opposite directions.

In figure 2.16 the geometry of PET acquisition is shown. The signal that reaches the photomultiplier is amplified and sent to a pulse height analyzer (PHA) that decides if the received photons belongs to the energy window chosen for the acquisition, with center 511 keV. The narrower this window is, the greater the number of events rejected by the PHA because classified as resulting from Compton interaction. The PHA is connected to a time-coincidence circuit that allows it to decide if two accepted events, within the selected energy window, are individual events or a coincidence. For determining the type of the event, the system uses a time window: once the PHA has accepted the single event, if the system registers another event inside the time window (amplitude of some nanoseconds), the two events are classified as a coincidence, otherwise they are classified as single. An event classified as a coincidence is assigned to the decay of one of the radioisotopes located on the line joining the two scintillators involved in the coincidence (line of response, LOR). This approach is also called *electronic collimation*, compared to the *physical collimation* in SPECT imaging.

The condition of simultaneity is more accurate in the case of a narrower time window. At the same time an excessive reduction of its width leads to a reduction of system efficiency. Events are accepted within a time interval selected by taking into

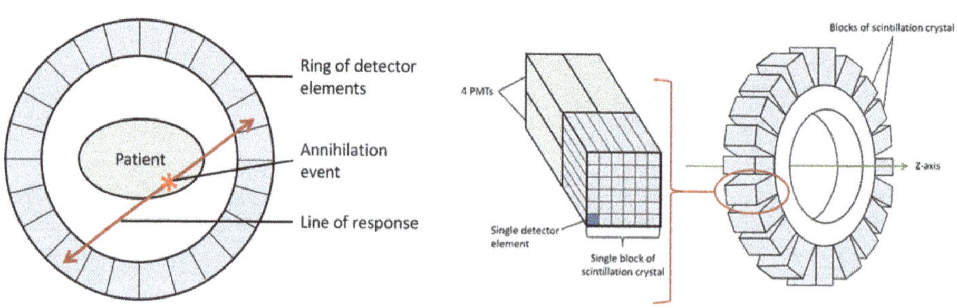

Figure 2.16. Figure illustrating PET geometry and detection system. Copyright Radiology Cafe ® 2011–2018.

account the flight time of the two annihilation photons, the signal collection time and the time required by the electronics.

PET data can be due to different types of coincidence since its response (ready events) can be expressed as the sum of three kinds of events: true coincidence, scatter and random coincidence. We speak of true coincidence when two photons deriving from an annihilation e^+ e^- leave the body without interacting and are revealed within the time window. Similarly, a scatter coincidence originates from a single annihilation event e^+ e^- but one or both photons undergo a Compton diffusion process as they pass through the body. Finally, we speak of a random event, when two photons are revealed in coincidence even if they do not derive from the annihilation of a single positron but, for example, from two different events of annihilation. In this case the system identifies a LOR that is not related to the position of the two decays. Events of this type are completely random and they cause an increase in the noise level and a loss of contrast in the reconstructed image. The three different types of coincidence are shown in figure 2.17.

In PET scanners several detector rings, similar to that shown in figure 2.16, are used. In such a way, two modes of acquisition are possible: one 2D and one 3D. The first differs from the second by the presence of collimator septa, built with shielding materials, that absorb the photons with a direction lying on inclined planes with respect to those of the detector rings. In this way only the photons coming from the planes formed by each ring are detected and the reconstruction can be made slice by slice in a 2D way. In a 3D acquisition, coincidences coming from different slices are also counted. The two acquisition modes are schematically illustrated in figure 2.18. The most recent PET scanner can pass from 2D to 3D configuration, using removable septa. An example of a PET 3D image is shown in figure 2.19. In modern PET–CT scanners, a low resolution CT image is acquired in the same session together with the PET image for improving the assessment of the localization of the injected radio-pharmaceutical.

In conventional PET coincidence electronics is used uniquely for determining the LOR of an annihilation event. The time-of-flight (TOF)-PET goes one step further, and tries to determine approximately the position of the annihilation event on the LOR by measuring the difference in arrival times. Developments on a TOF system

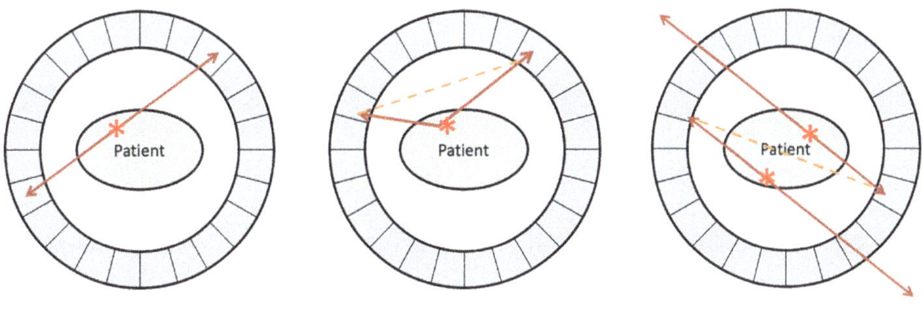

Figure 2.17. Different types of PET coincidences—from left to right: true coincidence; coincidence between a true event and a scattered photon; coincidence between two different events. The corresponding LORs are shown with dashed lines. Copyright Radiology Cafe® 2011–2018.

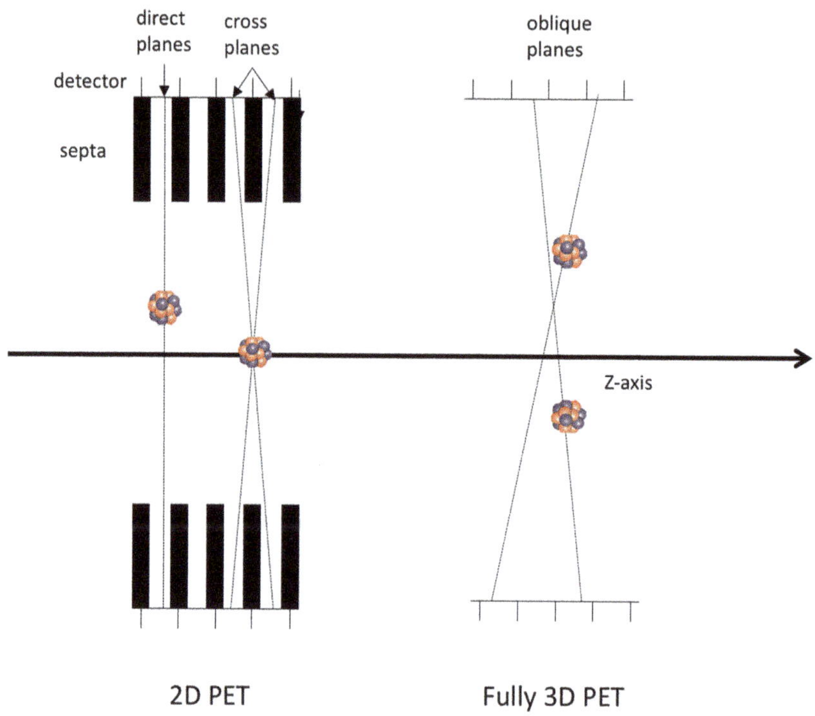

Figure 2.18. Axial view of multi-ring PET acquisition. On the left the 2D PET acquisition scheme is shown. If the collimator septa are removed (right panel) coincidences coming from different slices are counted and we have a fully-3D PET acquisition.

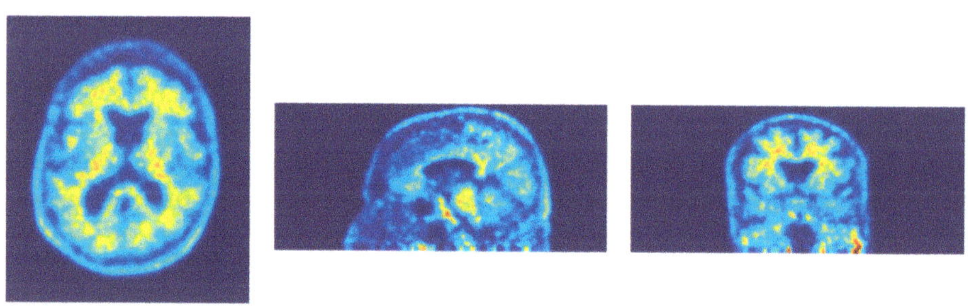

Figure 2.19. Axial, sagittal and coronal slices of an amyloid PET examination for Alzheimer's disease assessment.

started around 2000. The combination of fast photo-multipliers, scintillators with high density, modern electronics, and faster computing power for image reconstruction have made it possible to introduce this principle in clinical TOF-PET systems.

2.3 Astronomy

In 1609 Galileo Galilei turned to the sky the 'cannocchiale' manufactured by himself, the first example of a telescope. He observed the Moon, the Jupiter satellites, called by him *Medicean Stars*, and the hundreds of stars not visible with the naked eye and composing the Milky Way. His observations were reported in *Sidereus Nuncius*, published on March 13, 1610. It was the first published scientific report based on observations made through a telescope. Since then, observational astronomy has made steady advances due to improvements in telescope technology. One of the major advances was the replacement of refracting telescopes with reflecting ones, whose basic principle was introduced by Isaac Newton.

A traditional division of observational astronomy is given in terms of the observed region of the electromagnetic spectrum.

- **Optical astronomy** is the part of astronomy that uses optical components (mirrors, lenses and solid state detectors) to observe light from near infrared to near ultraviolet wavelengths. Visible-light astronomy (using wavelengths that can be detected by the eye, about 400–700 nm) falls in the middle of this range.
- **Infrared astronomy** deals with the detection and analysis of infrared radiation (this typically refers to wavelengths longer than the detection limit of silicon solid state detectors, about 1 μm wavelength). The most common tool is the reflecting telescope equipped with a detector sensitive to infrared wavelengths.
- **Radio and millimeter astronomy** detects radiation of millimeter to decameter wavelength. The receivers are similar to those used in radio broadcast transmission but much more sensitive.
- **High-energy astronomy** includes x-ray astronomy, gamma-ray astronomy, and extreme ultraviolet (UV) astronomy.

In addition to electromagnetic radiation, modern astrophysicists can also make observations by means of neutrinos, cosmic rays or gravitational waves.

Optical and radio astronomy can be performed with ground-based observatories, because the atmosphere is relatively transparent at these wavelengths. For this reason it is possible to realize extremely large telescopes, especially for radio astronomy: the second largest one is the Arecibo radio-radar telescope, sitting in a mountain cavity in Puerto Rico; the largest one is the Chinese five-hundred-meter aperture spherical telescope, or FAST, completed in 2016.

The largest optical reflecting telescopes are usually located at high altitudes to minimize the absorption and distortion caused by Earth's atmosphere. Examples are the 10.4 m Gran Telescopio Canarias (GTC) on the island of La Palma, the two 10 m telescopes of the Keck Observatory, near the summit of Manua Kea, Hawaii and the large binocular telescope on Mount Graham, Arizona, consisting of two 8.4 m mirrors which can be combined to form a Fizeau interferometer with a possible resolution of a 22.8 m telescope.

The atmosphere is opaque at the wavelengths used by x-ray astronomy, gamma-ray astronomy, UV astronomy and (except for a few wavelength 'windows') far infrared astronomy; therefore, observations must be carried out mostly from balloons or space observatories. Powerful gamma rays can, however, be detected by the large cascade of ionized particles and electromagnetic radiation they produce, and the study of cosmic rays is a rapidly expanding branch of astronomy (using Cherenkov arrays).

In this book we only consider optical and near infrared images observed by means of ground-based telescopes. The path of the electromagnetic radiation emitted from a source up to the detector, passes through the Earth's atmosphere and the telescope optics. Astronomical imaging and microscopic imaging have many aspects in common even if they treat images with completely different scales.

2.3.1 Electromagnetic sources

A *black body* is an object that absorbs all the light falling on it, with no reflecting power. Moreover, a black body in thermal equilibrium, i.e. at constant temperature, emits light very efficiently without any gaps or breaks in the brightness, with power and spectrum depending on its temperature (thermal spectrum). Though no object is a perfect black body, most stars, planets, moons and asteroids are near enough to being black bodies, so that they produce spectra very similar to a perfect thermal spectrum. The wavelength corresponding to the peak of the spectrum is given by the *Wien displacement law*, i.e. $\lambda_{\max} = \frac{b}{T}$, where T is the temperature and b a constant of proportionality, called the Wien displacement constant and approximately given by 2900 μm K, if the wavelength is measured in micrometers and the temperature in Kelvin.

This concept explains why there are different colors of stars. Red stars are cooler, and they emit most radiation in the red wavelengths. A hotter star like our sun emits most radiation in the yellow/green part of the spectrum. We don't see green stars because stars with peak wavelength in the green also emit a lot of radiation in the red and blue part of the spectrum. Our eyes combine all of these colors and we see white in this case. Even hotter stars and other objects emit most radiation in the blue, ultraviolet or even x-ray and gamma-ray part of the spectrum. Objects like these appear blue to our eyes. Much cooler objects like planets and humans emit most radiation in the infrared while cooler objects emit microwaves and radio waves.

Modern optical telescopes observe light sources in different bands using pass-band filters. Filters are named by letters defining the central wavelength and the bandwidth:

	λ_{eff}	$\Delta\,\lambda$
Ultraviolet		
U	365 nm	66 nm
Visible		
B	445 nm	94 nm
V	551 nm	88 nm
R	658 nm	138 nm
Near infrared		
J	1220 nm	213 nm
H	1630 nm	307 nm
K	2190 nm	390 nm
L	3450 nm	472 nm
M	4750 nm	460 nm

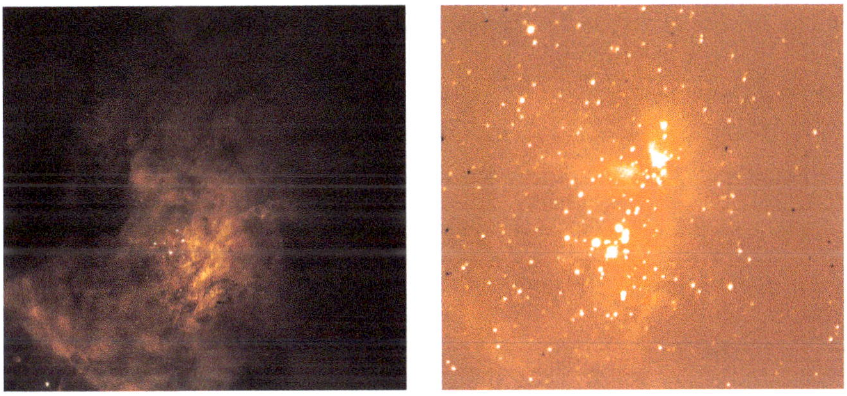

Figure 2.20. Two images of the Orion nebula observed at 650 nm (left panel) and 2500 nm (right panel). Since the extended dust clouds present in the nebula are not transparent at visible wavelengths while are transparent at microwave wavelengths, in the second image many more stars are visible.

In figure 2.20 we show two images of the Orion Nebula observed at different wavelengths, demonstrating the importance of observations at different wavelengths.

Another important parameter characterizing a celestial source is its *magnitude*. The apparent magnitude is a measure of its photon flux detectable from the Earth. The magnitude value is corrected in such a way to obtain the brightness that the object would have if the Earth were devoid of atmosphere. The definition is such that the lower the brightness of the celestial object the larger is its magnitude. Generally, magnitude is measured in the visible domain but other regions of the electromagnetic spectrum can be used. Indeed, a celestial source may emit electromagnetic radiation on a wide spectrum, with different intensities in different bands, so that it may be useful to define a magnitude for each band.

The apparent magnitude m_b for the observation band b is defined by:

$$m_b = -2.5 \log_{10} F_b + C \tag{2.16}$$

where F_b is the observed flux density with a filter b and C is the reference flux (zero-point) for that filter.

However, since an extremely luminous object may appear very weak if it is at a great distance, the apparent magnitude does not indicate the intrinsic luminosity of the celestial object. Actually the flux decreases with the square of the distance. This is expressed with the concept of absolute magnitude, i.e. the apparent magnitude of the object if it were at a distance of 10 parsec (1 parsec=3.26 light-years). In order to compute the absolute magnitude M given the apparent magnitude m, it is necessary to remember that the brightness of an object is inversely proportional to the square of its distance. It follows that the difference between the apparent magnitude m and the absolute magnitude M of an object will be expressed by the following formula:

$$M - m = -5 \log_{10}\left(\frac{d}{10}\right) \tag{2.17}$$

where d is the object distance (expressed in parsec). We are, in fact, comparing the brightness of the object in its real position with that it would have if it were 10 parsecs away.

2.3.2 Optical telescope and diffraction limited imaging

An optical telescope is a telescope that collects and focuses light, mainly from the visible or near infrared part of the electromagnetic spectrum; it creates a magnified image for direct view, or for making a photograph, or for collecting data through electronic image detectors.

There are three primary types of optical telescope:

- refractors, which use lenses (dioptrics),
- reflectors, which use mirrors (catoptrics),
- catadioptric telescopes, which combine lenses and mirrors.

The light collecting power of a telescope and its ability to resolve small details are directly related to the diameter (or aperture) of its objective (the primary lens or mirror that collects and focuses the light). The larger the objective, the more light the telescope collects and the finer details it resolves.

For this reason, nowadays all the telescopes used for scientific observations are reflectors which consist of a concave large mirror (the primary), defining the aperture of the telescope and focusing the radiation coming from the sky in a plane located above the primary. A second mirror (the secondary) is located before (Cassegrain telescope with concave mirror) or after (Gregorian telescope with convex mirror) the focal plane of the primary and reflects back the radiation with a focal plane located below the primary. For this reason the primary has a central hole allowing the passage of the radiation. The detector is located in the focal plane of the secondary.

A telescope, regardless of the details of its structure, contains an optical element whose pupil limits the beam of the incident light. It defines the aperture of the telescope which is the diameter of the light collecting region, assuming that this region has a circular geometry. For a refracting telescope, the aperture is the diameter of the objective lens while for a reflecting telescope it is the diameter of the primary mirror. For ground-based telescopes, increasing the aperture is often the easiest way to improve observations of faint objects. However, larger telescopes become more susceptible to the small-scale fluctuations (turbulence or seeing) in the Earth's atmosphere.

Using Fourier optics [9], the image of the pupil of the telescope in the image plane can be modeled as the Fourier transform of the function which is 1 over the pupil and 0 elsewhere. However, we must take into account that the typical emission processes that regulate the production of radiation in an astronomical object, generate the electromagnetic field in distinct points with statistically independent phase relationships. This reason, together with the hypothesis that the object has angular dimensions greater than the ratio between the observation wavelength and the telescope diameter [9], allows an extended source to be considered as spatially incoherent. The property allows the intensity distribution in the image plane to be calculated as the sum of the intensities produced by the individual points of the object. Therefore, the image of a point source is the modulus square of the Fourier transform indicated above.

This image can be called the PSF of the telescope in the case where the image of a point source is invariant by translation (isoplanatic system). In reality, an optical system hardly satisfies this property over its entire field-of-view; however, it is always possible to divide the object's field into sufficiently small regions (isoplanatic zones) in which the PSF can be considered invariant with respect to translation.

In the case of perfect optics of the system and absence of external disturbances (for example atmospheric turbulence), the PSF of a circular pupil (as in the case of microscopy) has a circular symmetry and is given by

$$h(r) = \left(\frac{2J_1\left(\frac{\pi D r}{\lambda}\right)}{\frac{\pi D r}{\lambda}} \right)^2 \tag{2.18}$$

where D is the aperture of the telescope as defined above, λ the central wavelength of the observation band and r the distance of a point of the image plane from the optical axis of the telescope. When these hypotheses are satisfied we say that the system is diffraction limited.

Refractor telescopes produce images with a great clarity and for this reason they are used by amateur astronomers who observe the elusive details of the planets, often not very contrasted. The dimensions of the optical telescopes range from a few centimeters in diameter for amateur low-end telescopes to several meters for the large telescopes of astronomical observatories. The large pupils are the undisputed domain of reflecting telescopes. Beyond a certain size, in fact, the lenses become so expensive and heavy that their use is technically and economically impracticable.

In general, it is possible to compute the PSF, corresponding to each one of the different pupils shown in figure 2.21 and corresponding to existing telescopes, just by computing the corresponding Fourier transform. Point spread functions (PSFs) corresponding to different shaped pupils are shown in figure 2.22.

The effect of diffraction blurring in Fourier domain is given by the pupil's autocorrelation function, i.e. the optical transfer function (OTF), that, in the circular case, is given by the Fourier transform of $h(r)$, defined in equation (2.18)

$$\hat{h}(\omega_n) = \frac{2}{\pi}\left(\arccos \omega_n - \omega_n\sqrt{1 - \omega_n^2}\right) \tag{2.19}$$

where $\omega_n = 2\pi\frac{\lambda\nu}{D}$, ν being the angular space frequency, λ the wavelength and D the pupil diameter.

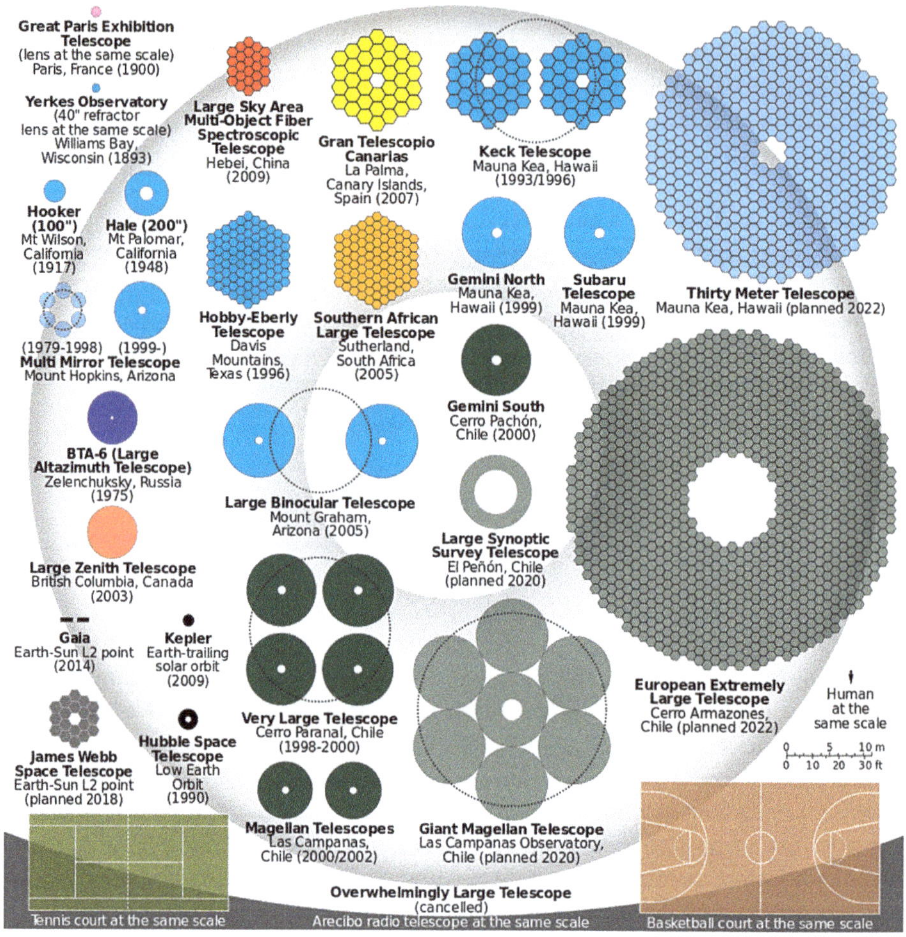

Figure 2.21. Diameter and shape of the primary mirrors of the largest ground-based telescopes. This El telescopio mas grande del mundo image has been obtained by the authors from the Wikimedia website where it was made available [by Cmglee] under a CC BY-SA 3.0 licence. It is included within this article on that basis. It is attributed to Cmglee.

2.3.3 Resolution limit

As in microscopy, one can use the Rayleigh resolution limit for discriminating two point sources coming from different angles, i.e. the two point-like sources cannot be resolved if their incidence directions are separated by an angle smaller than this limit. A large optical telescope may have a great angular resolution, but astronomical seeing and other atmospheric effects greatly reduce this limit: as a consequence a very large telescope may have the same resolution as a mirror with a diameter of few decimeters. Obviously it has much more sensitivity thanks to its large collection area.

The Rayleigh resolution or diffraction limit, R, of a telescope can usually be approximated by

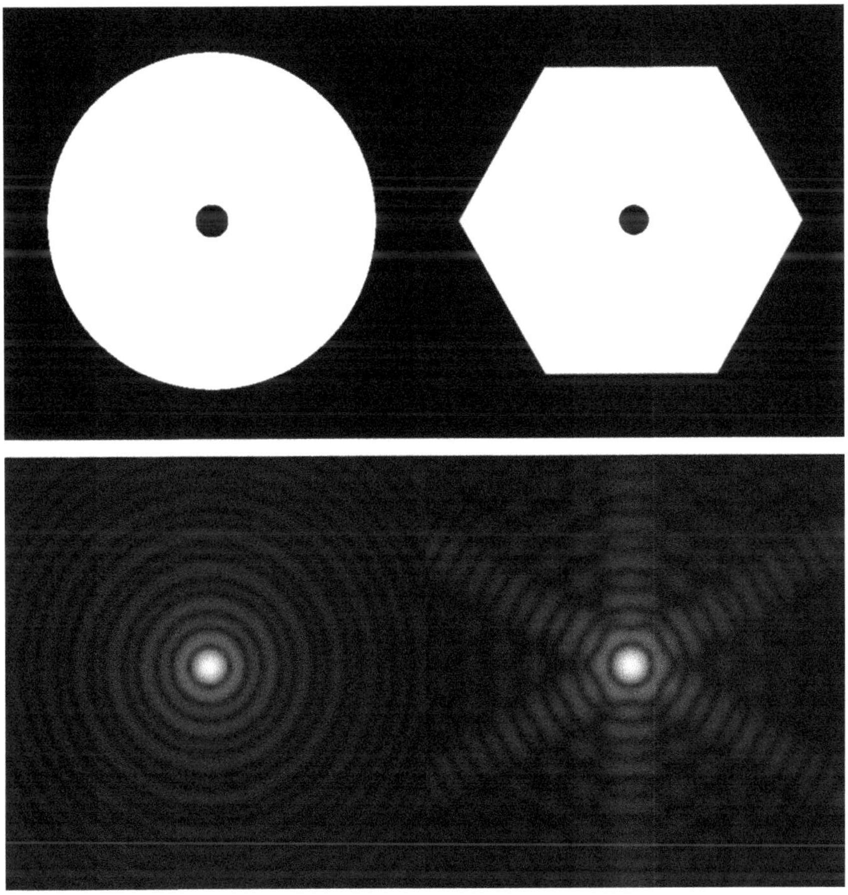

Figure 2.22. Comparison of the point spread functions (PSFs) of two differently shaped primary mirrors. The top row shows the mirrors and bottom row shows the cores of the corresponding PSFs, represented with a square-root gray scale.

$$R = 1.22\frac{\lambda}{D} \tag{2.20}$$

where D is the diameter of the telescope and λ the central wavelength of the observation band. The resolution of a telescope in equation (2.20) is expressed in *radians* and, usually, astronomers measure it in *arc-seconds*.

The Rayleigh criterion explains why, in astronomy, it is important to manufacture very large mirrors; in this way it is not only possible to collect much more light from faint sources (proportional to D^2) but it is also possible to improve angular resolution (inversely proportional to D) if the telescope is equipped with a good adaptive optic (AO) system (see section 2.3.4). It is also important to remark that resolution is proportional to the wavelength; therefore, it is better in the visible range than in the infrared range, even if infrared observation can provide more

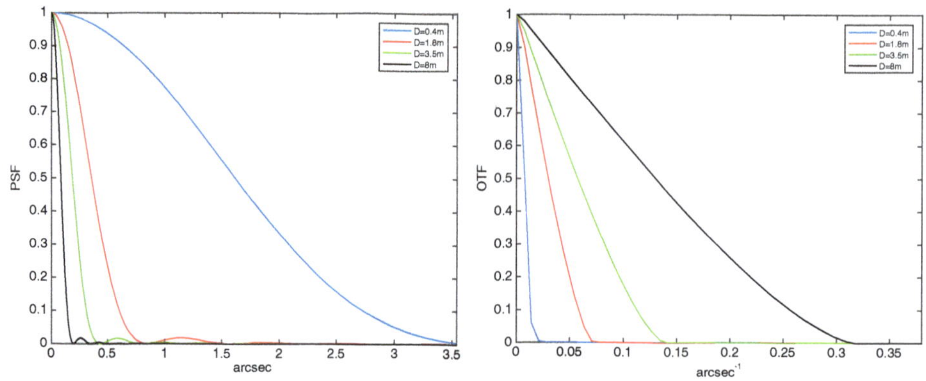

Figure 2.23. PSFs on the left and OTFs on the right, for different diameters of circular mirrors of a telescope, the diameter values ranging from 0.4 m up to 8 m at a wavelength of 2200 nm.

information on the physical processes characterizing the source. These limits can be reached by correcting the effects of atmospheric turbulence.

In figure 2.23 the PSFs and the corresponding OTFs are plotted in the case of telescopes with circular mirrors of different diameters, ranging from 0.4 m up to 8 m; the PSFs are calculated at a wavelength of 2200 nm.

2.3.4 Atmospheric turbulence

In spite of the increase in the diameter of the pupils of the telescopes and of the continuous technological progress in the quality of optics and detectors, Earth based astronomical observations inevitably suffer from the degradation due to the presence of the atmosphere. The turbulence present in its densest layers (up to a height of about 20 km) causes random variations of the optical paths making stars twinkle; moreover, by zooming on a star we see that the short-exposure images are affected by speckle.

The integration of many realizations of turbulence, necessary to increase the signal-to-noise ratio due to the small number of photons coming from celestial sources, causes a loss of components with high spatial frequency present in the image. The net result is that the limit in the resolution of terrestrial telescopes is about 1 s of arc, regardless of the diameter of the telescope (as long as it is larger than some decimeters), for images acquired with long exposures. This effect is called *seeing*.

AO is a technique to compensate in real time the wavefront deformations introduced by the Earth's atmosphere [2]. Compensation is possible by inserting into the optical beam of the telescope a device (typically a deformable mirror) that is capable of introducing a distortion of the wavefront equal and opposite to that generated by atmospheric turbulence. This correction must be performed in a fast cycle in order to track in real time the temporal evolution of the turbulence itself [19]. The scheme of an AO system is shown in figure 2.24. In figure 2.25 the image of a given sky region acquired with HST is compared with the image of the same region observed with one of the 8.4 m mirrors of the large binocular telescope. We recall that HST is equipped with a 2.5 m mirror and that, being a space telescope, it does not suffer for atmospheric turbulence.

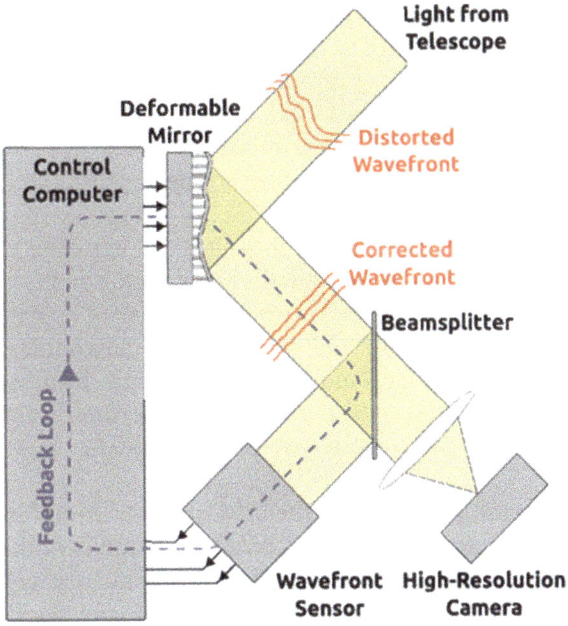

Figure 2.24. A typical scheme of an adaptive optics system. Reproduced from F Rigaut 2015 *Publ. Astron. Soc. Pacif.* 127 958. © 2015. The Astronomical Society of the Pacific. All rights reserved. Printed in U.S.A.

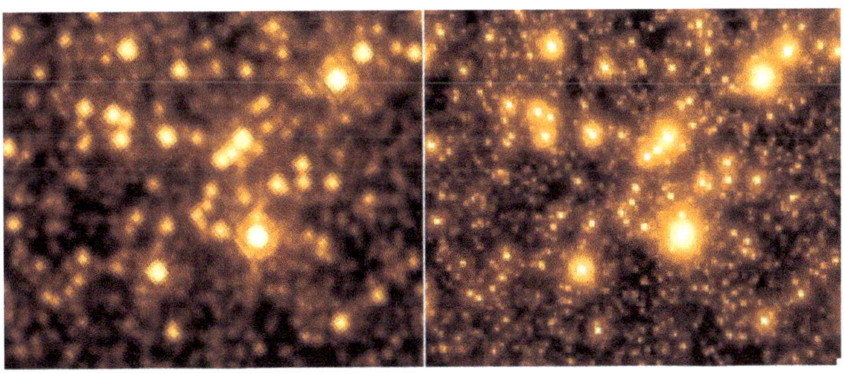

Figure 2.25. A central region of the globular cluster M92 at 1.6 μm as observed with the Hubble space telescope (left) and the LBT (large binocular telescope) in adaptive mode (right). It is immediately clear that the resolution and depth achieved with LBT surpasses those of the Hubble image (courtesy of LBTO). Reproduced from http://www.media.inaf.it/2012/01/23/ottica-adattiva-sotto-i-riflettori/ Courtesy of Media INAF.

2.3.4.1 Large binocular telescope (LBT)

LBT is located at about 3300 m on Mount Graham, Arizona. The choice of location caused considerable controversy from environmentalists and the San Carlos Apache Tribe, who claimed that the mountain is sacred.

LBT is a very innovative telescope, consisting of two 8.4 m primary mirrors on the same alt-azimuth mount with a center-to-center distance of 14.4 m; the two secondary mirrors are already equipped with the best existing AO system. Moreover, the particular structure of LBT allows observation modes not possible with traditional telescopes. One of these modes is based on Fizeau interferometry thanks to an instrument called an LBT interferometer (LBTI).

LBTI combines the light from the twin telescope mirrors to make high-resolution measurements of stars and galaxies and measures the emission from dust orbiting nearby stars. LBTI studies the formation of solar systems and is capable of directly detecting giant planets outside our solar system. Because of its unique geometry and relatively direct optical path, the LBTI offers science capabilities that are different from other interferometers. By combining the beams from separate telescope mirrors, according to a special optical concept (Fizeau interferometry), it provides high-resolution images of faint objects over a wide field-of-view, in the 1–20 μm wavelength range.

Since resolution is not uniform over the field-of-view, several images of the same scientific object must be acquired with different orientations of the baseline (i.e. the line joining the centers of the two mirrors). Then they must be suitably processed in order to get a unique image with a uniform resolution over the field.

In figure 2.26 the PSF and the corresponding OTF of LBTI interferometer are shown. The images are calculated at a wavelength of 2200 nm. Both images are represented using a square-root gray scale.

2.3.5 Detectors and data reduction

An important fact to be taken into account when describing the acquisition of astronomical images is the corpuscular nature of light. A rigorous treatment of the problem would require a quantum approach, but for the purposes of this book, the semi-classic approach is sufficient. It is structured in two phases: the propagation of light is treated in a classic way while the corpuscular aspect is introduced in the interaction radiation-detector.

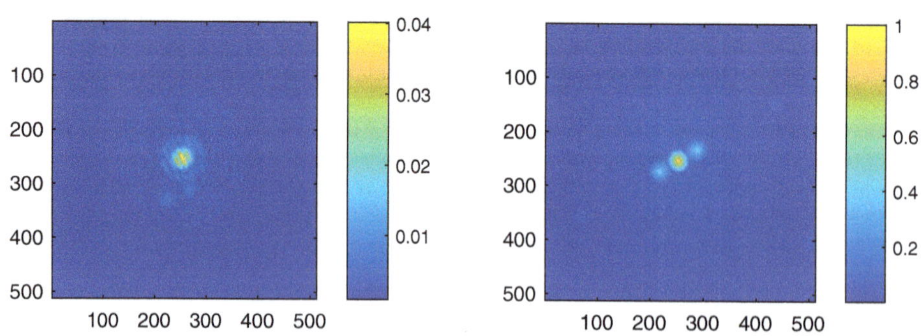

Figure 2.26. A frame of a clip video illustrating the PSF and the corresponding OTF of LBTI Fizeau interferometer. The images are calculated at a wavelength of 2200 nm. Both images are represented using a square-root gray scale. During the video one can observe the rotation of the baseline and the corresponding coverage of the frequency plane. Video available at http://iopscience.iop.org/book/978-0-7503-1437-4.

Most astronomical detectors in use today with telescopes of scientific observatories, as well as with several amateur telescopes, are cameras consisting of charged coupled devices (CCD). CCDs have revolutionized modern astronomy. They provided their first astronomical image in 1975 when scientists from the Jet Propulsion Laboratory imaged the planet Uranus at a wavelength of 890 nm (near infrared) using a prototype version of a CCD made by Texas Instruments Corporation as part of a development project for NASA spacecraft missions [12]. During the past 20 years, huge progress has been made in their manufacturing process allowing lower noise and best overall efficiency. In addition, larger format devices are produced and the read-out times are much shortened.

The usefulness of a detector is very often determined by the inherent amount of noise, i.e. the noise generated within the device itself. Indeed let us now look at the main sources of noise in a CCD, their importance and statistical distribution [20].

- Photon shot noise is inherently due to the photonic nature of the light incident on the detector. It has a Poisson distribution and often represents the main source of noise.
- Dark current shot noise derives from the quantization of the current generated in the CCD when the incident light is completely blocked. It has a Poisson distribution but it is not very important in the detectors used in astronomy, since these are adequately cooled.
- Fixed pattern noise represents the non-uniformity of the pixels (due to construction defects) detectable even in the absence of lighting.
- Photon response non-uniformity is similar to fixed pattern noise but concerns the characteristics of non-homogeneity of the response of the various pixels. Both are small in high quality CCDs used in astronomy.
- Read-out noise is electronic amplifier noise, usually quoted in electrons. It does not depend on exposure time. It is random with a Gaussian distribution.
- Quantization noise is due to the number of bits used for representing the datum of a single pixel.

Typical sources observed in astronomy are very weak, the night sky is not totally dark. Sky background increases with increasing wavelength. In order to obtain a clean image the following data reduction procedure is applied, starting from the raw data:

- acquisition of a set of short-time frames, to avoid the detector saturation, and for each frame:
    - removal of instrumental signatures, like dark current and field curvature
    - masking of unwanted signals, like cosmic rays, stellar halos and satellite tracks
- co-addition of corrected frames.

An example of this procedure is shown in figure 2.27: the calibration of an astronomical CCD image consists in removing the bias and thermal contribution (dark image, i.e. an image generated in the CCD when the incident light is completely blocked) and dividing the resultant image by the flat-field image

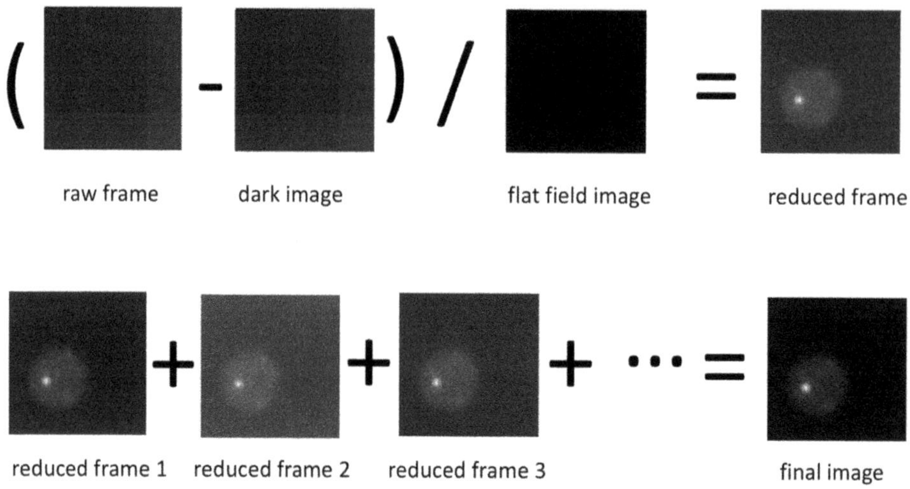

Figure 2.27. Data reduction scheme applied to IO (Jupiter's moon) frames, taken using LBTI the night of 24 Dec 2013 [6] (courtesy of LBTO).

(i.e. the image of a uniform object) in order to standardize the response of each image pixel.

The flat-field frame is a map of CCD sensitivity to light. It is an image of a uniform object such as twilight sky or a sheet of opal glass attached to the inside of the observatory dome. Chip sensitivity, vignetting and dust all appear as variations in the sensitivity of the CCD itself: division by flat-field image will remove these defects. The acquisition of the raw flat-field image itself must be corrected by dark image subtraction.

References

[1] Anger H O 1964 Scintillation camera with multichannel collimator *J. Nucl. Med.* **5** 515–31

[2] Beckers J M 1993 Adaptive optics for astronomy: principles, performance and applications *Annu. Rev. Astron. Astrophys.* **31** 13–62

[3] Bertero M, Boccacci P, Brakenhoff G J, Malfanti F and van der Voort H T M 1990 Three-dimensional image restoration and super-resolution in fluorescence confocal microscopy *J. Microsc.* **157** 3–20

[4] Born M and Wolf E 1999 *Principles of Optics* 7th edn (Cambridge: Cambridge University Press)

[5] Boulfelfel D, Rangayyan R M, Hahn L J and Kloiber R 1992 Use of the geometric mean of opposing planar projections in pre-reconstruction restoration of SPECT images *Phys. Med. Biol.* **37** 1915–29

[6] Conrad A *et al* 2015 Spatially Resolved M-band Emission from Io's Loki Patera Fizeau Imaging at the 22.8 m LBT *Astron. J.* **149** 175

[7] Diaspro A (ed) 2002 *Confocal and Two-Photon Microscopy: Foundations, Application and Advances* (New York: Wiley-Liss)

[8] Formiconi A R 1998 Geometrical response of multihole collimators *Phys. Med. Biol.* **43** 3359–79

[9] Goodman J W 1968 *Introduction to Fourier Optics* (New York: McGraw-Hill)
[10] Hell S W, Stelzer E H K, Lindek S and Cremer C 1994 Confocal microscopy with an increased detection aperture: type-b 4pi confocal microscopy *Opt. Lett.* **19** 222–4
[11] Hell S W and Wichmann J 1994 Breaking the diffraction resolution limit by stimulated emission: Stimulated-emission-depletion fluorescence microscopy *Opt. Lett.* **19** 780–2
[12] Howell S B 2006 *Handbook of CCD Astronomy. Cambridge Observing Handbooks for Research Astronomers* (Cambridge: Cambridge University Press)
[13] Huisken J, Swoger J, Del Bene F, Wittbrodt J and Stelzer E H K 2004 Optical sectioning deep inside live embryos by selective plane illumination microscopy *Science* **305** 1007–9
[14] Hutton B F 2014 The origins of SPECT and SPECT/CT *Eur. J. Nucl. Med. Mol. Imaging* **41** 3–16
[15] Klar T A and Hell S W 1999 Subdiffraction resolution in far-field fluorescence microscopy *Opt. Lett.* **24** 954–6
[16] Müller T, Schumann C and Kraegeloh A 2012 STED microscopy and its applications: new insights into cellular processes on the nanoscale *Chem. Phys. Chem.* **13** 1986–2000
[17] Neil M A A, Juskaitis R and Wilson T 1997 Method of obtaining optical sectioning by using structured light in a conventional microscope *Opt. Lett.* **22** 1905–7
[18] Nuyts J, De Man B, Fessler J A, Zbijewski W and Beekman F J 2013 Modelling the physics in iterative reconstruction for transmission computed tomography *Phys. Med. Biol.* **58** R63
[19] Rigaut F 2015 Astronomical adaptive optics *Publ. Astron. Soc. Pacif.* **127** 1197
[20] Romanishin W 2014 *An Introduction to Astronomical Photometry Using CCDs* (Scotts Valley, CA: CreateSpace Independent Publishing)
[21] Sheppard C J R, Xiason G, Min G and Maitreyee R 2006 Signal-to-noise ratio in confocal microscopes *Handbook of Biological Confocal Microscopy* 3rd edn ed J B Pawley (Berlin: Springer) pp 442–51
[22] Van Audenhaege K, Van Holen R, Vandenberghe S, Vanhove C, Metzler S D and Moore S C 2015 Review of SPECT collimator selection, optimization, and fabrication for clinical and preclinical imaging *Med. Phys.* **42** 4796–813
[23] Vicidomini G, Bianchini P and Diaspro A 2018 STED super-resolved microscopy *Nat. Methods* **15** 173

IOP Publishing

Inverse Imaging with Poisson Data

From cells to galaxies

Mario Bertero, Patrizia Boccacci and Valeria Ruggiero

Chapter 3

Mathematical modeling

In the previous chapter we discussed examples of very different imaging systems based on the detection of the electromagnetic radiation emitted by very different physical objects. These imaging systems provide indirect or degraded information about the source of the detected radiation. This information can be used to obtain, by data processing, an image of the unknown source (the object). To this purpose a mathematical model of the imaging process is required. The previous descriptions clarify that it may be quite hard to provide an accurate model. However, in practice, even a rather crude model can be sufficient. The crucial point is to introduce an operator which maps the set of the unknown sources into the set of the detected images. The reconstruction of the unknown sources requires the inversion of this operator and it follows that this inversion is an ill-posed problem in the sense of Hadamard. This is the first point discussed in this chapter.

The second point is the analysis of the detection system and data acquisition process. The detection has two main effects: data sampling and data noise. Thanks to the previous modeling, the first issue is strictly related to the well-known Wittaker–Shannon sampling theorem. The second issue requires the introduction of Poisson probability distribution since this is the probabilistic description of the main source of noise, namely photon counting. We also introduce the concept of observation time which is basic for discussing the noise level on the data.

3.1 Imaging system and forward problem

The imaging systems described in the previous chapter are based on the following principle: the object to be imaged emits electromagnetic radiation which, by means of physical devices, is conveyed to a detection system where it is measured; since the radiation is weak, the detection must be based on photon counting. In the source domain one can assume that the spatial (or angular) distribution of the sources (fluorocromes, radio-pharmaceuticals, astronomical objects, etc) is roughly proportional to the spatial (angular) distribution of the emitted radiation so that, if one is

doi:10.1088/2053-2563/aae109ch3

able to recover this distribution from the detected radiation, one gets an image of the source distribution. Therefore the key point is to find a model describing, with a sufficient approximation, the propagation of the electromagnetic radiation through the imaging system.

A model satisfying the previous requirement is usually called a *forward model*: given the source distribution, it allows one to compute the radiation arriving on the detection system. Since the detection of this radiation is, in several instances, described in terms of photons, a completely discrete model looks to be the most natural approach and this is the way followed by the majority of authors. However, in this way, a basic feature of the image reconstruction problem, which consists in estimating the sources from the detected radiation, does not clearly appear. This feature is numerical instability, a discrete manifestation of the *ill-posedness*, an infinite-dimensional phenomenon of the image reconstruction problem.

Therefore, in this chapter we first attempt a formulation of the forward model in an infinite-dimensional setting, i.e. in terms of functions; this approach leads to the introduction of the so-called *forward operator*. To this purpose, the use of a continuous description of the emitted radiation can be roughly justified by the fact that photons are bosons so that an arbitrary number of them can occupy the same quantum state; if their number is sufficiently large, their propagation can be described in terms of an electromagnetic wave. More precisely, a monochromatic electromagnetic wave if all the photons have the same energy. Another possible approach consists in representing photons in terms of particles and to this purpose the concept of intensity of a Poisson process will be introduced. This point will be discussed in chapter 9.

It is obvious that we are obliged to introduce other very crude approximations for describing parts of the physical system, for instance lenses or collimators or mirrors, so that the model can appear unrealistic or, if you prefer, very abstracted. Moreover, we also consider approximations leading to a linear forward operator (and in many cases to an affine-linear one). Linearity is important because it introduces a great simplification in the mathematical analysis of the ill-posedness of the inversion of the forward operator and, consequently, of the numerical instability of the inversion of the model deriving from the discretization of the infinite-dimensional one. Moreover, the latter provides a hint to the choice of appropriate data sampling.

The first step is to define input and output of the imaging system. As indicated above, the signal arriving at the entrance of the imaging system can be described by some electromagnetic radiation roughly proportional to the distribution of the emitting sources, with wavelengths depending on the particular problem. For instance wavelengths in the visible spectrum in the case of microscopy, in the visible and near-infrared spectrum in the case of astronomy, in the x-ray domain for transmission tomography and in the γ-ray spectrum in the case of SPECT and PET.

The input radiation is emitted by a source which is described by a density: in the case of fluorescence microscopy, the density of the fluorocromes in the cell volume; in the case of astronomy, the angular density of the sources in the sky; in the case of emission tomography, the density of the radio-isotopes metabolized by the human body. Then the purpose of the modeling of the imaging system is to establish a relationship between the density of the source, denoted by x and the intensity of the radiation arriving on the detection system, denoted by y. The two functions in general depend on variables defined on different domains, since the domains of source and image do not coincide and, in general, a one-to-one mapping cannot be established between their points. However, this mapping can be established in the first case we consider in the next section while, in the second case, the variables of object and image domain have a completely different meaning.

3.1.1 Microscopy and astronomy

Image formation in microscopy and astronomy is described by a mathematical model with the same structure if the imaging system is isoplanatic, i.e. the image of a point source does not depend on the location of the source. The main difference is that, while in astronomy images are always 2D, in microscopy they can be 2D or 3D. Indeed, in the case of optical sectioning (see section 2.1.2), one can obtain a volume image by putting together the different planar images obtained by focusing the microscope on different planes of the object volume.

In both applications an optical system is used, i.e. a device consisting of optical components (lenses, mirrors, beam splitters, etc). By applying Fourier optics and the scalar theory of diffraction it can be shown that there exists a linear relationship between the input and the output intensity [16].

Let us first consider **confocal microscopy** and let us assume magnification 1, so that the domain of the object, which is the density of fluorocromes, and that of the image are essentially the same subset S of 3D space; more precisely, thanks to geometrical optics one can establish a one-to-one correspondence between the points of the object domain and those of the image domain.

Let us introduce a variable $\vec{r}$, with components $\{r_1, r_2, r_3\}$, which is the position of a point of S in a coordinate system with the origin in the center of S and the r_3-axis coinciding with the optical axis of the system. The coordinate r_3 of a point is called its axial coordinate while r_1, r_2 are called its lateral coordinates. Then, let us indicate with $h(\vec{r})$ the image of a source located in the origin with a source given by a δ-distribution. The function $h(\vec{r})$ is the PSF of the imaging system discussed in section 2.1.2.1 (see figure 2.6); isoplanatism means that, if the same source is located in $\vec{r}'$, then its image is given by $h(\vec{r} - \vec{r}')$. Finally, if we take into account that the source with density x can be considered as the superposition of point sources with intensity $x(\vec{r}')d\vec{r}'$, thanks to the linearity of the system, we obtain that the image is given by

$$y(\vec{r}) = (Lx)(\vec{r}) \tag{3.1}$$

where L is a linear integral operator given by

$$(Lx)(\vec{r}) = \int_S h(\vec{r} - \vec{r}')x(\vec{r}')d\vec{r}', \ \vec{r} \in S; \tag{3.2}$$

therefore the input–output relationship is described by a convolution operator. If the source (object) to be imaged is completely contained in S, then we can extend the integral to $\mathbf{R}^q$ ($q = 3$ in the present case) by setting to zero the integrand outside S.

Given a function z, we recall that the Fourier transform (FT) of z is defined by

$$\hat{z}(\vec{\omega}) = \int_{\mathbf{R}^q} z(\vec{r})e^{-i<\vec{\omega},\vec{r}>}d\vec{r}, \tag{3.3}$$

where $< .,. >$ denotes the scalar product in $\mathbf{R}^q$. Then, thanks to the *convolution theorem*, the equation (3.1) is equivalent to the following relationship between Fourier transforms

$$\hat{y}(\vec{\omega}) = \hat{h}(\vec{\omega})\hat{x}(\vec{\omega}). \tag{3.4}$$

The model holds true also in the case of 2D images acquired with a confocal or STED microscope and the relationship between Fourier transforms is still valid if the object is completely contained into S so that the integral can be extended to $\mathbf{R}^q$, with $q = 2$.

The forward operator in **astronomy** is also described by a convolution operator in $\mathbf{R}^q$, with $q = 2$, if we associate to each direction in the sky the point in the image plane given by geometrical optics. In this way the forward operator is essentially the relationship between the image x given by geometrical optics and the image y given by physical optics, which takes into account wave propagation and diffraction effects. It is obvious that now $\vec{r}$ has only two components $\{r_1, r_2\}$ as in the case of planar microscopic images.

We recall the main properties of the PSF $h(\vec{r})$ in the 2D case. The extension to the 3D case is easy.

- The intensity PSF of an optical instrument is band-limited, i.e. its FT, also called OTF (see section 2.3.2), has a bounded support $\mathcal{B}$, called the *band* of the optical instrument. In the case of a microscope with circular lenses or a telescope with circular mirrors $\mathcal{B}$ is a disc with a radius proportional to the diameter of the lens/mirror and inversely proportional to the wavelength of the observed radiation (see equation (2.19)).

  Indeed, in such a case, the amplitude PSF is the FT (in suitably normalized variables) of the circular pupil of the instrument (we recall that we call pupil the planar region defined by the entrance of an optical instrument). Therefore, the intensity PSF, which is the square modulus of the amplitude PSF (see section 2.1.2.1), is also band-limited; its band $\mathcal{B}$ is also a disc with a radius which is twice the radius of the pupil. The analysis can be extended to the case of pupils with arbitrary shapes, by taking into account the relationship between the band of a function and that of its square modulus.

  For a simple demonstration of the band-limiting of the 3D intensity PSF in the case of confocal microscopy we refer, for instance, to [6].

  The property of band-limiting has important consequences since it implies that the PSF is an infinitely differentiable function, more precisely an analytic function, and, therefore, cannot have a bounded support. Anyway it can have lines (in the 2D case) or surfaces (in the 3D case) where it is zero. It implies also that the noise-free image is a band-limited and analytic function.

- The PSF is usually normalized as follows

$$\int_{\mathbf{R}^q} h(\vec{r})d\vec{r} = 1 \Rightarrow \int_{\mathbf{R}^q} y(\vec{r})d\vec{r} = \int_{\mathbf{R}^q} x(\vec{r})d\vec{r}, \tag{3.5}$$

so that this condition implies that the total intensity (or total flux) of the image coincides with that of the object. Moreover this normalization condition implies that the PSF can also be considered as a probability distribution.

Optical imaging systems are seldom isoplanatic over the whole image field, but it is usually possible to divide the field into regions within which the system is approximately space-invariant. Therefore, in all cases, even if the function x can be irregular (with jumps, point sources, complex structure, etc) its image y is very smooth. This fact has important consequences which are discussed in the next sections. The problem of space-variant PSF is discussed in section 8.5, taking into account the differences between microscopy and astronomy.

The PSFs described above are aberration free, can be called ideal PSFs and, in the case of circular pupils, are given by the Airy pattern discussed in the previous chapter. Different kinds of aberrations introduce, in general, modifications of the ideal PSFs. These can be due to aberrations in the optical components of the instruments. However, in the case of telescopes, the main source of aberration is atmospheric turbulence, while, in the case of microscopes, it is the presence of inhomogeneities in the refractive index of the specimen. The result is to produce an effective band considerably smaller than the band of the optical instrument. However, as we discussed in section 2.3.4, in modern ground-based telescopes a new technology, called *adaptive optics* (AO) [35], allows one to compensate, at least partially, for the effect of atmospheric turbulence; recent advances in this techniques allows one to get a PSF close to the diffraction limited one [15]. Mathematical problems motivated by AO technology are discussed in [13]. A similar technology has been introduced also in microscopy [8] for a partial compensation of the effect of spatial variations of the refractive index of the specimen.

The important concept of *resolution limit*, related to the shape of the PSF is already discussed in chapter 2. The recently published book of De Villers and Pike [12] is completely dedicated to the discussion of this concept and its application and extension to several domains of applied science. In section 2.1.2.1 we introduced the classic concept of *Rayleigh resolution limit*. Even if this criterion can be considered as a rule of thumb, based on perception by the human visual system, as also clearly stated by Rayleigh ([11], p 14), in section 3.3 we show that it has also a deep mathematical meaning with important practical consequences.

The model given in equation (3.2) is a linear one. However, in the following we will find that, in order to take advantage of the non-negativity of object and image, it is convenient to take into account an image perturbation which can be described as an image background; in this case the model becomes a linear affine one

$$y(\vec{r}) = (Lx)(\vec{r}) + b(\vec{r}), \tag{3.6}$$

where $b(\vec{r})$ is a known function (very often a given constant). In microscopy it can arise from auto-fluorescence, inadequate removal of fluorescent staining material, offset levels of the detector gain or other electronic sources [37]. Its accurate measurement is important for obtaining sensible image reconstruction [36]. In astronomy the background is due to both external radiation generated by sky emission and internal radiation generated by the detector [30]. It can be estimated by a preliminary processing of the detected images.

3.1.2 Emission tomography

We first consider an idealized model of SPECT imaging. As explained in section 2.2.3.1, the SPECT imaging system contains a physical collimator consisting of holes in a slab made of heavy material, the holes being aligned along parallel lines, each one defining a section of the body to be imaged. The effect of a hole is to collect radiation coming from a given direction. In practice, since the aperture of the hole is small but not zero and the thickness of the slab is finite, the acceptance region is a cone with a small aperture angle. We neglect this point (which must be taken into account in a discrete model) and we assume that each hole accepts only photons emitted by radio-isotopes lying on the straight line defined by the axis of the hole. Moreover, we neglect the aperture of the hole as well as the distance between holes forming a line of the collimator. In other words, we assume that the holes of the collimator are extremely close, narrow and long, so that each hole defines a straight line crossing the body.

If we also neglect scatter and absorption effects, then each hole collects all the photons arriving from the radio-isotopes located on its axis and this number is roughly proportional to the number of radio-isotopes located on this line; let x be their density.

To introduce the model, let us consider a plane orthogonal to the rotation axis of the system, identified by a line of holes of the collimator. Let us introduce in this plane a coordinate system, fixed with the body, with origin in the intersection of the plane with the rotation axis. For a given rotation angle ϕ of the collimator, we introduce a new coordinate system, with the same origin and axes defined by the couple of vectors $\vec{\theta}$, $\vec{\theta}^{\perp}$, where $\vec{\theta} = (\cos\phi, \sin\phi)$. We denote by s, t the coordinates of a point with respect to this system (see figure 3.1). Then, the points on the axis of a hole of the collimator have a fixed value of s. With these definitions the datum collected by the corresponding hole is given by

$$y_{\vec{\theta}}(s) = \int_{-\infty}^{+\infty} x(s\vec{\theta} + t\vec{\theta}^{\perp})dt. \tag{3.7}$$

This equation, with $\vec{\theta}$ fixed and s variable defines what is called a *projection in the direction* $\vec{\theta}$ of the density x. If x is constant within a disc with center in the origin, then the projections do not depend on $\vec{\theta}$.

The function of the two variables s, $\vec{\theta}$, defined by the previous equation, is called the *Radon transform* of x and given by

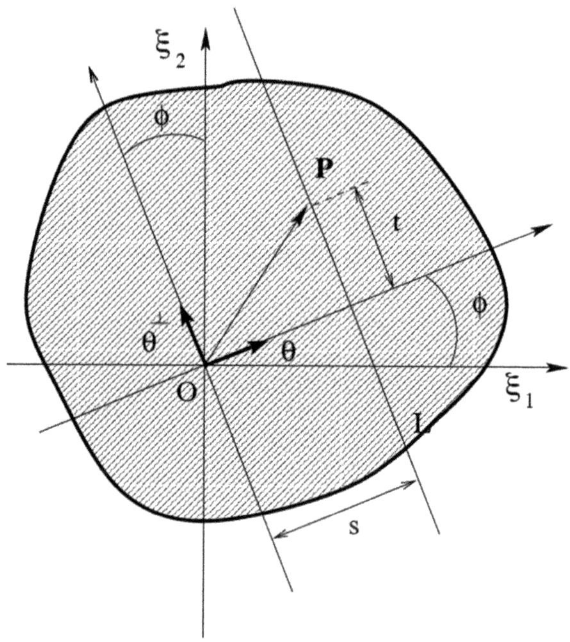

Figure 3.1. Coordinates used in the definition of a projection.

$$y(s, \vec{\theta}) = (Rx)(s, \vec{\theta}) = \int_{-\infty}^{+\infty} x(s\vec{\theta} + t\vec{\theta}^{\perp})dt, \tag{3.8}$$

where $y(s, \vec{\theta})$ is the data function and x the source distribution.

The Radon transform of x is related to the important concept of *sinogram*. This is the digital version of the Radon transform representation in the s, ϕ plane; its values are represented as gray levels and therefore, this is the image of a planar section of the body as provided by a SPECT scanner. If this section is contained in a disc with radius a then the sinogram is defined within the rectangle $|s| \leqslant a, \quad |\phi| \leqslant \pi$. A point in the sinogram domain corresponds to a straight line through the source and its value in that point is the value of the integral of the source density along the corresponding line. A picture of the sinogram of the section of a toy object, consisting of three disks, is shown in figure 3.2 and the picture explains the origin of the name sinogram.

Thanks to the symmetry of the Radon transform $(Ru)(s, \vec{\theta}) = (Ru)(-s, -\vec{\theta})$, it should be sufficient to consider only values of ϕ in the interval $[0, \pi)$. However, this symmetry does not hold in real scanners, as already discussed in section 2.2.3.1, because of attenuation and because a hole of the collimator accepts photons arriving from a cone and not from a straight line.

The main result, which is basic for the discussion of the next section, is the so-called *Fourier slice theorem* which states that the 1D Fourier transform of the projection of x in the direction $\vec{\theta}$ coincides with the 2D Fourier transform of x along the direction $\vec{\theta}$

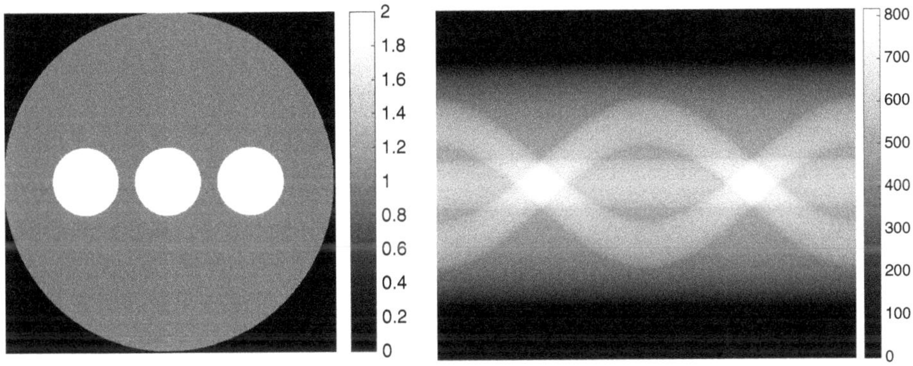

Figure 3.2. Example of sinogram. The source object (left panel) consists of three identical disks with uniform emission and the rotation axis is placed in the center of the figure. The corresponding sinogram (right panel), in the plane ϕ, s, consists of three bands corresponding to the three disks, showing the 'sinusoidal' behavior of the projections of the two lateral disks.

$$\hat{y}_{\vec{\theta}}(\omega) = \int_{-\infty}^{+\infty} y_{\vec{\theta}}(s)e^{-i\omega s}ds = \hat{x}(\omega\vec{\theta}). \tag{3.9}$$

This result provides the information content of one projection.

In the previous analysis we neglected the effects of scatter and attenuation. However in the modern SPECT–CT scanners, it is easy to compensate for the attenuation effect. The key point is the so-called *attenuated Radon transform* which, in terms of the previous notation can be written as follows

$$(R_{\mu}x)(s, \vec{\theta}) = \int_{-\infty}^{+\infty} e^{-\int_{t}^{+\infty} \mu(s\vec{\theta}+t'\vec{\theta}^{\perp})dt'} x(s\vec{\theta} + t\vec{\theta}^{\perp})dt, \tag{3.10}$$

where μ is the linear attenuation coefficient (see section 2.2.1). In figure 3.3 we show the effect of attenuation on the projection of a homogeneous cylinder with radius 1 cm and center in the origin of the coordinate system, so that all projections coincide; the projection computed without attenuation is compared with that where a constant attenuation is taken into account (the latter projection is computed by means of equation (3.10)).

As concerns PET scanners with detector rings separated by collimator septa, one ring defines a section of the body to be imaged. If we neglect again absorption and the other effects described in section 2.2.3.2, then each LOR (line of response) is a straight line through the body. Assuming very small detectors the response of each pair of detectors is basically the integral along the corresponding LOR of the source density. In this way we have a different parametrization of the straight lines through the body but, in principle, they can be rearranged to obtain a sinogram as in the case of SPECT so that we can apply the same mathematical analysis. A different model should be used if the septa are removed but the analysis of this case is beyond the scope of this book.

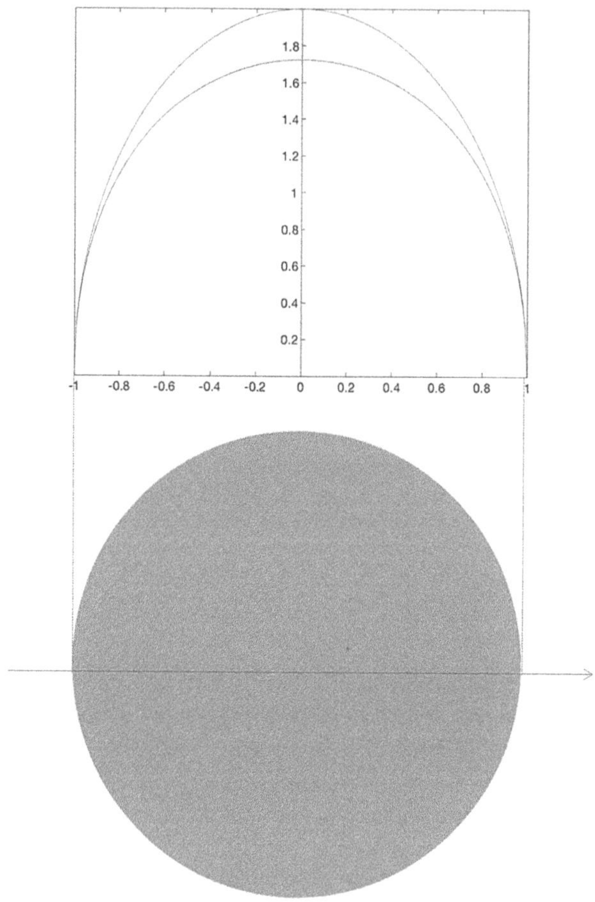

Figure 3.3. Effect of the attenuation on the projection of a uniform disc. In the upper panel, the green line is the projection in the absence of attenuation while the red line is the projection of the same disc with a constant attenuation.

3.2 Ill-posedness of the backward (inverse) problem

In general, the forward problem consists in computing the image y given the forward operator and the input object x. The operators L, R or R_μ are examples of forward operators. A feature, which is particularly evident in the case of a convolution operator, is that it transforms an irregular function x into a function y which is very smooth and this smoothing effect can be interpreted as a loss of information.

In the backward problem we have an exchange between the data and the unknown function: in the notations of the previous section, given y and the forward operator, determine x (or an approximation of x). The solution of this problem is given by the inverse of the forward operator, if the inverse operator exists; this is a particular case of a class of problems traditionally called *inverse problems*. A wide literature exists on them (see, for instance [5, 14, 19, 38]) and the methods for their approximate solution. Indeed, the main and specific feature of these problem is that they are *ill-posed* in the sense of Hadamard.

We recall that a problem is *well-posed* in the sense of Hadamard when its solution is unique, exists for any data function in a suitable space and depends continuously on the data with respect to the metric of the data space. It is obvious that the definition of well-posedness depends on the choice of the data space; we show in the following that, in the case of the inverse problems deriving from the forward models of the previous section, no sensible choice of data space exists such that the conditions of existence and continuity are satisfied. In other words the solution may not exist or may not depend continuously on the data. The last point means that, if for a given image y a solution exists, then a very small perturbation is sufficient for producing a new image y' for which no solution exists.

We first consider the inversion of the linear operator (3.2), a problem which is usually called **image deconvolution**. If we assume, for simplicity, that S coincides with $\mathbf{R}^2$ in the case of planar images or $\mathbf{R}^3$ in the case of volumetric images, then the relationship between the Fourier transforms of x and y is given by equation (3.4). Since we know that the PSF is band-limited with a band $\mathcal{B}$, then any non-zero function x, whose FT is zero on $\mathcal{B}$ and non-zero outside $\mathcal{B}$, produces an image which is zero everywhere (such a function is sometimes called an *invisible object*). We conclude that, given a band-limited image y, the solution of the inverse problem is not unique and the first condition of well-posedness is violated.

Anyway, assume that we only search for functions x whose FT is zero outside $\mathcal{B}$. Then, from equation (3.4), we can derive

$$\hat{x}(\vec{\omega}) = \begin{cases} \dfrac{\hat{y}(\vec{\omega})}{\hat{h}(\vec{\omega})}, & \vec{\omega} \in \mathcal{B} \\ 0, & \text{elsewhere} \end{cases} \tag{3.11}$$

for an arbitrary band-limited function y whose FT is zero outside $\mathcal{B}$. However, the PSF is integrable (it satisfies the normalization condition (3.5)) and therefore, thanks to the well-known Riemann–Lebesgue theorem, its FT is continuous and tends to zero for $|\vec{\omega}| \to \infty$. It follows that the PSF tends to zero at the boundary of $\mathcal{B}$. On the other hand, if y is a generic band-limited function, for instance an L^2 function (as one can assume by taking into account measurement errors), then $\hat{y}(\vec{\omega})$ can take non-zero values on the boundary of $\mathcal{B}$. In such a case, as follows from equation (3.11), $\hat{x}(\vec{\omega})$ may tend to ∞ at the boundary of $\mathcal{B}$, so that its inverse FT may not exist and the second condition of Hadamard is violated. The third condition is also violated thanks to general theorems of functional analysis since the previous result implies that the image of the operator L is not closed in the space of band-limited L^2 functions.

As concerns **Radon transform inversion**, from the Fourier slice theorem, equation (3.9), it is easy to derive an inversion formula by means of FT inversion in polar coordinates. The result is as follows

$$x(\vec{r}) = \frac{1}{2(2\pi)^2} \int_0^{2\pi} \left(\int_{-\infty}^{+\infty} |\omega| \hat{y}_\theta(\omega) e^{i\omega<\vec{\theta},\vec{r}>} d\omega \right) d\phi, \tag{3.12}$$

where $< .,. >$ denotes the scalar product in $\mathbf{R}^2$. For the convenience of the reader, in section 3.6.1 we give the derivation of this formula, which is basic for the so-called *filtered back-projection* (FBP), the inversion algorithm still frequently used in commercial machines [25]. Indeed the computation of the inversion formula can be performed in the following steps.

- Define the *filtered projections* as follows

$$G_{\vec{\theta}}(s) = \frac{1}{2\pi} \int_{-\infty}^{+\infty} |\omega| \hat{y}_{\vec{\theta}}(\omega)\, e^{i\omega s} d\omega, \tag{3.13}$$

  where $\hat{y}_{\vec{\theta}}$ is the 1D Fourier transform of the projection $y_{\vec{\theta}}$ defined in equation (3.7). An example of filtered projection is given in figure 3.4 where the projection of a uniform disc is compared with its filtered version; remark that the integral of a filtered projection is always zero.
- Define the back-projection operator as follows

$$(R^{\#}y)(\vec{r}) = \int_0^{2\pi} y(\langle \vec{r}, \vec{\theta} \rangle, \vec{\theta}) d\phi. \tag{3.14}$$

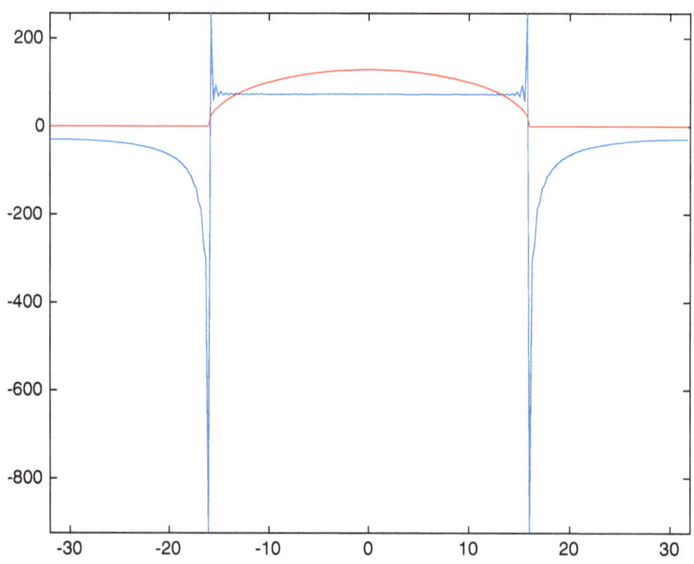

Figure 3.4. Comparison of the projection of a uniform and centered circle, before (red line) and after (blue line) filtering.

- The function x is obtained by applying the back-projection operator to the filtered projections

$$x = \frac{1}{4\pi} R^{\#} G \tag{3.15}$$

where $G(s, \vec{\theta}) = G_{\vec{\theta}}(s)$.

The ill-posedness of the Radon transform inversion is concentrated in the computation of the filtered projections which requires the differentiability of the projection y, a condition which is not satisfied if the projections are affected by experimental errors.

A way for overcoming this difficulty consists, for instance, in truncating the integral at a maximum frequency $\omega_{\max}$. This approach is equivalent to considering a band-limited version of the measured projections.

As concerns the inversion of the attenuated Radon transform, a simple and elegant formula is proved in [23].

Remark 3.1. *The previous simple analysis of ill-posedness does not apply to the case of bounded domains for image and source objects. However, in this case, the operators L and R are compact in suitable Hilbert spaces. Then the analysis can be performed in terms of singular functions of the compact operators. This analysis is provided to the interested reader in the supplementary material, section 3.6.4.*

3.3 Detection and data sampling

The band-limiting of the noise-free data function, which is a consequence of the properties of the imaging operators investigated in the previous section, has a beneficial effect when one must consider the problem of data sampling. Indeed, a basic property of a band-limited function is the possibility of its representation, without any loss of information, by means of its samples taken at equidistant points. This property is expressed by the so-called *sampling theorem* also called *Wittaker–Shannon theorem* [18].

Let us consider first the case of functions of one variable. Then, let y be a square-integrable function whose FT is zero outside the interval $[-\omega_{\max}, \omega_{\max}]$; this interval is the band $\mathcal{B}$ of the function and $\omega_{\max}$ is its *band-width*. Then, according to the sampling theorem, y can be represented as follows

$$y(r) = \sum_{k=-\infty}^{+\infty} y(r_k) \mathrm{sinc}\left[\frac{\omega_{\max}}{\pi}(r - r_k)\right], \tag{3.16}$$

where

$$\mathrm{sinc}(r) = \frac{\sin(r)}{r}, \quad r_k = \frac{\pi}{\omega_{\max}} k, \tag{3.17}$$

and the r_k are the sampling points. A simple derivation of the sampling theorem is given in section 3.6.2.

The distance between adjacent sampling points is given by $\delta_{samp} = \pi/\omega_{max}$ and is called *sampling* distance or also Nyquist distance. The Nyquist rate is the number of sampling points in the unit interval and therefore it is given by $\nu_{Nyquist} = 1/\delta_{samp} = \omega_{max}/\pi = 2\nu_{max}$, i.e. twice the maximum frequency associated with the function y.

A first important remark is that the sampling distance is strictly related to the Rayleigh criterion as derived from equation (2.18). Indeed, if in this equation we set $\omega_{max} = \pi D/\lambda$ we obtain that the radius of the first circle where the PSF is zero, which is precisely this criterion, is given by

$$\rho_1 = 1.22\frac{\lambda}{D} = 1.22\frac{\pi}{\omega_{max}}; \tag{3.18}$$

the additional factor 1.22 derives from the fact that this relationship is derived in the 2D case while in the 1D case equation (2.18) should be replaced by $h(r) = \mathrm{sinc}^2\left[\frac{\omega_{max}}{\pi}r\right]$.

The second important remark is that the sampling theorem does not hold if one uses a sampling distance $\delta' > \delta_{samp}$ but it holds true for any $\delta' \leqslant \delta_{samp}$. One says that in the first case the function is *under-sampled* while in the second case it is *over-sampled.* If the function is sampled at the Nyquist rate then the interpolation between the sampled values must be performed using the sampling functions appearing in equation (3.16) for obtaining a sensible representation of the function; but, if the function is over-sampled, then much simpler interpolation can be used and, in some cases, even a visual interpolation of the sampling points can provide a view of the function (remember that band-limited functions are quite smooth). For this reason in several situations the sampling theorem is used for designing a suitable over-sampling of the function (in general, 2–3 points within a sampling distance).

The extension to functions of two variables is easy if we assume that the support of the FT of the functions is interior to the square $\mathcal{B} = \{\vec{\omega} = \{\omega_1, \omega_2\} \,||\omega_1| \leqslant \omega_{max}, |\omega_2| \leqslant \omega_{max}\}$ Then, if $\vec{r} = \{r_1, r_2\}$ we have

$$y(\vec{r}) = \sum_{k,l=-\infty}^{+\infty} y(r_{1,k}, r_{2,l})\mathrm{sinc}\left[\frac{\omega_{max}}{\pi}(r_1 - r_{1,k})\right]\mathrm{sinc}\left[\frac{\omega_{max}}{\pi}(r_2 - r_{2,l})\right], \tag{3.19}$$

the sampling points being defined as in equation (3.17). In general, this expansion is not the most efficient one, in the sense that it does not require the minimum number of sampling points per area. More general sampling expansions can be used with sampling points forming a non-rectangular lattice [26]. The optimum choice depends on the shape of the band. However, this improvement, in general, is not required in imaging problems since data over-sampling may be desirable; moreover, a reduction of the number of sampling points implies also a reduction of the SNR.

In general, sampling is performed on a uniform grid in the image plane for 2D microscopic and astronomical imaging and in the image volume 3D for confocal imaging. If a CCD camera is used, each element of the camera defines a pixel of a 2D digital image. In the case of microscopy, as already mentioned in chapter 2, a 3D image is obtained by focusing the instrument at different depths and acquiring a 2D image for each depth by means of a scanning system. In confocal microscopy, the sampling rate is usually uniform along the two lateral directions, while, along the optical axis, it is smaller by about a factor 3, in order to accommodate the difference between lateral and axial resolution (see figure 2.6). As a consequence, a voxel can be viewed as a parallelepiped with a square basis. Finally, in emission tomography, if we focus on SPECT, a gamma camera should provide a sampling of the sinogram in agreement with the expected resolution in image reconstruction (based, for instance, on filtered back-projection).

3.4 Detection and data noise

A detection system consists of elements, such as photo-multipliers or CCD, each one covering a collecting surface. In a simplified model of the detection process, each element picks up an amount of energy coming from the imaging device in the form of an electromagnetic wave. If y represents the intensity of this wave, then the amount of energy which hits the detector element is given by the integral of y over the element surface, multiplied by the observation time τ. Finally, if ν is the frequency of the wave, then the previous energy divided by $h\,\nu (h$ = Planck's constant) is the number of photons arriving on the detector element (see also section 2.1.2.2).

More precisely, the previous quantity represents an average value of the number of detected photons. Indeed, this number is subject to fluctuations which are due to fluctuations in the emission-transmission process, quantum conversion process, etc. In other words, if we measure the same phenomenon in the same conditions but in subsequent time intervals, then, in each one of these intervals, the number of detected photons is not the same. This is the main source of *noise*, even if other sources are present due to intrinsic properties of the detection system or to defects of the detection system such as efficiency variation from one pixel to another one. A more accurate noise model in the case of a CCD is discussed at the end of this section.

We first consider the noise intrinsic to photon counting. This noise, which occurs also in electronic devices as a consequence of the discrete nature of electronic charge and is called *shot noise*, is also called *photon noise* in imaging. Its effect appears when the number of photons emitted by the source in a given time interval is not very large.

Assume that the following conditions are satisfied:

- the emission of one photon does not affect the probability of the emission of a second one;
- the rate at which photons are emitted is constant;
- two photons cannot be emitted at exactly the same time.

Under these conditions, if y_i is the number of photons captured by the ith detector with average value z_i, then y_i is the realization of an integer valued *Poisson random variable* (r.v.) Y_i with a probability distribution given by

$$\mathrm{P}_{Y_i}(y_i) = e^{-z_i}\frac{z_i^{y_i}}{y_i!}, \quad y_i = 0, 1, 2, \ldots. \tag{3.20}$$

In image reconstruction problems z_i must be derived from the model of the forward problem.

Let $\mathcal{I}$ be the index set characterizing the detector elements; then the discrete image $y = \{y_i\}_{i\in\mathcal{I}}$ is the realization of a multivariate r.v. (or random vector) $Y = \{Y_i\}_{i\in\mathcal{I}}$. Since, in general, it is reasonable to assume that the r.v.s Y_i are statistically independent, the probability distribution of y is given by

$$\mathrm{P}_Y(y) = \prod_{i\in\mathcal{I}} e^{-z_i}\frac{z_i^{y_i}}{y_i!}. \tag{3.21}$$

If z is given, then one can compute the probability of any possible realization of Y, i.e. the probability of any detected image corresponding to the noise-free image z. If, for a given value of i we have $z_i = 0$, then the corresponding random variable Y_i assumes the value 0 with probability 1.

We recall the main properties of a Poisson r.v.

- Mean and variance of Y_i coincide and are given by

$$\mu_i = E[Y_i] = z_i, \quad \sigma_i^2 = E[(Y_i - \mu_i)^2] = z_i, \tag{3.22}$$

  so that the standard deviation is given by $\sigma_i = \sqrt{z_i}$ and the relative standard deviation by $\sigma_i/\mu_i = 1/\sqrt{z_i}$. In other words, the fluctuations are large for large values of z_i while the relative fluctuations are small. Since the arrivals of photons occur at different times, it is obvious that, if we increase the time of observation in microscopy and astronomy or we increase the dose of radio-isotopes in emission tomography, we increase z_i and therefore we reduce the relative noise. This effect is illustrated in figure 3.5. The noise-free image (i.e. z), shown in the upper left panel, has the same morphology in the three cases of noisy images but, with respect to the first one (upper right panel) the two others are obtained after a rescaling of the test image respectively by a factor 5 and 10.
- The sum of two independent Poisson r.v. is again a Poisson r.v. with mean and variance which are the sum of the means and variances of the two r.v. It follows that, if we consider the sum of the r.v. associated to the different pixels, this is the total number of photons detected in the image domain and its mean and variance are given by

$$\mu = E\left[\sum_{i\in\mathcal{I}} Y_i\right] = \sum_{i\in\mathcal{I}} z_i, \quad \sigma^2 = E\left[\left(\sum_{i\in\mathcal{I}} Y_i - \mu\right)^2\right] = \sum_{i\in\mathcal{I}} z_i. \tag{3.23}$$

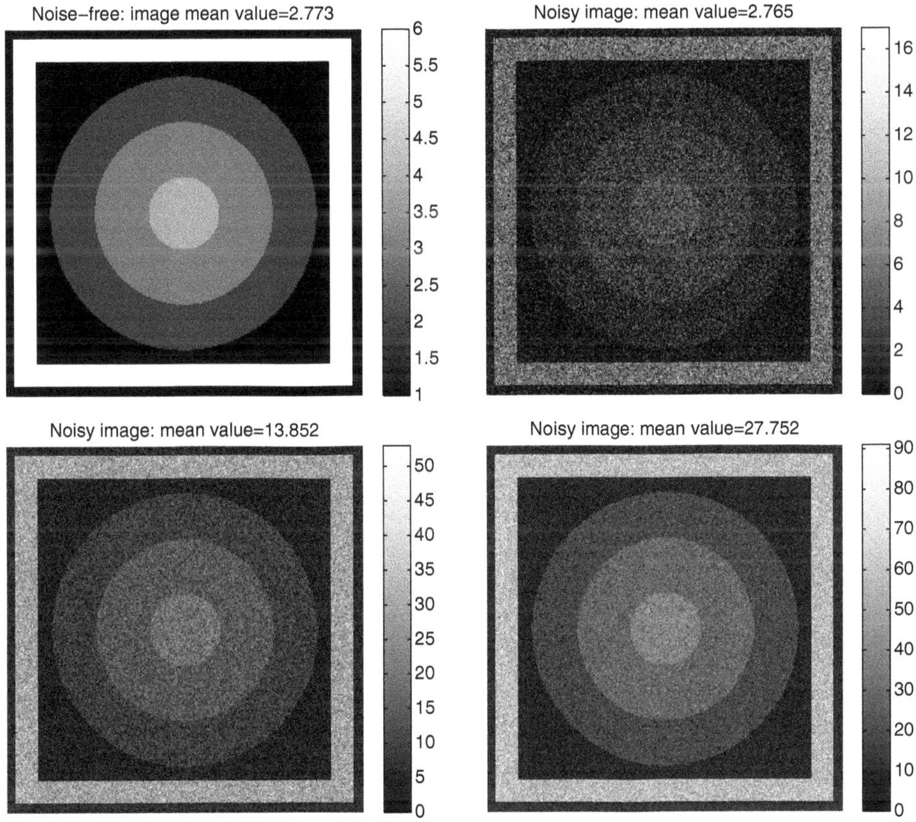

Figure 3.5. First row: left panel, a test image without noise; right panel, the test image corrupted by Poisson noise. Second row: left panel, the image corrupted by Poisson noise after a scaling by 5 of the test image; right panel, the image corrupted by Poisson noise after a scaling by 10 of the test image.

- For large values of z_i and large values of y_i, close to z_i in the sense that $|y_i - z_i|/z_i$ is small, using Stirling approximation of the factorial it can be shown that the Poisson distribution is approximated by a Gaussian distribution

$$e^{-z_i}\frac{z_i^{y_i}}{y_i!} \simeq \frac{1}{\sqrt{2\pi z_i}} e^{-\frac{(y_i - z_i)^2}{2z_i}}. \tag{3.24}$$

This approximation makes evident the difference between Poisson noise and additive Gaussian noise. Indeed, while the latter is signal independent the previous one depends on the signal. In figure 3.6 we show this difference by comparing the same test object corrupted with the two different kinds of noise. The test object is shown in the first row. In the left panel of the second row the test object is corrupted by additive Gaussian noise and by Poisson noise in the right panel. Finally, in the last row we show the difference between noisy and noise-free image. While in the case of additive Gaussian

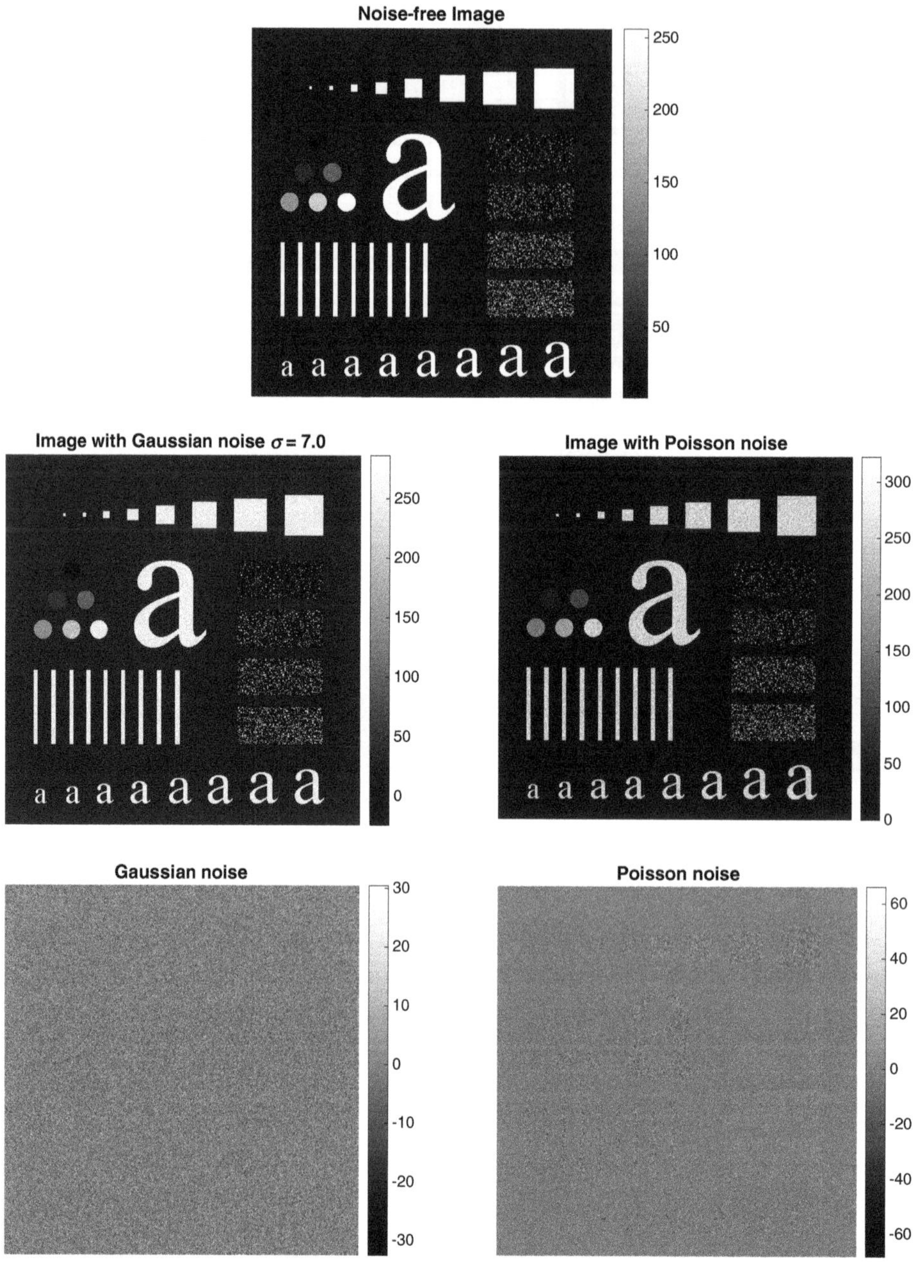

Figure 3.6. First row: the test object. Second row: the test object corrupted by additive Gaussian noise (left panel) and by Poisson noise (right panel). Third row: difference between noisy and noise-free image in the case of additive Gaussian noise (left panel) and in the case of Poisson noise (right panel).

noise this difference is independent of the noise-free image, in the case of Poisson noise the difference contains a reminiscence of the original image, i.e. the noise variance is not uniform over the image domain but is pixel-dependent. Therefore, even if, in the case of large number of photons, one could approximate the Poisson r.v. with a Gaussian r.v., its mean and variance are pixel-dependent and, in principle, unknown. This remark provides the basis of some approximations of the Poisson noise [2, 33] used in image processing, while a variance homogenization of the pixel values based on the Anscombe transform [1] is used for applying methods based on least-squares approaches [34].

- In the analysis of the results of image reconstruction methods it may be useful to consider the so-called **normalized residual**, defined as the difference between the noisy and the noise-free image divided by the standard deviation, i.e. the square root of the variance; more precisely

$$\text{Gauss:}\ R_i = \frac{y_i - z_i}{\sigma}, \quad \text{Poisson:}\ R_i = \frac{y_i - z_i}{\sqrt{z_i}}. \tag{3.25}$$

  As follows from the approximation given in equation (3.24), in both cases, one should obtain a map of white Gaussian noise with zero mean and variance 1. More precisely, in the case of Poisson noise this is true in the case of large numbers of photons. In figure 3.7 we show the maps of the normalized residuals (first line) and of the corresponding histograms (second line) in the case of the test object of figure 3.6.

The property of Poisson r.v. illustrated in figure 3.5 makes clear that it is important to introduce the concept of *observation time*. Indeed one can interpret the three test images as the result of three observations of the same object with different observation times, let us say 1 s for the first one, 5 and 10 s, respectively, for the two others. As the observation time increases, the absolute error in a given pixel, i.e. the difference between the noisy and the noise-free value, increases but the relative error decreases, so that the image becomes cleaner. Therefore, the explicit introduction of an observation time τ may be convenient for investigating the limit when the relative noise level tends to zero since it corresponds to τ tending to infinity.

To this purpose, for a given pixel i, one can introduce the Poisson r.v. $Y_i^{(\tau)}$ with parameter τz_i where z_i is the parameter corresponding to unit time. Since mean and variance are both given by τz_i, it follows that the relative standard deviation is given by $1/\sqrt{\tau z_i}$ and tends to zero as τ tends to infinity, even if the mean tends to infinity. To avoid this difficulty it is convenient to introduce the *temporally-normalized* Poisson r.v. [17] given by

$$\tilde{Y}_i^{(\tau)} = \frac{1}{\tau} Y_i^{(\tau)}. \tag{3.26}$$

Mean and variance of this r.v. are given, respectively, by z_i and z_i/τ so that the relative standard deviation is given by $1/\sqrt{\tau z_i}$. Then, when τ tends to infinity, the

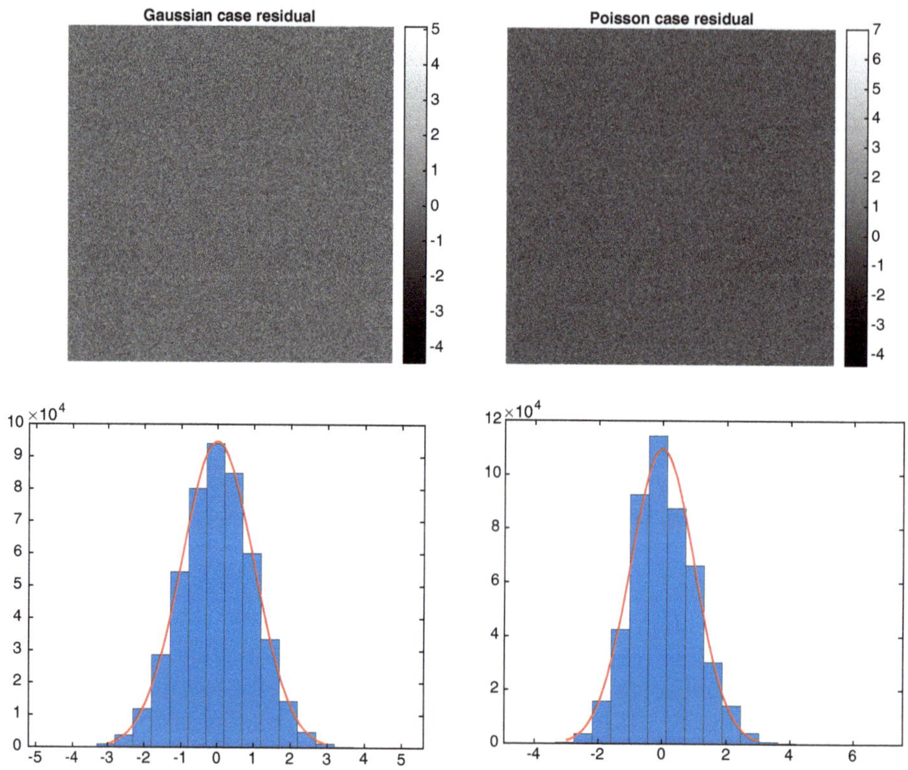

Figure 3.7. Upper line. Maps of the normalized residuals defined in equation (3.25): additive Gaussian noise (left panel) and Poisson noise (right panel). Lower line. Histograms of the two maps shown in the previous line. The test object used in this computation is shown in figure 3.6.

corresponding sequence of temporally-normalized Poisson r.v. concentrates around the mean value z_i; such a behavior can be interpreted as noise tending to zero.

Before concluding this section it is important to remark that Poisson noise is not the unique source of noise even if it is the most important one. As concerns photomultipliers the corrections to Poisson noise are discussed in [9]. As concerns a CCD an accurate description of noisy data is given in [30]. According to this model the detected data y_i are given by

$$y_i = y_i^{\mathrm{obj}} + y_i^{\mathrm{ext_{back}}} + y_i^{\mathrm{int_{back}}} + y_i^{\mathrm{dark}} + y_i^{\mathrm{bias}} + y_i^{\mathrm{ron}}, \tag{3.27}$$

where:

- y_i^{obj} is the number of photo-electron conversions due to the source radiation and it is Poisson distributed;
- $y_i^{\mathrm{ext_{back}}}$ is the number of photo-electrons due to external background and it is Poisson distributed;
- $y_i^{\mathrm{int_{back}}}$ is the number of photo-electrons due to internal background radiation from luminescent radiation on the CCD itself and it is Poisson distributed;

- y_i^{dark} is the number of thermo-electrons due to heat and it is Poisson distributed;
- y_i^{bias} is the number of electrons due to bias or fat zeroes and it is Poisson distributed;
- y_i^{ron} is the noise due to the amplifier through which the pixel value is read and it is a Gaussian random variable with mean m and variance σ^2; it is called *read-out-noise* (RON).

As we see, except the RON, all the other random variables are Poisson distributed. A mixed Poisson–Gaussian model can be used for dealing with these data [3, 4, 32]. However, the most simple RON compensation is obtained by adding the RON variance σ^2 to the data and the background and approximating the Gaussian distribution with a Poisson distribution with expected value σ^2 [31]. In this way data can be described by a unique Poisson distribution.

3.5 The discrete models

The sampling theorem provides a basis for the design of a detection system with a correct data sampling (and possible over-sampling) whenever the realization of this sampling can be obtained by means of the existing technology. For instance, in the case of a CCD camera possible limitations can be introduced by the size of a CCD.

For simplicity we consider the case where detection occurs in a plane. Then the sampling defines a set of sub-domains of the image domain S, let us say S_i, $i \in \mathcal{I}$, with $\cup_{i \in \mathcal{I}} S_i = S$. We also assume that each sub-domain is a small square defining what is called a *pixel* in microscopy and astronomy, and a *bin* in tomography. In the case of microscopy and tomography the volumetric image is obtained by combining different slices with a given distance between them. Then it may be convenient to introduce voxels by adding to the pixels of the slices a third dimension given by the distance between adjacent slices. Therefore, the index i may be an index or multi-index. It is a single integer number if pixels/voxels are numbered in a lexicographic order, but it can also be a pair of integer numbers (i_1, i_2) in the case of 2D images, or a triple of integer numbers (i_1, i_2, i_3) in the case of 3D images.

The purpose of the detectors is to collect and count (with a given efficiency) the photons which arrive from the source through the imaging system and hit the different detection elements. This process can be modeled if the data y_i are defined by integrating the continuous model given, for instance, in equation (3.6), over the area of the detector elements S_i. The inverse problem, i.e. the reconstruction of x given the data y_i, can also be treated in the framework of this semi-discrete model [7], i.e. discrete data y and source function x depending on continuous space variables. Since this approach is not widely considered in the case of Poisson data, in this book we mainly consider the completely discrete approach which is obtained by introducing a discrete representation of the unknown object. This approach requires a separate treatment of the cases of convolution and tomography.

- In microscopy and astronomy, if the PSF is space-invariant, data are modeled by a convolution operator (see equations (3.2) and (3.4)). In order to obtain the discrete version of equation (3.2) one can introduce this relationship into the definition of the discrete data by writing the integral over S as the sum of the integrals over the sub-domains S_j in the object plane

$$y_i = \int_{S_i} y(\vec{r})d\vec{r} = \int_{S_i} \left(\sum_{j\in\mathcal{J}} \int_{S_j} h(\vec{r} - \vec{r}\,')x(\vec{r}\,')d\vec{r}\,' \right) d\vec{r}. \tag{3.28}$$

Next, one can approximate the integrals over the (small) sub-domain S_i and S_j by the product of their measures $m(S_i)$ and $m(S_j)$, respectively, and the value of the integrands in their midpoints $\vec{r}_i$ and $\vec{r}_j$; one obtains

$$y_i = \sum_{j\in\mathcal{J}} H_{i,j}x_j; \; x_j = x(\vec{r}_j)m(S_j), \; H_{i,j} = h(\vec{r}_i - \vec{r}_j)m(S_i). \tag{3.29}$$

It is easy to recognize that, if we introduce a pair of indices for characterizing the pixels, then the matrix H is a block Toeplitz matrix with Toeplitz blocks. If both the object and the image are completely contained in the domain S, then the matrix can be replaced by a block circulant matrix with circulant blocks (for a discussion see, for instance, [5], section 2.7). In both cases matrix–vector product can be computed efficiently by means of the fast Fourier transform (FFT) (see, for instance, [38], chapter 5).

- As described in section 3.1.2 data acquired in emission tomography are the projections of a body section along lines identified by a projection angle and a distance from the rotation axis, see equation (3.7). Let us discretize the unknown body section with a grid of square pixels characterized by an index j which flows along the section in a lexicographical way; we can write x_j with $j = 1, \cdots, n$, n being the total number of pixels. On the other hand, the data acquired by the tomograph (raw data) are labeled by an index i which flows along the sinogram (one projection at a time) and we can write y_i with $i = 1, \cdots, Mm$ where m is the product between the number of lines per projections and the number of projections.

  Therefore, the forward problem is defined by a matrix H where the element $H_{i,j}$ provides the weight (contribution) of the unknown x_j to the data y_i. Indeed this matrix describes the tomograph and can be used to model many physical phenomena. In the simplest case H will contain only 1 and 0. If the value of $H_{i,j}$ is 1, it means that the cell x_j is on the ray which contributes to y_i. For further details see [20].

  This approach to the problem of reconstructing tomographic sections from their projections, allows us to take into account other physical phenomena such as attenuation, scattering and space-variant response of the system. In any case, the matrix H will be very sparse.

As already indicated, we use the index or multi-index j for labeling the pixels/voxels of the source object and we denote as $\mathcal{J}$ the set of its values. Moreover, we denote as

x_j the sampled values of the function x; then, if we take also into account the discretization of the background which appears in equation (3.6), in terms of the data y_i the discrete model of the forward problem is given by

$$y = H\,x + b, \tag{3.30}$$

where y is the vector with components y_i, x the vector with components x_j and b the vector with components b_i; all vectors x, y, b have non-negative components. In the following, for a generic vector h which has at least one positive component we will write $h \geqslant 0$ while, if all its components are positive, then we will write $h > 0$.

Since H and b are given, the solution of the inverse problem is reduced to the solution of the linear equation (3.30), i.e. compute x given y. However, the data y appearing at this stage of the analysis should be considered as exact data while, in practice, the measured data will provide approximations of the y_i. The approximation is due to several kinds of errors, the most important being that due to noise, discussed in the previous section. With an abuse of notation we denote also by y_i the real data and with these approximate data we can at most obtain an approximation of the x_j. Therefore, the basic problem, in this preliminary formulation, amounts to the solution of the linear system resulting from equation (3.30) with approximate data y.

As concerns the matrix H, which derives from the discretization of the linear operator, we assume that it satisfies the following conditions

$$H_{i,j} \geqslant 0; \ \sum_{i\in\mathcal{I}} H_{i,j} > 0, \ \forall j \in \mathcal{J}; \ \sum_{j\in\mathcal{J}} H_{i,j} > 0, \ \forall i \in \mathcal{I}; \tag{3.31}$$

in other words there exists at least one non-zero entry on each row and column of H.

In the following, in some instances, it is convenient to assume that the normalization conditions

$$\sum_{i\in\mathcal{I}} H_{i,j} = 1, \ \forall j \in \mathcal{J} \tag{3.32}$$

are satisfied. As stated in the seminal paper of Shepp and Vardi [29], these conditions do not involve a loss of generality of the results obtained by means of them, but make their interpretation simpler. For instance, if we set $y = Hx$, then, as one can easily verify, they imply equality of the ℓ_1 norms of x and y for x, $y \geqslant 0$. Then the interpretation of this relationship is that *the total number of photons is the same in the source object and in the corresponding image*, or also that *the source object and the image have the same total flux*; this is the physical meaning of the normalization conditions. Another important consequence is that the elements of each column can be interpreted as a probability distribution, i.e. $H_{i,j}$ is the probability that a photon, emitted in the pixel/voxel j of the object, reaches the pixel/voxel i of the detector.

If the imaging matrix does not satisfy the normalization condition, it can be introduced by a renormalization of the columns of the matrix. However, this procedure may not be convenient in some cases, for instance tomographic imaging. On the other hand, if the imaging matrix is approximated by the cyclic convolution of the object with a periodic and non-negative *point spread function* (PSF), $h = \{h_i\}_{i \in \mathcal{I}}$, i.e. $Hx = h * x$, then the normalization conditions are satisfied if

$$\sum_{i \in \mathcal{I}} h_i = 1, \tag{3.33}$$

a condition required in the everyday use of deconvolution in microscopy and astronomy.

Let us assume that $m = \#\{I\}$ and $n = \#\{J\}$ and that discrete image and object have been ordered in such a way that y is a vector of length m while x is a vector of length n; therefore $H \in \mathbf{R}^{m \times n}$. Let us remark that, in microscopy and astronomy, one can assume $m = n$, while in emission tomography one can have $m > n$ or $m < n$. We denote p the rank of H, $p \leqslant \min\{m, n\}$.

If H, y, b are known, the first point is to discuss the solution of the simple linear algebraic equation

$$H\,x = y - b. \tag{3.34}$$

A very simple discussion is provided by the system of singular values and vectors of H, i.e. the set of triples $\{\sigma_k, u^{(k)}, v^{(k)}\}_{k=1}^{p}$ which provide the representation (3.56) of H (see section 3.6.3). Then, if we indicate as $y_s = y - b$ the data after background subtraction, equation (3.34) becomes

$$Hx = \sum_{k=1}^{p} \sigma_k (x^T v^{(k)}) u^{(k)} = \sum_{k=1}^{p} (y_s^T u^{(k)}) u^{(k)} + y_s^{\perp} = y_s, \tag{3.35}$$

where $y_s^{\perp}$ denotes the component of y_s perpendicular to the range of H. This form of the linear equation makes evident the following properties which will be used in the following.

- The solution does not exist if $y_s^{\perp} \neq 0$.
- If $y_s^{\perp} = 0$ but $p < n$ the solution is not unique. On the other hand, if $p = n$, then there exists a unique solution given by

$$x^* = \sum_{k=1}^{p} \frac{y_s^T u^{(k)}}{\sigma_k} v^{(k)}. \tag{3.36}$$

The important point is the stability of the previous solution. Since the imaging matrix derives, in general, from the discretization of a compact operator whose singular values accumulate to zero (see section 3.6.4) it is obvious that, except in the case of a very coarse discretization, the last singular value σ_p is much smaller than the largest one, i.e. σ_1. The quantity

$$\kappa(H) = \frac{\sigma_1}{\sigma_p} \tag{3.37}$$

is called the *condition number* of the matrix H.

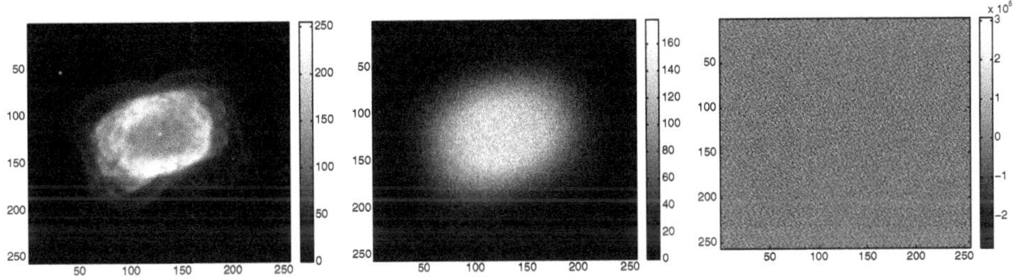

Figure 3.8. Left panel: the source object; middle panel: simulation of the detected image; right panel: the corresponding solution of the linear equation.

This number depends on the problem, on the data sampling and, in general, on the discretization of the problem: the finer the discretization, the larger the condition number. It can take extremely high values. Its relevance derives from the fact that, if δy_s is a small perturbation of the data y_s, then, as shown by a simple calculation (see, for instance, [5], p 82), the variation on the solution δx^* is bounded by

$$\frac{\|\delta x^*\|}{\|x^*\|} \leqslant \kappa(H)\frac{\|\delta y_s\|}{\|y_s\|}, \tag{3.38}$$

so that, even a very small variation of the data can produce an extremely large variation of the solution of the linear system. The previous inequality is optimal in the sense that equality can be reached.

Since our data are randomly distributed, it is obvious that it is quite difficult, in practice impossible, to obtain a reasonable solution by this approach. We can also conclude, with a reasonable certainty, that *the solution will never be non-negative*, a statement which will be used in the next chapter.

We show an example in figure 3.8, where the source object is the image, observed with the Hubble space telescope (HST), of the planetary nebula NGC7027, located around 3000 light-years away in the Cygnus constellation. The source object is convolved with an ideal PSF and perturbed by Poisson noise after addition of a constant background. While the values of the image are non-negative and smaller than about 170, the values of the solution of the linear equation lie approximately in the interval between $(-2.5 \times 10^5, 3 \times 10^5)$; moreover, no detail of the structure of the source object is visible.

Remark 3.2. *All the simulations presented in this book are obtained using the following procedure.*

- *The object is supposed to be limited in space and is convolved with the PSF, using FFT, in the case of microscopy or astronomy. For tomographic simulation the free-noise image is obtained by means of the projection operator, using the Matlab® radon function.*
- *The noise-free image, obtained from the previous step, is corrupted with Poisson noise using the Matlab® imnoise function. Through this operation each pixel value of the noise-free image is replaced by the value generated by a Poisson*

distribution with mean given by the previous pixel value. The noise level depends on the scaling of the object. (See the discussion in section 3.4.)

We conclude this section by quoting a famous and fundamental statement of Cornelius Lanczos [21], p 132: *a lack of information cannot be remedied by any mathematical trickery*. Indeed, the ill-conditioning of the imaging matrix derives from the ill-posedness of the imaging operator and this derives from the fact that some information is lost in the transmission of the image from the input to the output of the imaging system. This loss of information is very evident, for instance, in the case of microscopy or astronomy, where the output is given by the convolution product of the input with the PSF of the system which is band-limited. Hence the output image is very smooth even if the input object is very irregular. In other words, all the spatial frequencies of the object outside the band of the PSF are lost.

One can try to compensate this loss of information with some *a priori* information on the object to be reconstructed. This is the golden rule in the treatment of image reconstruction or, in general, of inverse problems. It is the basic point in the formulation of regularization methods, Bayesian methods, constrained solutions and so on. In particular, in the case of statistical methods, one can also introduce part of the information available in the statistics of the data. This is the approach considered in this book which is devoted to the image reconstruction methods based on the assumption that the data satisfy Poisson statistics.

3.6 Supplementary material

3.6.1 Filtered back-projection

In this section we derive the inversion formula of the Radon transform as given by the filtered back-projection, equation (3.12). The Fourier slice theorem, whose content is given in equation (3.9), shows a relationship between the 1D Fourier transform of a projection and the 2D Fourier transform of the unknown function x in polar coordinates. Therefore, the starting point is the FT inversion in this form

$$x(\vec{r}) = \frac{1}{(2\pi)^2} \int_0^{2\pi} \left(\int_0^{+\infty} \omega \hat{x}(\omega\vec{\theta}) e^{i\omega<\vec{\theta},\vec{r}>} d\omega \right) d\phi, \tag{3.39}$$

where, again, $\vec{\theta} = (\cos\phi, \sin\phi)$.

If we observe that $\hat{x}(\omega\vec{\theta})$ is a periodic function of ϕ with period 2π, by replacing ϕ with $\phi + \pi$, i.e. $\vec{\theta}$ with - $\vec{\theta}$, we obtain

$$x(\vec{r}) = \frac{1}{(2\pi)^2} \int_0^{2\pi} \left(\int_0^{+\infty} \omega \; \hat{x}(-\omega\vec{\theta}) e^{-i\omega<\vec{\theta},\vec{r}>} d\omega \right) d\phi. \tag{3.40}$$

From this expression, by means of a further change of variable which consists in replacing ω with $-\omega$, we get

$$x(\vec{r}) = \frac{1}{(2\pi)^2} \int_0^{2\pi} \left(\int_{-\infty}^{0} |\omega| \hat{x}(\omega\vec{\theta}) e^{i\omega<\vec{\theta},\vec{r}>} d\omega \right) d\phi. \tag{3.41}$$

By adding equations (3.39) and (3.41) and using the Fourier slice theorem, equation (3.9), we obtain equation (3.12).

3.6.2 Sampling theorem

In this section we provide a derivation of the sampling theorem in the case of functions of one variable and we discuss some of its properties.

Let $x(r)$ be a function whose FT has a support interior to the interval $\mathcal{B} = [-\omega_{\max}, \omega_{\max}]$; then from the inverse FT we obtain

$$x(r) = \frac{1}{2\pi} \int_{-\omega_{\max}}^{+\omega_{\max}} \hat{x}(\omega) e^{-ir\omega} d\omega. \tag{3.42}$$

If we assume that $\hat{x}$ is a square-integrable function of ω then x is an analytic square-integrable function of r.

We can represent $\hat{x}$ on the interval $[-\omega_{\max}, \omega_{\max}]$ by means of its Fourier series, convergent in $L^2(\mathcal{B})$, hence also in $L^1(\mathcal{B})$, i.e.

$$\hat{x}(\omega) = \sum_{k=-\infty}^{+\infty} \hat{x}_k e^{i\frac{\pi}{\omega_{\max}}k\omega}, \tag{3.43}$$

$$\hat{x}_k = \frac{1}{2\pi} \int_{-\omega_{\max}}^{+\omega_{\max}} \hat{x}(\omega) e^{-i\frac{\pi}{\omega_{\max}}k\omega} = x\left(\frac{\pi}{\omega_{\max}}k\right). \tag{3.44}$$

By inserting this series in equation (3.42), integrating term-by-term and taking into account that

$$\int_{-1}^{1} e^{i\omega r} d\omega = \frac{\sin r}{r}, \tag{3.45}$$

we obtain the sampling expansion (3.16), the series being uniformly convergent.

The interpolation between the sampling points r_k is performed by the sampling functions

$$S_{\omega_{\max}}(r - r_k) = \operatorname{sinc}\left[\frac{\omega_{\max}}{\pi}\left(r - \frac{\pi}{\omega_{\max}}k\right)\right], \tag{3.46}$$

which have very interesting properties. The function associated with the sampling point r_k is 1 at r_k and zero at the other sampling points $r_{k'}$, $k' \neq k$. Moreover the sampling functions associated with different sampling points are orthogonal in $L^2(\mathbf{R})$, the space of square-integrable functions

$$\int_{-\infty}^{+\infty} S_{\omega_{\max}}(r - r_k) S_{\omega_{\max}}(r - r_{k'}) = \frac{\pi}{\omega_{\max}} \delta_{kk'} \tag{3.47}$$

and they form an orthogonal basis in the subspace of the square-integrable band-limited functions with band-width ω_{max}. This property implies also the following formula for the scalar product of two functions x and z having the same band $\mathcal{B}$

$$\int_{-\infty}^{+\infty} x(r)z(r)dr = \sum_{k=-\infty}^{+\infty} x(r_k)z(r_k)\frac{\pi}{\omega_{max}}, \tag{3.48}$$

which shows that the scalar product of the two functions can be computed from their samples using the trapezoidal rule.

3.6.3 Singular value decomposition (SVD) of a matrix

Consider the imaging matrix $H \in \mathbf{R}^{m\times n}$ which is, in general, non-square, and, if it is square, non-symmetric. Therefore, a spectral representation of the matrix cannot be obtained in terms of eigenvalues and eigenvectors; the alternative is provided by the singular values and singular vectors and the corresponding representation, called *singular value decomposition* (SVD), is a very useful tool for discussing properties of the matrix.

To this purpose, let us consider the two matrices

$$\bar{H} = HH^T \in \mathbf{R}^{m\times m}, \;\; \tilde{H} = H^TH \in \mathbf{R}^{n\times n}. \tag{3.49}$$

The following result holds true.

Lemma 3.1. *The matrices $\bar{H}$, $\tilde{H}$ are symmetric and positive semi-definite. Moreover, the following relationships hold true between null spaces*

$$\mathcal{N}(\bar{H}) = \mathcal{N}(H^T), \;\; \mathcal{N}(\tilde{H}) = \mathcal{N}(H) \tag{3.50}$$

so that the two matrices have the same rank p of H.

Proof. It is obvious that the two matrices are symmetric as well as positive semi-definite. As concerns the relationship between null spaces, let us prove the second; the proof of the first is similar. If $Hv = 0$ then also $H^THv = 0$ so that $\mathcal{N}(H) \subset \mathcal{N}(\tilde{H})$. On the other hand, if $H^THv = 0$ then $0 = (H^THv)^Tv = (Hv)^T(Hv)$, hence $Hv = 0$ so that $\mathcal{N}(H) \subset \mathcal{N}(\tilde{H})$; the equality between the two sets follows. Since $\dim\mathcal{N}(H) = n - p$, if p is the rank of H, we obtain that H^TH has the same rank. □

The lemma implies that the two matrices can be diagonalized and their non-null eigenvalues are positive. Since both matrices have the same rank p, both matrices have exactly p positive eigenvalues, while the null eigenvalue has multiplicity $m - p$ for the matrix $\bar{H}$ and $n - p$ for the matrix $\tilde{H}$. If $m \neq n$, at least one of the two matrices has the null eigenvalue and precisely the matrix which has the largest dimension.

Proposition 3.1. *The matrices $\bar{H}$ and $\tilde{H}$ have exactly the same positive eigenvalues with the same multiplicity.*

Proof. Let us consider $\tilde{H}$ and let us order its positive eigenvalues, each one counted as many times as its multiplicity, in such a way that $\sigma_1^2 \geqslant \sigma_2^2 \geqslant \cdots \geqslant \sigma_p^2$. Moreover, let us denote as $v^{(k)}$ the eigenvector associated with σ_k^2, these eigenvectors being selected in such a way that $(v^{(k)})^T v^{(j)} = 0$ if $k \neq j$.

To each eigenvector $v^{(k)}$, of length n, let us associate a non-null vector $u^{(k)}$, of length m, given by

$$u^{(k)} = \frac{1}{\sigma_k} H v^{(k)}; \tag{3.51}$$

it is an eigenvector of $\hat{H}$ associated with the eigenvalue σ_k^2 as follows from the computation

$$HH^T u^{(k)} = \frac{1}{\sigma_k} H(H^T H) v^{(k)} = \sigma_k H v^{(k)} = \sigma_k^2 u^{(k)}. \tag{3.52}$$

These eigenvectors are linearly independent because they form an orthonormal system as follows from the computation

$$(u^{(k)})^T u^{(j)} = \frac{1}{\sigma_k \sigma_j} (H^T H v^{(k)})^T v^{(j)} = \frac{\sigma_k}{\sigma_j} (v^{(k)})^T v^{(j)} = \delta_{k,j}. \tag{3.53}$$

Since the matrix $\hat{H}$ has exactly p linearly independent eigenvectors, we have obtained, by construction, the set of these eigenvectors. □

From equation (3.51) and from the same equation multiplied by H^T we obtain the set of coupled equations

$$Hv^{(k)} = \sigma_k u^{(k)}, \quad H^T u^{(k)} = \sigma_k v^{(k)}, \tag{3.54}$$

which is called by Lanczos [21] *shifted eigenvalue problem.*

The positive numbers σ_k are called the *singular values* of the matrix H (or H^T) while the vectors $u^{(k)}$, $v^{(k)}$ are called its *singular vectors.* Sometimes the set of the triples $\{\sigma_k, u^{(k)}, v^{(k)}\}_{k=1}^p$ is called the *singular system* of the matrix. From our construction it follows that each singular value is counted as many times as its multiplicity and that they are ordered to form a non-increasing set $\sigma_1 \geqslant \sigma_2 \geqslant \cdots \geqslant \sigma_p$.

From equation (3.54) it follows that the vectors $v^{(k)}$ are orthogonal to the null space of H while the vectors $u^{(k)}$ are orthogonal to the null space of H^T. Therefore, the $v^{(k)}$ form an orthonormal basis in $\mathcal{R}(H^T) = \mathcal{N}(H)^\perp$ while the $u^{(k)}$ form a basis in $\mathcal{R}(H) = \mathcal{N}(H)^\perp$. It follows that

$$Hx = \sum_{k=1}^{p} ((Hx)^T u^{(k)}) u^{(k)}, \tag{3.55}$$

and, from the second of the shifted equations (3.54)

$$Hx = \sum_{k=1}^{p} \sigma_k (x^T v^{(k)}) u^{(k)}; \tag{3.56}$$

similarly

$$H^T y = \sum_{k=1}^{p} \sigma_k (y^T u^{(k)}) v^{(k)}. \tag{3.57}$$

If we introduce the $p \times p$ diagonal matrix Σ whose diagonal elements are the singular values σ_k, the isometric $m \times p$ matrix U whose columns are the vectors $u^{(k)}$ and the isometric $n \times p$ matrix V whose columns are the vectors $v^{(k)}$, then the previous equations can be written in the concise form

$$H = U^T \Sigma V, \quad H^T = V^T \Sigma U, \tag{3.58}$$

which is the usual form of SVD in numerical analysis.

3.6.4 SVD of a compact operator in Hilbert spaces

Let X, Y be separable Hilbert spaces and $L : X \to Y$ a compact operator from X in Y. As usual we denote as $L^* : Y \to X$ its adjoint operator. The operators $\tilde{L} = L^*L : X \to X$ and $\hat{L} = LL^* : Y \to Y$ are compact and selfadjoint operators in X and Y, respectively. Moreover, they are also positive semi-definite. The basic tool for introducing the SVD of a compact operator is the following spectral theorem for compact selfadjoint operators [28].

Proposition 3.2. *Let $T : X \to X$ be a selfadjoint compact operator in the separable Hilbert space X. Moreover, let us assume that T is not a finite rank operator. Then the following properties hold true:*

- *the eigenvalues of T are real;*
- *each eigenvalue has finite multiplicity;*
- *the sequence of the eigenvalues λ_k has 0 as unique accumulation point;*
- *by counting each eigenvalue as many times as its multiplicity, they can be ordered to form a sequence such that $|\lambda_1| \geqslant |\lambda_2| \geqslant \cdots$; moreover, one can add the eigenvalue $\lambda_0 = 0$ if the null space of the operator is not trivial; its multiplicity is possibly infinity;*
- *if one associates an eigenfunction u_k to each eigenvalue λ_k in such a way that they form an orthonormal system, then the set of all eigenfunctions of T, including those associated with 0, form an orthonormal basis in X;*
- *the following spectral representation holds true*

$$Tx = \sum_{k=1}^{\infty} \lambda_k < x, u_k >_X u_k, \tag{3.59}$$

where $<.,.>_X$ denotes the scalar product in X.

This result allows us to construct the singular system of the compact operators $L, \ \ L^*$ as in the case of a matrix. Indeed we can start with the eigenvalue problem of the operator $\tilde{L} = L^*L$

$$L^*Lv_k = \sigma_k^2 v_k. \tag{3.60}$$

The eigenfunctions v_k form an orthonormal basis in the orthogonal complement of $\mathcal{N}(L^*L) = \mathcal{N}(L)$, hence in $\mathcal{R}(L^*)$. Next, one can associate with each v_k a function $u_k \in Y$ through the shifted eigenvalue problem

$$Lv_k = \sigma_k u_k, \ \ L^*u_k = \sigma_k v_k, \tag{3.61}$$

and show that the u_k are eigenfunctions of LL^* associated with the eigenvalue σ_k^2 and that they are orthonormal. Therefore, they form an orthonormal system in Y. One can prove that they form a basis in $\mathcal{R}(L)$ by showing that a vector $y \in \mathcal{R}(L)$ orthogonal to the u_k is zero. Indeed, let $y = Lx$ and $0 = (Lx, u_k)_Y = (x, L^*u_k)_X = \sigma_k(x, v_k)_X, \ \ \forall k$; it follows that $x \in \mathcal{N}(L)$ so that $y = 0$. The singular value decomposition of the operators $L, \ \ L^*$ follows

$$Lx = \sum_{k=1}^{\infty} \sigma_k < x, v_k >_X u_k \tag{3.62}$$

$$L^*y = \sum_{k=1}^{\infty} \sigma_k < y, u_k >_Y v_k. \tag{3.63}$$

From these representations it follows that the ranges of $L, \ L^*$ are not closed in $Y, \ X$ respectively. Indeed, a function y belongs to the range of L if and only if it satisfies the following condition, called the *Picard condition* in honor of the paper of E Picard on the SVD of integral operators [27]

$$\sum_{k=1}^{\infty} \frac{|<y, u_k>_Y|^2}{\sigma_k^2} \leqslant \infty. \tag{3.64}$$

This condition implies that the problem of the inversion of the compact operator L from Y to X is ill-posed in the sense of Hadamard. Indeed, the problem of solving the linear equation $y = Lx$ does not have a solution for any $y \in Y$ but only for y satisfying the Picard condition.

We consider a couple of examples of compact operators in the framework of the modeling of section 3.1. The first is a convolution operator, as that defined in equation (3.2), with S a bounded domain of $\mathbf{R}^q$ (with $q = 2$ or 3) while the domain of y can be $\mathbf{R}^q$. Let us assume that the PSF, which satisfies condition (3.5), is also a function of $L^2(\mathbf{R}^q)$, then we have

$$\int_S \left(\int_{\mathbf{R}^q} |h(\vec{r} - \vec{r}\,')|^2 d\vec{r}\,' \right) d\vec{r} = m(S) \int_{\mathbf{R}^q} |h(\vec{r})|^2 d\vec{r} < \infty, \tag{3.65}$$

where $m(S)$ is the finite measure of S. Therefore, the convolution operator is a Hilbert–Schmidt integral operator (i.e. it belongs to the class of integral operators investigated by Picard [27]). This is a well-known class of compact operators with a finite trace $\mathrm{Tr}(L^*L)$, which is given by equation (3.65) and is related to the singular values by

$$\mathrm{Tr}(L^*L) = \sum_{k=1}^{\infty} \sigma_k^2, \tag{3.66}$$

so that the series of the squares of the singular values is convergent.

A second example is provided by the Radon transform defined in equation (3.8) when we consider a space of functions with a given bounded support. Without loss of generality we can assume that the support is interior to the disc of radius one. Then, if we set

$$w(s) = \sqrt{1 - s^2}, \tag{3.67}$$

the Radon transform of a function $x \in L^2(D)$, where D is the unit disc, can be written as follows

$$(Rx)(s, \vec{\theta}) = \int_{-w(s)}^{w(s)} x(s\vec{\theta} + t\vec{\theta}^{\perp})dt. \quad |s| \leqslant 1. \tag{3.68}$$

A simple property of the functions in the range of the operator R is obtained by applying Schwarz inequality to the integral in this equation; we obtain

$$|(Rx)(s, \vec{\theta})| \leqslant 2w(s) \int_{-w(s)}^{w(s)} |x(s\vec{\theta} + t\vec{\theta}^{\perp})|^2 dt, \tag{3.69}$$

so that

$$\int_{-1}^{1} w^{-1}(s)|(Rx)(s, \vec{\theta})|^2 \leqslant 2 \int_D |x(\vec{r})|^2 d\vec{r}. \tag{3.70}$$

This inequality implies that an appropriate space for the data functions $y = Rx$ is the weighted Hilbert space $Y = L^2(w^{-1}; [-1, 1])$ whose norm is defined by

$$\|y\|^2 = \int_0^{2\pi} d\phi \int_{-1}^{1} \frac{ds}{w(s)} |y(s, \vec{\theta})|^2; \tag{3.71}$$

indeed, if $X = L^2(D)$, then R is a bounded operator from X in Y since

$$\|Rx\|_Y \leqslant \sqrt{4\pi}\,\|x\|_X. \tag{3.72}$$

It can be shown that the previous inequality is tight so that $\|R\| = 4\pi$.

With the previous choice of X, Y it has been shown that the Radon transform is a compact operator by a direct construction of its singular system [11, 22]. The result proved in [11] is very general since it concerns the Radon transform in n-dimensional Euclidean spaces. The singular functions are expressed in terms of orthogonal polynomials and they provide orthonormal basis in X and Y. The general result is also reported in [24] while the case of functions of two variables is discussed in [5]. The SVD for the problem with a finite number of projections is given in [10].

We do not report these results in this book; the interested reader can consult the indicated references.

References

[1] Anscombe F J 1948 The transformation of Poisson, binomial and negative-binomial data *Biometrika* **35** 246–54

[2] Bardsley J M 2008 Stopping rules for a nonnegatively constrained iterative method for ill-posed Poisson imaging problems *BIT Numer. Math.* **46** 651–64

[3] Benvenuto F, La Camera A, Theys C, Ferrari A, Lantéri H and Bertero M 2008 The study of an iterative method for the reconstruction of images corrupted by Poisson and Gaussian noise *Inverse Probl.* **24** 035016

[4] Benvenuto F, La Camera A, Theys C, Ferrari A, Lantéri H and Bertero M 2012 Corrigendum: The study of an iterative method for the reconstruction of images corrupted by Poisson and Gaussian noise *Inverse Probl.* **28** 069502

[5] Bertero M and Boccacci P 1998 *Introduction to Inverse Problems in Imaging* (Bristol: Institute of Physics)

[6] Bertero M, Boccacci P, Malfanti F and Pike E R 1994 Super-resolution in confocal scanning microscopy: V. Axial super-resolution in the incoherent case *Inverse Probl.* **10** 1059–77

[7] Bertero M, De Mol C and Pike E R 1985 Linear inverse problems with discrete data:I-General formulation and singular system analysis *Inverse Probl.* **1** 300–30

[8] Booth M J 2007 Adaptive optics in microscopy Philos *Trans. R. Soc.* A **365** 2828–43

[9] Candy B H 1985 Photomultiplier characteristics and practice relevant to photon counting *Rev. Sci. Instrum.* **56** 183–93

[10] Caponnetto A and Bertero M 1997 Tomography with a finite set of projections: singular value decomposition and resolution *Inverse Probl.* **13** 1191–205

[11] Davison E M 1981 A singular value decomposition for the Radon transform in n-dimensional Euclidean space *Numer. Funct. Anal. Optim.* **3** 321–40

[12] de Villers G and Pike E R 2017 *The Limits of Resolution* (Boca Raton, FL: CRC Press)

[13] Ellerbroek B L and Vogel C R 2009 Inverse problems in astronomical adaptive optics *Inverse Probl.* **25** 063001

[14] Engl H W, Hanke M and Neubauer A 1996 *Regularization of Inverse Problems* (Dordrecht: Kluwer)

[15] Esposito E *et al* 2010 First light AO (FLAO) system for LBT: final integration, acceptance test in Europe, and preliminary on-sky commissioning results, *Proc. SPIE 7736, Adaptive Optics Systems II*; (15 July 2010) 773609

[16] Goodman J W 1968 *Introduction to Fourier Optics* (New York: McGraw-Hill)

[17] Hohage T and Werner F 2016 Inverse problems with Poisson data: statistical regularization, theory, applications and algorithms *Inverse Probl.* **32** 093001

[18] Jerri A J 1977 The Shannon sampling theorem-its various extensions and applications: a tutorial review *Proc. IEEE* **65** 1565–96

[19] Kaipio J and Somersalo E 2005 *Statistical and Computational Inverse Problems* (Berlin: Springer)

[20] Kak A C and Slaney M 2001 *Principles of Computerized Tomographic Imaging* (Philadelphia, PA: SIAM)

[21] Lanczos C 1961 *Linear Differential Operators* (London: Van Nostrand)

[22] Louis A K 1984 Orthogonal function series expansions and the null space of the Radon transform *SIAM J. Math. Anal.* **15** 621–33

[23] Natterer F 2001 Inversion of the attenuated Radon transform *Inverse Probl.* **17** 113–9

[24] Natterer F and Wübbeling F 2001 *Mathematical Methods in Image Reconstruction* (Philadelphia, PA: SIAM)

[25] Pan X, Sidky E and Vannier M 2009 Why do commercial CT scanners still employ traditional, filtered back-projection for image reconstruction? *Inverse Probl.* **25** 123009

[26] Petersen D P and Middleton D 1962 Sampling and reconstruction of wave-number-limited functions in N-dimensional Euclidean spaces *Inf. Control* **5** 279–323

[27] Picard E 1910 Sur un théoreme général relatif aux équations intégrales de prèmiere espèce et sur quelques problèmes de physique mathématique *Rend. Circ. Mat. Palermo* **29** 79–97

[28] Reed M and Simon B 1972 *Methods of Modern Mathematical Physics I: Functional Analysis* (New York: Academic)

[29] Shepp L A and Vardi Y 1982 Maximum likelihood reconstruction for emission tomography *IEEE Trans. Med. Imaging* **1** 113–22

[30] Snyder D L, Hammoud A M and White R L 1993 Image recovery from data acquired by a charge-coupled-device camera *J. Opt. Soc. Am.* A **10** 1014–23

[31] Snyder D L, Helstrom A D, Lanterman C V, Faisal M and White R L 1995 Compensation for read-out noise in CCD images *J. Opt. Soc. Am.* A **12** 272–83

[32] Snyder D L, Helstrom C W, Lanterman A D, Faisal M and White R L 1994 Compensation for read-out noise in HST image restoration *The Restoration of HST Images and Spectra II* ed R J Hanish and R L White (Baltimore: The Space Telescope Science Institute) pp 139–54

[33] Staglianó A, Boccacci P and Bertero M 2011 Analysis of an approximate model for Poisson data reconstruction and a related discrepancy principle *Inverse Probl.* **27** 125003

[34] Stark J L, Murtagh F and Bijaoui A 1998 *Image Processing and Data Analysis: The Multiscale Approach* (Cambridge: Cambridge University Press)

[35] Tyson R K 2015 *Principles of Adaptive Optics* (Boca Raton, FL: CRC Press)

[36] van Kempen G M P and van Vliet L J 2000 Background estimation in nonlinear image restoration *J. Opt. Soc. Am.* A **17** 425–33

[37] van Kempen G M P, van Vliet L J, Verveer P J and van der Voort H T M 1997 A quantitative comparison of image restoration methods for confocal microscopy *J. Microsc.* **185** 354–65

[38] Vogel C R 2002 *Computational Methods for Inverse Problems* (Philadelphia, PA: SIAM)

Chapter 4

Statistical approaches in a discrete setting

In this chapter and in the subsequent ones we consider the discrete inverse problems resulting from the discretization of linear problems which are ill-posed in the sense of Hadamard. Therefore, in order to reduce the uncertainty of the solution, a re-formulation of the problem is required, by taking into account all available information: data statistics, constraints on the unknown solution, such as non-negativity, possible smoothness properties, etc. Since we are considering Poisson data, the maximum likelihood approach proposed by Shepp and Vardi is the natural starting point. Due to the instability of the maximum likelihood solutions in specific cases, the Bayesian approach is introduced and its regularization properties are discussed. Analogies with Tikhonov regularization theory are outlined.

4.1 Maximum likelihood approach and data-fidelity function

In sections 3.2 and 3.5 we discussed the typical difficulties arising in the solution of image reconstruction problems. The classic approach for dealing with these difficulties is provided by Tikhonov regularization theory (see, for instance, [11]) which is a variational approach based on the minimization of a penalized and/or constrained least-squares functional. In the discrete case, using our notation, it can be written as follows

$$f_0(x; y) = \|Hx - y_s\|^2, \tag{4.1}$$

and it provides a measure of the misfit between the computed data Hx and the measured and background subtracted data y_s. It can be called the *data-fidelity function.*

A probabilistic approach to regularization theory (see, for instance, [5], chapter 7) shows that this data-fidelity function arises in a quite natural way if one considers a *maximum likelihood* (ML) approach to the image reconstruction problem when data are perturbed by additive Gaussian white noise, hence an assumption not applicable to the problems considered in this book.

doi:10.1088/2053-2563/aae109ch4

To introduce in a simple way the ML approach to image reconstruction with Poisson data, we first consider a simple problem which is basic in section 4.3.

As we know, if we know the expected values z_i of the numbers of photons which can be detected in the image pixels/voxels $i \in \mathcal{I}$, then we can compute the probability of any possible realization y of the r.v. Y by means of equation (3.21). On the other hand, if the parameters z are unknown while a realization y of Y is given, then the problem arises of estimating z from the given y. The ML approach is based on the so-called *likelihood function*, i.e. the function of z defined by

$$\mathcal{L}_y(z) = \prod_{i \in \mathcal{I}} e^{-z_i} \frac{z_i^{y_i}}{y_i!}. \tag{4.2}$$

Note that, somehow, we exchange the role of the two sets of variables: while the probability distribution is a function of y given z, the likelihood is a function of z given y. Moreover, due to the meaning of z, this function must be considered as restricted to the non-negative orthant. Then the ML-estimates of z are the non-negative arrays which maximize the likelihood function. Even if the solution of this constrained problem is trivial and given by $z = y$, as proved in proposition 4.1, it provides a useful introduction to the much more difficult problem of image reconstruction.

The problem is simplified if we consider the negative logarithm of the likelihood and we observe that the set of the minimizers of the negative logarithm coincides with the set of the maximizers of the likelihood. By taking the negative logarithm of $\mathcal{L}_y(z)$ and rearranging the terms depending only on y, hence independent of z (we add the terms $-\ln y_i! + y_i \ln y_i - y_i$), we obtain the following function

$$-\ln \mathcal{L}_y(z) \to \sum_{i \in \mathcal{I}} \left\{ y_i \ln \frac{y_i}{z_i} + z_i - y_i \right\} := \mathrm{KL}(y; z) \tag{4.3}$$

where $y_i \ln y_i = 0$ if $y_i = 0$. $\mathrm{KL}(y; z)$ is the so-called *generalized Kullback–Leibler* (KL) *divergence*. Its values are finite and real on its effective domain, i.e. the set $\{z \,|z_i > 0 \text{ for } y_i > 0;\ z_i \in \mathbf{R} \text{ for } y_i = 0\}$, even if, as already remarked, we are only interested in studying its behavior on the non-negative orthant $z \geqslant 0$.

This function was first introduced by Kullback and Leibler in probability theory [21]; it is a divergence because it is non-negative on the non-negative orthant and is zero if and only if $z = y$; it is not a metric distance because is not symmetric with respect to the exchange of the two variables and does not satisfy triangle inequality.

If we introduce the two sets of indices $\mathcal{I}_0 = \{i \in \mathcal{I}\colon y_i = 0\}$ and $\mathcal{I}_1 = \{i \in \mathcal{I}\colon y_i > 0\}$ so that $\mathcal{I} = \mathcal{I}_0 \cup \mathcal{I}_1$, then the KL divergence can be written explicitly as the real-valued extended function given by

$$\mathrm{KL}(y; z) = \sum_{i \in \mathcal{I}_0} z_i + \sum_{i \in \mathcal{I}_1} \left\{ y_i \ln \frac{y_i}{z_i} + z_i - y_i \right\} \tag{4.4}$$

on its effective domain and ∞ if $z_i \leqslant 0$, $i \in \mathcal{I}_1$. We collect the main properties of the KL divergence in the following proposition.

Proposition 4.1. *For any given value of $y \geqslant 0$, the KL divergence is a convex, coercive and non-negative function of z on the non-negative orthant $\{z \geqslant 0\}$; it is strongly convex if and only if $y > 0$. In all cases it has a unique minimizer $z = y$ on the non-negative orthant. Moreover, it is differentiable on its effective domain.*

Proof. Since the KL divergence is separable, the result follows from an analysis of the different terms. A term corresponding to $y_i = 0$ is just z_i, which is obviously convex, coercive and non-negative on the non-negative orthant where the minimum is at $z_i = 0$. On the other hand, if $y_i > 0$, then the corresponding term is $y_i \ln \frac{y_i}{z_i} + z_i - y_i$ and again one can easily verify that it is convex, coercive and non-negative with a minimum at $z_i = y_i$. Non-negativity follows from the elementary inequality $\ln u \leqslant u - 1,\ u > 0$. Moreover, the terms with indices in $\mathcal{I}_1$ are not only convex but also strongly convex, so that strong convexity of $\mathrm{KL}(y; z)$ follows from the condition $y > 0$.

Finally, the gradient and the Hessian can be easily computed on the positive orthant (contained in the effective domain of the function). The gradient is given by

$$\frac{\partial \mathrm{KL}(y; z)}{\partial z_i} = 1, \ \ i \in \mathcal{I}_0; \ \frac{\partial \mathrm{KL}(y; z)}{\partial z_i} = 1 - \frac{y_i}{z_i}, \ \ i \in \mathcal{I}_1 \tag{4.5}$$

while the Hessian is a diagonal matrix with diagonal elements which are 0 for $i \in \mathcal{I}_0$, while, for $i \in \mathcal{I}_1$, are given by

$$\frac{\partial^2 \mathrm{KL}(y; z)}{\partial^2 z_i} = \frac{y_i}{z_i^2}. \tag{4.6}$$

The Hessian is positive definite if and only if $y > 0$. □

In figure 4.1 we plot the level sets of two KL divergences depending on two variables and restricted to the non-negative quadrant, one with the minimum point on the boundary, the other with the minimum point interior to the quadrant. It is obvious that the first one is defined also outside the quadrant, in particular for $z_2 < 0$ while the second one is defined only inside the quadrant. Therefore, in the first case, the minimum point is a constrained one.

If we now model the unknown expected values z using the model of the forward problem introduced in section 3.3, equation (3.30), i.e. $z = Hx + b$ with H and b given while the source object x is unknown, the negative logarithm of the likelihood is a function of x which can be called the *data-fidelity function* for Poisson data; it replaces the data-fidelity function (4.1), which applies to Gaussian noise, and is given by

$$f_0(x; y) := \mathrm{KL}(y; Hx + b), \tag{4.7}$$

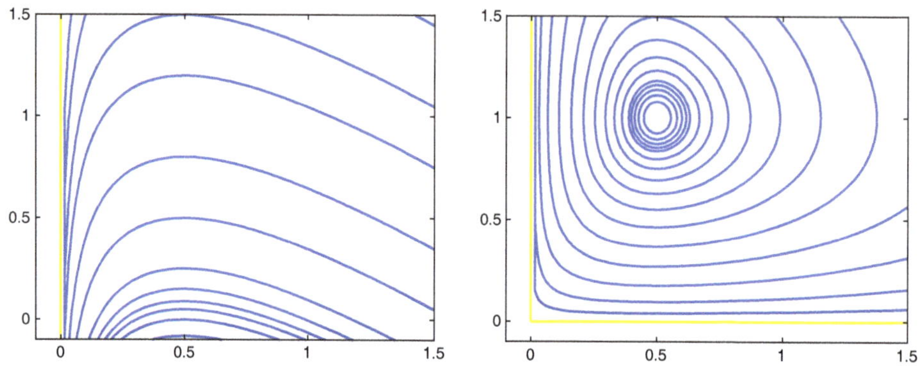

Figure 4.1. Contour maps of KL divergence, depending on two variables, with different values of y. In the left panel the values of y are (0.5,0) while in the right panel they are (0.5,1). Obviously these are also the coordinates of the minimum points on the non-negative quadrant. In the first case it lies on the boundary while in the second it is interior to the non-negative quadrant.

or, in a more explicit form

$$f_0(x; y) = \sum_{i\in\mathcal{I}_0}(Hx + b)_i + \sum_{i\in\mathcal{I}_1}\left\{y_i \ln\frac{y_i}{(Hx + b)_i} + (Hx + b)_i - y_i\right\}. \tag{4.8}$$

This is a real-valued extended function; the effective domain consists of all points x where the value of $f_0(x; y)$ is finite, while its value is ∞ if for some value of x one of the components $(Hx + b)_i$, with $i \in \mathcal{I}_1$, is zero. However, thanks to the physical meaning of x, $f_0(x; y)$ is restricted to the non-negative orthant. Hence, we denote as $\mathcal{D}(f_0)$ the set of the points $x \geqslant 0$ where f_0 is finite, i.e. the intersection of the effective domain with the non-negative orthant. The following proposition summarizes the main properties of the data-fidelity function.

Proposition 4.2. *If H satisfies the conditions* (3.31) *and $b > 0$, then, for any given value of $y \geqslant 0$, the data-fidelity function $f_0(x; y)$ is a convex, non-negative and coercive function of $x \geqslant 0$. It is also strongly convex if $y > 0$ and the null space of H is trivial. Moreover, it is differentiable on $\mathcal{D}(f_0)$ and, if $b > 0$, it is Lipschitz continuously differentiable on the non-negative orthant, with Lipschitz constant L such that*

$$L \leqslant \frac{\|y\|_\infty}{b^2}\|H\|_2^2. \tag{4.9}$$

Proof. As in the case of the KL divergence we can first consider the different terms. Since each one is the composition of a linear affine function with a convex one, it is also convex. Moreover, the terms corresponding to positive y_i are strongly convex if the null space of H is trivial because, in such a case, we have $Hx_1 \neq Hx_2$ if $x_1 \neq x_2$. The non-negativity follows as in the case of the KL divergence.

As concerns coercivity, under the assumptions of the proposition, let us remark that $f_0(x; y)$ can also be written as follows

$$f_0(x; y) = \|Hx\|_1 + \|b\|_1 - \|y\|_1 + \sum_{i \in \mathcal{I}_1} \left\{ y_i \ln \frac{y_i}{(Hx + b)_i} \right\}; \tag{4.10}$$

then, if H satisfies the normalization condition (3.32), we have $\|Hx\|_1 = \|x\|_1$ and coercivity is obvious. In the other cases, with H satisfying conditions (3.31), we can write

$$v_j = \sum_{i \in \mathcal{I}} H_{i,j} > 0, \quad \|Hx\|_1 = v^T x \geqslant v_{\min} \|x\|_1, \tag{4.11}$$

where $v_{\min} = \min_j v_j > 0$. Therefore, $\|Hx\|_1$ is always coercive.

As concerns differentiability of f_0 on $\mathcal{D}(f_0)$, it is easy to verify that the gradient and Hessian are given by

$$\begin{aligned} \nabla_x f_0(x; y) &= H^T \mathbf{1} - H^T \frac{y}{Hx + b} \\ \nabla_x^2 f_0(x; y) &= H^T \mathrm{diag}\left(\frac{y}{(Hx + b)^2} \right) H, \end{aligned} \tag{4.12}$$

where the quotient of vectors is component-wise and $\mathbf{1}$ is the vector in the image space with all the components equal to 1. Moreover, the Hessian is a positive definite matrix if $y > 0$ and the null space of H is trivial. If the matrix H satisfies the normalization condition of equation (3.32), then $H^T \mathbf{1} = \mathbf{1}$.

Finally, since $\nabla_x^2 f_0(x; y)$ is positive semi-definite on the non-negative orthant, the Lipschitz constant L is equal to the largest eigenvalue of $\nabla_x^2 f_0(x; y)$ over this set

$$L = \sup_{x \geqslant 0} \sup_{\|z\|_2 = 1} z^T H^T \mathrm{diag}\left(\frac{y}{(Hx + b)^2} \right) Hz.$$

Since H is non-negative, the supremum over $x \geqslant 0$ is attained at $x = 0$. Therefore, we simply need to bound the largest eigenvalue of the symmetric matrix $H^T \mathrm{diag}\left(\frac{y}{b^2} \right) H$. Using the properties of matrix norms, we obtain the inequality (4.9) (see [18], lemma 1). This result is relevant in the analysis of some optimization methods. □

These results imply that minimizers of $f_0(x; y)$, which will be denoted as x^*, exist in $\mathcal{D}(f_0)$, are global and form a convex set. They can also be defined as the solutions of the constrained optimization problem

$$x^* = \arg \min_{x \geqslant 0} f_0(x; y). \tag{4.13}$$

This point of view allows the possibility of considering additional constraints, as we briefly discuss at the end of this section.

It is interesting to investigate properties of these solutions to understand if they can provide a sensible solution to the image reconstruction problem. A few of these properties are collected in the following proposition.

Proposition 4.3. *The following properties of the solutions of problem* (4.13) *hold true.*

- x^* *is unique if* $y > 0$ *and the null space of* H, $\mathcal{N}(H)$ *is trivial.*
- *If* $b > 0$, *then* $x^*=0$ *is not a solution if* $y \geqslant b$ *and, for at least one value of* $i \in \mathcal{I}$, $y_i > b_i$.
- *If* $x^*>0$, *i.e. the minimizer is interior to the non-negative orthant, then it is a solution of the equation* $Hx^* + b = y$.

Proof. As concerns the first point, it is obvious that, if the stated conditions are satisfied, then the Hessian of $f_0(x; y)$, equation (4.12), is positive definite and the function is strongly convex; hence the minimizer is unique.

As concerns the second point, the gradient in $x = 0$ is given by

$$\nabla_x f_0(0; y) = H^T\left(\mathbf{1} - \frac{y}{b}\right), \tag{4.14}$$

and therefore has non-positive components if the stated condition is satisfied; it follows that $x = 0$ cannot be a minimizer. We remark that the condition has a precise physical meaning: it implies that the image must contain a 'signal' besides the background.

The proof of the third point is not so direct as the proof of the two others. The minimizers of the data-fidelity function satisfy the first order optimality conditions which, in the case $b > 0$, are simply given by the Karush–Khun–Tucker (KKT) conditions

$$x^*\left(H^T\mathbf{1} - H^T\frac{y}{Hx^*+b}\right) = 0, \quad x^* \geqslant 0, \quad H^T\mathbf{1} - H^T\frac{y}{Hx^*+b} \geqslant 0, \tag{4.15}$$

where quotient and product of vectors are intended component-wise. The second and third condition imply that the first one is equivalent to

$$(x^*)^T\left(H^T\mathbf{1} - H^T\frac{y}{Hx^*+b}\right) = 0, \tag{4.16}$$

or also

$$(Hx^*)^T\left(\mathbf{1} - \frac{y}{Hx^*+b}\right) = 0. \tag{4.17}$$

If $x^* > 0$ (so that $Hx^* > 0$), this equation implies

$$\mathbf{1} - \frac{y}{Hx^*+b} = 0, \tag{4.18}$$

i.e. x^* is a solution of the equation $Hx^* + b = y$. □

The last point of this proposition has interesting consequences. Indeed, as we know from the discussion of section 3.5, the solution of the linear equation may not exist and, if it exists, is not positive. It follows that the relationship $Hx^* + b = y$ cannot be true and therefore all the minimizers of $f_0(x; y)$ lie on the boundary of the non-negative orthant. Therefore, the minimizers must have zero-valued components; in practice they have many of these null components, i.e. they are presumably sparse objects. This effect is sometimes called *checkerboard effect* [22] or also *night-sky effect* [3]. It implies that the minimizers of the data-fidelity function, i.e. the solutions of the ML problem, in general, are not good estimates of the unknown source object. Apparently, the ML approach has not provided a solution to the difficulties related to the solution of the linear equation.

An example of the checkerboard effect is shown in figure 4.2 (where the source object and the corresponding image are those of figure 3.8) and is compared with the solution of the linear equation. It is clear that the checkerboard effect here appears as a night-sky effect: the non-zero values are smaller than 1.6×10^4 and their density is higher in the region of the nebula. These properties imply that the ML solutions do not provide a significant improvement with respect to the solution of the linear equation, which is shown again for direct comparison.

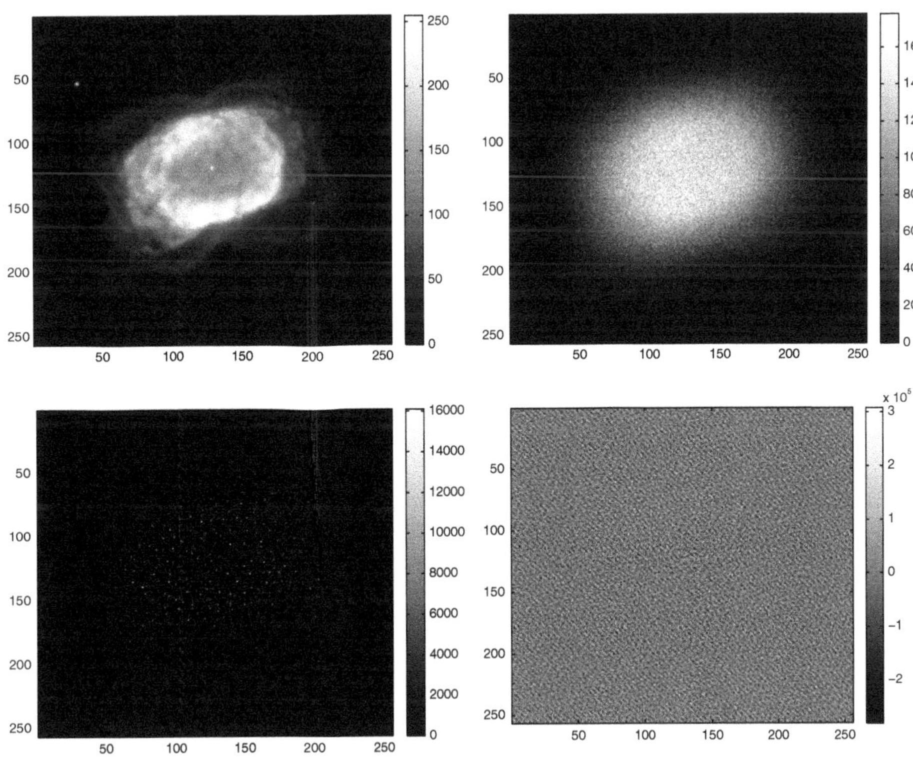

Figure 4.2. First row: the source object (left panel) and the corresponding detected image (right panel). Second row: the minimizer of the data-fidelity function (left panel) and the solution of the linear equation (right panel).

However, this example suggests that this property is not always a defect. For instance, in astronomy, if the problem is to reconstruct a star cluster, then the minimizers of the data-fidelity function can provide a preliminary solution giving the locations of the different stars. In figure 4.3 we show the result obtained in the case of a toy example. The object is a binary consisting of two stars with equal intensity, set to 10^4, over a background of 100. The object is convolved with the diffraction limited PSF and is perturbed with Poisson noise. The result is shown in the left panel of figure 4.3. The minimizer is computed by pushing to convergence one of the iterative algorithms described in the next chapters and is shown in the right panel of the same figure. The reconstruction is essentially free of artifacts, as highlighted by the log scale used for the representation of the result.

The sparsity of the minimizers may be related to the fact that, if $b = 0$, they satisfy a sparsity constraint. Indeed, from the first KKT-condition of equation (4.15), by summing with respect to j and exchanging the sums with respect to j and i (the latter appearing in the computation of H^T), we obtain

$$\sum_{i \in \mathcal{I}} (Hx^*)_i = \sum_{i \in \mathcal{I}} \frac{(Hx^*)_i}{(Hx^*)_i + b_i} y_i. \tag{4.19}$$

In the case $b = 0$ this relationship becomes

$$\sum_{i \in \mathcal{I}} (Hx^*)_i = \sum_{i \in \mathcal{I}} y_i, \tag{4.20}$$

and its interpretation is easy: the flux of the computed image associated with the minimizer x^* coincides with the data flux, a condition which is certainly desirable in the case of a quantitative analysis of the reconstructed source object. Finally, if the matrix H satisfies the normalization condition (3.32), then x^* satisfies the condition

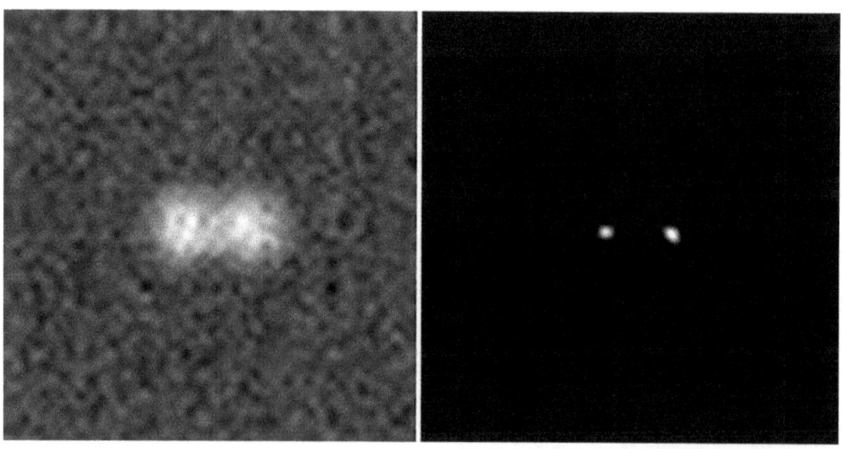

Figure 4.3. The blurred and noisy image of a binary star (left) and the corresponding minimizer of the KL divergence (right). Both images are represented in a log scale. Reproduced from M Bertero *et al* 2009 *Inverse Problems* **25** 123006. © IOP Publishing. Reproduced with permission. All rights reserved.

$$\sum_{j\in\mathcal{J}} x_j^* = \sum_{i\in\mathcal{I}} y_i, \tag{4.21}$$

i.e. all minimizers satisfy a constraint which can justify their sparsity.

This remark clarifies that, in addition to non-negativity, it can be important to introduce an additional constraint in the minimization of the data-fidelity function when $b > 0$ as well as in the minimization of the penalized data-fidelity function introduced in the next section, namely a constraint which guarantees that the computed and the detected image contain the same number of photons. In conclusion, in the following we will always consider constrained minimization and we will indicate as C the convex set defined by the constraints. Therefore, C can be the non-negative orthant or the subset of the non-negative orthant defined by

$$C = \left\{ x \geqslant 0 : \sum_{i\in\mathcal{I}} (Hx + b)_i = \sum_{i\in\mathcal{I}} y_i \right\}. \tag{4.22}$$

We remark that this set is compact since from the bound (4.11) and the previous constraint we obtain $\|x\|_1 \leqslant c/v_{\min}$ where c is the constant

$$c = \sum_{i\in\mathcal{I}} (y_i - b_i) \tag{4.23}$$

which must be positive if the image contains a signal above the background.

4.2 Bayesian regularization

The analysis developed in the previous section implies that an ML solution always exists and, in many cases, it is also unique. Therefore, in a strict sense, the ML problem is well-posed in the sense of Hadamard since continuity always holds true in a discrete problem. In spite of that, in the large majority of image reconstruction problems, the ML solutions are not sensible (see figure 4.2). Obviously, the reason is that, since the problem is the discrete version of an ill-posed problem, information about data statistics combined with the non-negativity of the solution is not sufficient to stabilize the problem; additional information on the solution is required.

The numerical instability of the ML solutions was soon recognized and an application of the method of *sieves*, proposed by Grenander [16], was used for obtaining stable solutions in emission tomography [26]. In the language of inverse and ill-posed problems this is essentially a *mollifier* method. In other words some kind of regularization is introduced, even if, in the present context, this term does not have the same precise meaning it has in Tikhonov regularization theory.

According to the general philosophy of the theory of inverse and ill-posed problems, sensible solutions can be obtained if we introduce additional *a priori* information on the unknown source object. Since we are in a statistical setting, because we are considering the ML approach to the image reconstruction problem, then also this prior information must be formulated in a statistical context. The way is the general *Bayesian paradigm* proposed by Geman and Geman [13], subsequently applied to SPECT imaging [14].

In a Bayesian approach, the unknown object x is also considered as a realization of a multi-valued r.v., let us say X, and the *a priori* information is encoded into a given probability distribution $P_X(x)$ of this r.v., the so-called prior probability distribution, or simply the *prior*.

If the probability distribution (3.21), with the model $z = Hx + b$, is viewed as a conditional probability of Y for a given value of X, i.e. $P_Y(y; x) = P_Y(y|X = x) := P_Y(y|x)$, then Bayes formula ([24], chapter 2) provides the conditional probability of X for a given value y of Y

$$P_X(x|y) = \frac{P_Y(y|x)P_X(x)}{P_Y(y)}. \tag{4.24}$$

If we insert in this equation the detected value of y, we obtain the so-called *posterior probability distribution* of X

$$P_y^X(x) = \frac{\mathcal{L}_y(x)P_X(x)}{P_Y(y)}. \tag{4.25}$$

By definition, a *maximum a posteriori* (MAP) estimate of the unknown source object is any maximizer of the posterior probability distribution. The set of these estimates coincides with the set of the minimizers of the negative logarithm of this function; by neglecting an irrelevant term which depends only of y and rearranging the other constants as in equation (4.3), we obtain the function

$$-\ln P_y^X(x) \to f_0(x; y) - \ln P_X(x), \tag{4.26}$$

with $f_0(x; y)$ defined as in equation (4.7). In other words the Bayesian approach is equivalent to penalize the data-fidelity function with the negative logarithm of the prior.

The most frequently used priors are the so-called Gibbs priors which are given in terms of Gaussian functions. If we take into account that all variables we consider are non-negative, they can be expressed in the following general form

$$P_X(x) = \frac{\chi_+(x)}{Z} e^{-\beta f_1(x)}, \tag{4.27}$$

where Z is a normalization constant and $\chi_+(x)$ is the characteristic function of the non-negative orthant, which is 1 on the orthant and 0 elsewhere. The function $f_1(x)$ is a non-negative function, usually called the energy function or potential function; in this book it will be called *regularization function*. Moreover, the parameter β is a positive parameter, in general called hyper-parameter, which will be called *regularization parameter*. The denominations we use are coherent with those used in Tikhonov regularization theory.

If we insert equation (4.27) in equation (4.26), we obtain, except for the constant $\ln Z$

$$-\ln P_y^X(x) = f_0(x; y) + \beta f_1(x) + \iota_+(x), \tag{4.28}$$

where $\iota_+(x)$ is the indicator function of the non-negative orthant, i.e. the function which is zero on the orthant and ∞ outside.

The previous analysis indicates that the MAP estimates are the non-negative minimizers of the function

$$f_\beta(x;\, y) = f_0(x;\, y) + \beta f_1(x), \tag{4.29}$$

which can be called the Bayesian objective function, i.e. the MAP estimates, indicated as x^*_β are solutions of the constrained minimization problem

$$x^*_\beta = \arg\min_{x \geqslant 0} f_\beta(x;\, y). \tag{4.30}$$

The problem of existence and uniqueness of the solution arises.

We give a few examples of frequently used regularization functions.

- **Tikhonov regularization of order 0 (T0)** i.e. the square of the ℓ_2 norm of x

$$f_1(x) = \frac{1}{2} \sum_{j \in \mathcal{J}} x_j^2. \tag{4.31}$$

- **Tikhonov regularization of order 1 (T1)** i.e. the square of the ℓ_2 norm of the modulus of the discrete gradient of x

$$f_1(x) = \frac{1}{2} \sum_{j \in \mathcal{J}} |\nabla x|_j^2. \tag{4.32}$$

- **Tikhonov regularization of order 2 (T2)**, i.e. the ℓ_2 norm of the discrete Laplacian of x

$$f_1(x) = \frac{1}{2} \sum_{j \in \mathcal{J}} (\Delta x)_j^2. \tag{4.33}$$

- **Sparsity regularization in pixel space**, i.e. the ℓ_1 norm of x (remember that we consider non-negative vectors)

$$f_1(x) = \frac{1}{2} \sum_{j \in \mathcal{J}} x_j. \tag{4.34}$$

- **Hyper-surface (HS) regularization**

$$f_1(x) = \sum_{j \in \mathcal{J}} \left(\sqrt{|\nabla x|_j^2 + \delta^2} - \delta \right), \tag{4.35}$$

where δ is a thresholding parameter. An interesting comment about this regularization function is the following. If we consider the components x_j as the samples of a differentiable function $x(s)$, then, in the case $\delta = 1$, the sum in equation (4.35) (without the term $-\delta$) is the discretization of the integral of the graph of this function. In this context, if δ is different from 1, then it has the meaning of a sampling distance.

- **Total variation (TV) regularization**, i.e. the ℓ_1 norm of the modulus of the discrete gradient of x, which is obtained from the HS regularizer by setting $\delta = 0$

$$f_1(x) = \sum_{i \in \mathcal{I}} |\nabla x|_i. \tag{4.36}$$

- **Frame-analysis regularization**, i.e. the ℓ_1 norm of the coefficients of x on some wavelet basis or tight frame, given by a transform P applied to x

$$f_1(x) = \sum_{i \in \tilde{\mathcal{I}}} |(Px)_i|. \tag{4.37}$$

In order to show how quantitative *a priori* information on the MAP estimates, expressed by means of a Gibbs prior, represents a qualitative information on the unknown source object, in figure 4.4 we show realizations of the priors corresponding to the first, fourth and third regularization function listed above, namely T0 (or white noise) prior, sparsity prior (or impulse noise), both with positivity constraint,

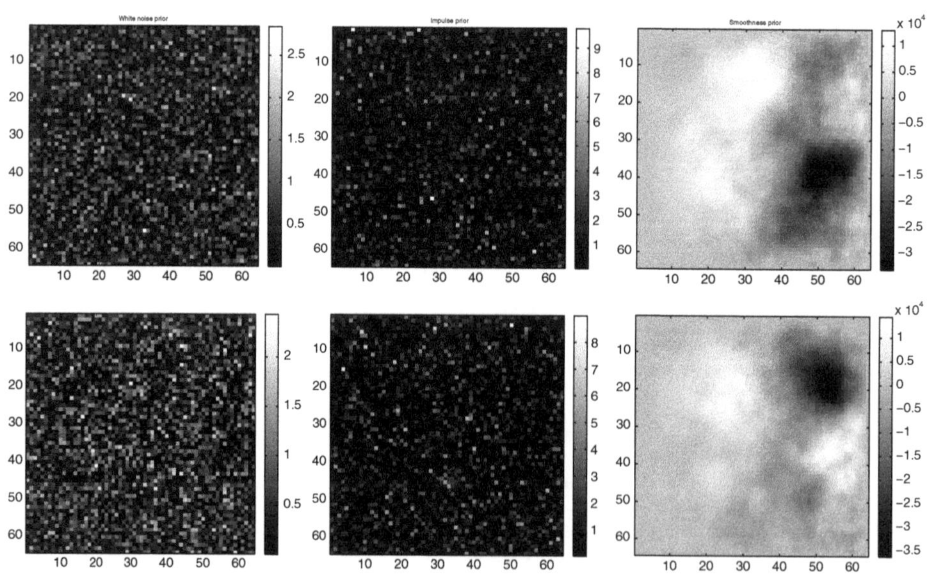

Figure 4.4. Two examples of realizations of three priors: white noise, i.e. T0 regularization (left panels), impulse prior, i.e. sparsity regularization in pixel space (middle panels) and Gaussian smooth prior, i.e. T2 regularization (right panels). In the first two cases a non-negativity constraint is used.

and T2 prior (all with $\beta = 1$). These realizations are computed using the method indicated in [19], section 3.3. It is evident that T0 regularization implies a boundedness of the solution, sparsity regularization implies the existence of localized sources while T2 regularization implies smoothness of the solution.

All the regularization functions listed above are convex and convexity is, in general, a requirement for $f_1(x)$, since, in the case of a non-convex function the analysis of the minimizers is much more complex (for instance, possible existence of local minimizers, saddle points etc). In the convex case, thanks to convexity and coercivity of $f_0(x; y)$ for $x \geqslant 0$ (proposition 4.2), non-negative MAP estimates exist. If $f_0(x; y)$ or $f_1(x)$ is strongly convex, then also $f_\beta(x; y)$ is strongly convex and therefore the MAP estimate is unique.

If $f_0(x; y)$ is not strongly convex, then only T0 regularization is strongly convex, hence providing strong convexity of $f_\beta(x; y)$ and uniqueness of the minimizer. All the other regularization functions are only convex. However, it is possible to prove [7] that in the case of the regularization functions T1, T2 and HS, the null space of the Hessian of $f_\beta(x; y)$ is trivial, even if $f_0(x; y)$ is not strongly convex. Therefore, $f_\beta(x; y)$ is strongly convex and the MAP estimate is unique in all these cases. The interested reader can find the proof of this result in section 4.4.

We conclude with a few comments on HS regularization which, in our opinion, is particularly interesting. First, if we consider the components x_j as the samples of a function $x(s)$, with sampling distance δ, then the HS function is (except for a constant term) the discrete version of the graph of $x(s)$. Therefore, the minimization of the corresponding objective function $f_\beta(x; y)$ is related to the problem of the search of a solution with minimal graph surface.

Moreover, the HS regularization can be considered as a kind of 'interpolation' between T1 and TV regularization. Indeed, if in a region (defined by a subset of values of the index j) $|\nabla x|_j$ is small with respect to δ then, in that region, the corresponding terms of the HS function are approximated by $|\nabla x|_j^2/(2\delta)$ (indeed $\delta\sqrt{1 + |\nabla x|_j^2/\delta^2} - \delta \simeq \delta(1 + |\nabla x|_j^2/(2\delta^2)) - \delta$) and therefore, in that region, HS provides a smooth regularization; on the other hand, if in another region $|\nabla x|_j$ is large because of the existence of edges, then it is approximated by $|\nabla x|_j$, i.e. HS behaves as TV regularization in that region. In other words, in domains where x is smooth the minimizers provided by HS are also smooth (T1 regularization), without the annoying cartoon effects which appear in the case of TV regularization; otherwise, in domains where edges of x appear, these can be correctly reproduced as in the case of TV regularization. Obviously this nice behavior depends on the choice of δ and this can be a critical point in the application of HS regularization.

Finally, a very interesting feature of HS regularization is that the minimizer of the corresponding objective function $f_\beta(x; y)$ is unique, as proved in section 4.6.1, and can be computed very efficiently, since $f_\beta(x; y)$ is differentiable (see section 4.6.1).

Obviously it is possible to introduce other regularization functions with similar properties, namely differentiability and interpolation between T1 and TV. One example used in existing software is what we call MISTRAL regularization (see, section 5.5.3) which can be derived from the function $v(s) = s - \delta \ln(1 + \frac{s}{\delta})$ by setting $s = |\nabla|_j$ and summing over j; another one is provided by the function $v(s) = \ln(\cosh \frac{s}{\delta})$. These regularization functions, however, do not have a simple interpretation as HS.

4.3 Denoising problems

In several applications the number of detected photons per pixel is small. Such a situation arises for instance in medical imaging when a low dose is administered to the patient; other examples are found in microscopy or in astronomy if one observes a very distant and faint object. A simulated example of a very noisy image of a test object is shown in the upper right panel of figure 3.5. In these cases it may be convenient to design methods able to reduce the noise for improving the quality of the image; they are denoted *denoising methods.*

Several methods have been considered for different kinds of noise. Since in this book we consider data perturbed by Poisson noise, we describe variational methods specific for this kind of noise and derived from the maximum likelihood—Bayes approach described in the previous sections for image reconstruction. Indeed image denoising can be considered as a particular case of these problems when the imaging matrix is simply the unit matrix I.

If we denote now as x the estimate of the noise-free image we are searching for, then the data-fidelity function is given by

$$f_0(x; y) = \mathrm{KL}(y; x), \tag{4.38}$$

with the KL divergence restricted to the non-negative orthant. We already know that on this domain the function has a unique minimizer given by $x^* = y$. Hence the ML solution of the denoising problem is trivial and it is necessary to introduce *a priori* information through a Bayesian approach for obtaining non-trivial solutions.

Without repeating the analysis of the previous section, we consider the solutions of the following minimization problem

$$x_\beta^* = \arg\min_{x \geqslant 0} \{f_0(x; y) + \beta f_1(x)\} \quad := f_\beta(x; y), \tag{4.39}$$

as the solution of the denoising problem. Again $f_1(x)$ comes from the negative logarithm of the prior and expresses in a quantitative way qualitative properties of the noise-free image.

It is important to remark that some priors providing interesting results in the image reconstruction problem, in the case of denoising lead to solutions which are not sensible. The first case is that of the ℓ_1 norm. Indeed, if $f_1(x) = \|x\|_1$, we easily find that $x_\beta^* = (1 + \beta)^{-1}y$. The second case is that of the ℓ_2 norm. Indeed, if $f_1(x) = \frac{1}{2}\|x\|_2^2$, it is still possible to compute the minimizer since $f_\beta(x; y)$ is separable with respect to the components of x. By elementary computations, solving a second degree algebraic equation, we find

$$\left(x_\beta^*\right)_i = \frac{1}{2\beta}\left\{\sqrt{1 + 4\beta y_i} - 1\right\}, \quad i \in \mathcal{I}. \tag{4.40}$$

It is easy to check that for $\beta = 0$ we re-obtain the ML solution $x^* = y$ while for β tending to infinity x_β^* tends to zero.

Again the previous solution is not interesting. For the denoising problem one needs priors which correlate the values of adjacent pixels such as the smooth priors, the edge-preserving priors or the frame-analysis priors indicated in the previous section.

An application of denoising to tomography is discussed in [12].

4.4 Selection of the regularization parameter

Given the regularization function $f_1(x)$, a very important issue is the selection of the value of the regularization parameter β since the reliability of the reconstruction strongly depends on its value. This problem has been the subject of a large number of papers in the case of Tikhonov regularization theory. To our knowledge the literature is not so wide in the case of the Bayesian regularization of Poisson data inversion.

In the case of data perturbed by additive Gaussian noise, the data-fidelity function $f_0(x; y)$ is the square of the metric distance between detected and computed data as given in equation (4.1). Then, if we denote again as x_β^* the minimizer of the regularized function $f_\beta(x; y) = f_0(x; y) + \beta f_1(x)$, the choice of the regularization parameter is claimed by the well-known *Morozov discrepancy principle* [11]. If we assume that the data y are perturbed by additive white Gaussian noise with variance σ^2, then for the discrete problem, modeled as in section 3.5, the value of β is given by the unique solution of the equation

$$\|Hx_\beta^* + b - y\|^2 = \sigma^2 m, \tag{4.41}$$

where m is the number of data; the left-hand side of this equation is the expected value of the right-hand side when x_β^* is replaced by the exact solution $\bar{x}$. This recipe is frequently used in practice even if it can over-smooth the solution [5].

Before discussing a possible extension of this principle to the Poisson case, we point out that, in general, much about the choice of β can be learned from numerical experiments on simulated data. Indeed in such a case we start with a synthetic source object $\bar{x}$, we generate the corresponding noise-free image using a given imaging matrix and we perturb the result with Poisson noise (see remark 3.2). Therefore, under these conditions we know exactly the object $\bar{x}$ which should be estimated by the reconstruction method. If x_β^* is the result of a regularization scheme, i.e. the (possibly) unique minimizer of a given regularized function $f_\beta(x; y)$, then the choice of β relies on a comparison between $\bar{x}$ and x_β^*. Incidentally, we remark that in the computation of x_β^* the same imaging matrix used for generating the data is also used; such a procedure is sometimes called *inverse crime* because one does not take into account possible errors on the imaging matrix which certainly arise in practical applications.

The comparison between a synthetic object and the corresponding reconstruction requires the introduction of some *figure of merit* (FOM) able to quantify the difference between the two objects. To this purpose one can consider for instance their metric distance, which, as for all FOMs, depends on both the regularization function and the regularization parameter β (for simplicity we indicate only the dependence on β); if $||.||$ denotes as usual the ℓ_2 norm, then we have

$$\rho(\beta) = \frac{||x_\beta^* - \bar{x}||}{||\bar{x}||}, \tag{4.42}$$

a quantity which can be called *relative root mean square* (rms) error.

Taking into account the particular structure of the data-fidelity function for Poisson data, one can also use as a FOM the KL divergence between x_β^* and $\bar{x}$, i.e.

$$\mathrm{KL}(\beta) = \sum_{j\in\mathcal{J}} \left\{ \bar{x}_j \ln \frac{\bar{x}_j}{\left(x_\beta^*\right)_j} + \left(x_\beta^*\right)_j - \bar{x}_j \right\}, \tag{4.43}$$

which, in this context, looks more natural than the metric distance. For both FOMs we expect that they have at least one minimum as functions of β; however we are not aware of a proof of this behavior in the context we are considering even if, in general, a minimum is found in numerical experiments.

A quite different measure of similarity between the two source objects is provided by the *structural similarity* (SSIM) index introduced in [28]. It is given by the following quantity

$$\mathrm{SSIM}(\beta) = \frac{(2\mu_1\mu_2 + c_1)(2\sigma_{12} + c_2)}{(\mu_1^2 + \mu_2^2 + c_1)(\sigma_1^2 + \sigma_2^2 + c_2)}, \tag{4.44}$$

where $\mu_1,\ \sigma_1^2$ are mean and variance of $\bar{x}$, $\mu_2,\ \sigma_2^2$ are those of x_β^* while $\sigma_{1,2}$ is the covariance of the two objects; it depends on β as well as $\mu_2,\ \sigma_2^2$. The constants $c_1,\ c_2$ are given by $c_1 = (k_1 L)^2$ and $c_2 = (k_2 L)^2$, where L is the dynamic range of the pixel values and, by default, $k_1 = 0.01,\ k_2 = 0.03$. A remarkable property of SSIM is that it is symmetric with respect to the exchange of the two objects and becomes 1 when the two objects coincide. Therefore, in this case one must search for a value of β such that SSIM is close to 1 as far as possible.

We conclude this discussion of the numerical experiments by remarking that, to check the statistical goodness of the reconstructions provided by the different regularization functions $f_1(x)$ and the different FOMs, one can look at the normalized residuals introduced in section 3.4, equation (3.2), which can now be written as follows (again, we indicate only the dependence on β)

$$R_i(\beta) = \frac{y_i - \left(Hx_\beta^* + b\right)_i}{\sqrt{\left(Hx_\beta^* + b\right)_i}}. \tag{4.45}$$

As discussed in [25] in the case of white Gaussian noise (see the definition in equation (3.25)), if x_β^* is replaced by the exact solution $\bar{x}$ then the corresponding normalized residual is essentially a realization of white Gaussian noise with zero mean and variance 1. Therefore, if x_β^* is a reconstruction of $\bar{x}$ free of artifacts, then the corresponding residual should have a similar behavior, otherwise it will exhibit spatial correlations. An application to a problem with Poisson data is discussed in [1].

In the case of real data it is obvious that we cannot use the FOMs we have just introduced because we only know the detected image. A formal counterpart of Morozov principle, based on the data-fidelity function (4.4), is proposed in [30] for the denoising problem and in [6] for the deblurring problem. It rests upon the following result, also proved in [30] (see, also the corrigendum [31]).

Lemma 4.1. *Let Y_λ be a Poisson r.v. with expected value λ and consider the function of Y_λ defined by*

$$F(Y_\lambda) = 2\left\{ Y_\lambda \ln\left(\frac{Y_\lambda}{\lambda}\right) + \lambda - Y_\lambda \right\}. \tag{4.46}$$

Then the following asymptotic estimate of the expected value of $F(Y_\lambda)$ holds true

$$E[F(Y_\lambda)] = 1 + O\left(\frac{1}{\lambda}\right), \quad \lambda \to +\infty. \tag{4.47}$$

For the convenience of the reader we give a slightly simplified proof of this lemma in section 4.6.2. Moreover, in figure 4.5 we give the plot of the function $E[F(Y_\lambda)]$ showing that the function approaches rapidly the asymptotic value and is >1 for $\lambda > 1$.

To introduce the criterion, let us consider a simulated image of an object $\bar{x}$ and let us associate to each pixel/voxel i the quantity

$$D_i(\bar{x}) = 2\left\{ y_i \ln \frac{y_i}{(H\bar{x} + b)_i} + (H\bar{x} + b)_i - y_i \right\}. \tag{4.48}$$

Since this is a simulated image, y_i is precisely a realization of a Poisson random variable with expected value $(Hx + b)_i$. If this value is sufficiently large, then, thanks to lemma 4.1, the expected value of the r.v. with realization $D_i(x)$ is approximately 1. It follows that, if we associate to each pixel the quantity (4.48), the values of these quantities associated with different pixels fluctuate around 1 so that their sum is approximately given by the number of pixels/voxels of the image, provided this number is also sufficiently large.

The same phenomenon should happen in the case of a real image if:

- data statistics is well approximated by Poisson statistics;
- a good approximation of imaging matrix H and background b is available;
- the reconstructed object x_β^*, for some value of β, is close to the true one $\bar{x}$.

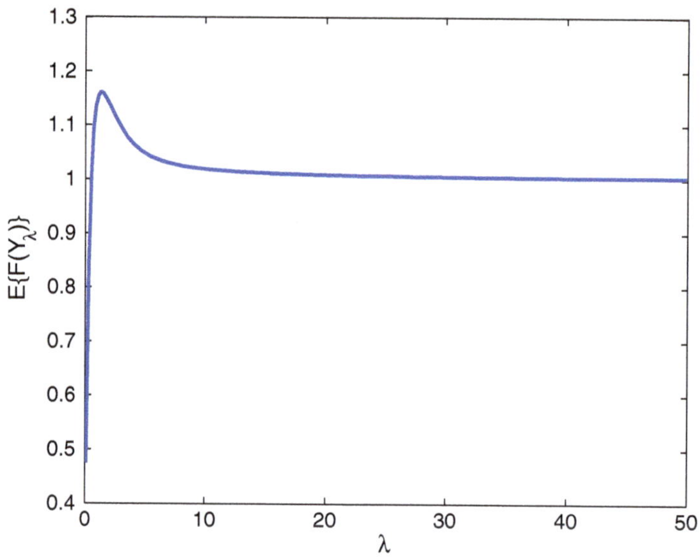

Figure 4.5. Plot of $E[F(Y_\lambda)]$ as a function of λ. The asymptotic value of 1 is already approached between 10 and 20, i.e. for a moderate average number of photons. Reproduced from [31].

If these conditions are satisfied, let us define as *discrepancy function for Poisson data* the following quantity

$$D_y(x) = \frac{2}{\#\mathcal{I}} f_0(x; y), \tag{4.49}$$

which is proportional to the data-fidelity function. Then, a *discrepancy principle for Poisson data*, in the vein of Morozov discrepancy principle, can be formulated as follows: *search for a value of* β *such that*

$$D_y(x_\beta^*) := D_y(\beta) = 1. \tag{4.50}$$

The function $D_y(\beta)$ is well defined if x_β^* is unique. However, the problem of the existence and uniqueness of a solution of equation (4.50) arises and is not trivial. An insight on this problem can be given by the following lemma proved in [6]. The proof is reported in section 4.6.3.

Lemma 4.2. *If the function* $f_\beta(x; y)$ *is strictly convex, coercive and differentiable on the set* C *of the feasible solutions, then* $f_0(x_\beta^*; y)$ *and* $f_1(x_\beta^*)$ *are respectively an increasing and a decreasing function of* β.

This lemma implies that, if a solution of equation (4.50) exists, then it is unique. For existence the requirement is $D_y(0) < 1$ and $lim_{\beta\to\infty} D_y(\beta) > 1$. The problem of existence is discussed in [6].

Remark 4.1. *As remarked above, the criterion (4.50) is strictly related to the assumption that the data satisfy Poisson statistics; therefore, it may not be useful if data are manipulated in such a way that this assumption does not hold true. In addition, the number of photons must be sufficiently large and a good approximation of the imaging matrix is also required. Therefore, the choice of the value 1 in equation (4.50) may not be the most convenient one. In general one can replace this equation by* $D_y(\beta) = 1 + \eta$ *where* η *is a small positive or negative number. The results of existence and uniqueness of the solution discussed in* [6] *can also be extended to this case.*

For comparing the performance of the discrepancy principle with respect to that of the three FOMs defined in equations (4.42)–(4.44) we consider two 256×256 test objects, frequently used in the literature on image reconstruction and denoted, respectively, as **cameraman** and **spacecraft**.

For the cameraman we consider values between 0 and 1000 and, for the spacecraft, values between 0 and 2550 (remark that in [32] these images are called respectively, *Scameraman* and *Sspacecraft*; we refer to this paper for a detailed numerical analysis of the discrepancy principle). The cameraman image is convolved with a Gaussian PSF with standard deviation of 1.3; on the other hand, the spacecraft, after the addition of a background set to 10, is convolved with a PSF which simulates that taken by a ground-based telescope (downloaded from www.mathcs.emory.edu/nagy/RestoreTools/index.html). Finally, the convolved images are perturbed with Poisson noise. The results are shown in figure 4.6.

A comment about the main differences between the two images. The spacecraft is convolved with a PSF which is much broader than that used in the case of the cameraman. This is evident by comparing the two images in the right panels: all the fine details of the spacecraft object are lost in the convolved and noisy image while a detailed shape of the cameraman is still recognizable. Another consequence of the different PSFs is that, even if the range of pixel values of the spacecraft object is broader than that of the cameraman, this difference is lost in the convolved images since the range is [19, 938] for the cameraman and [5, 1135] for the spacecraft so that Poisson noise perturbation is comparable in the two cases.

Thanks to the structure of the two test objects, a reconstruction based on edge-preserving regularization seems quite natural. Therefore, we consider the HS regularization defined in equation (4.35) with $\delta = 10^{-4} \max_i(y_i)$; since in this case the objective function is differentiable, it is possible to compute the minimizers of $f_\beta(x; y)$, for several different values of β, with one of the fast minimization methods described in chapter 6, more precisely SGP (see section 6.3.1). Then, using the computed x_β^*, we plot the values of the four FOMs defined above. The results in the case of the cameraman are shown in figure 4.7 while the results for the spacecraft are shown in figure 4.9. Note that in these plots we use a log scale for the values of β, $\rho(\beta)$ and $KL(\beta)$, while we use a linear scale for the values of $D_y(\beta)$ and SSIM.

A few comments on these results, noting that, for a comparison of the results provided by the different FOMs we use the value of the relative rms error, even if we know that its minimum value is given by the minimization of $\rho(\beta)$; the reason is that

Figure 4.6. Left panels—the test objects: cameraman (upper panel) and spacecraft (lower panel). Right panels—the corresponding blurred and noisy images.

this quantification of the error is more familiar than others. It does not mean that reconstructions provided by the minimization of $\rho(\beta)$ are superior to the reconstructions obtained with other criteria.

We consider first the cameraman. In this case, the discrepancy function $D_y(\beta)$ takes the value 1 for $\beta = 1.245 \times 10^{-2}$ with a relative rms of 11.5%; its values range between 0.986 and 1.030 for $\beta \in [1.0 \times 10^{-2}, 1.8 \times 10^{-2}]$. In this interval, the values of $\rho(\beta)$ range between 0.114 and 0.118, while its minimum value of 11% is obtained for $\beta = 3.3 \times 10^{-3}$, so that the reconstruction provided by the discrepancy principle should be slightly smoother than that corresponding to the minimum relative rms error. We remark that such a feature is similar to that of the Morozov discrepancy principle in the standard Tikhonov regularization theory. In the same interval of values of β, the values of $KL(\beta)$ range between 3.426×10^4 and 3.624×10^4 while 3.301×10^4 is the minimum value observed at $\beta = 5.8 \times 10^{-2}$ and corresponding to a relative rms error of 11.1%. Finally, the values of SSIM in the selected interval range between 0.775 and 0.765, while the maximum value 0.780 is reached for

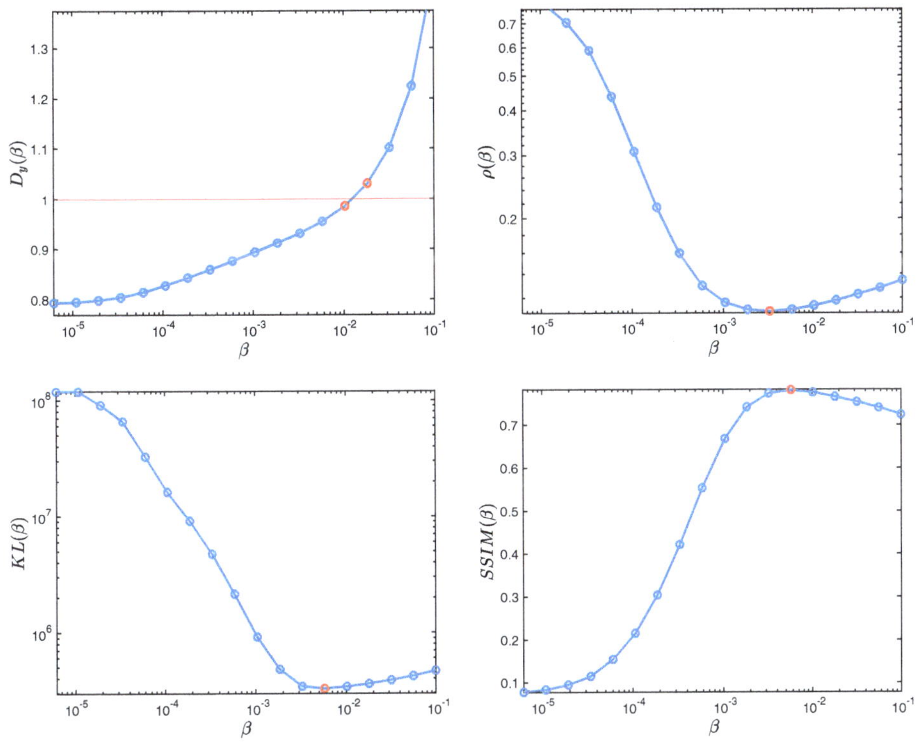

Figure 4.7. Plot of the FOMs as functions of β in the case of the cameraman: discrepancy function (upper left); relative rms (upper right); KL (lower left); SSIM (lower right).

$\beta = 5.8 \times 10^{-2}$, the same value of the minimum of $KL(\beta)$. In figure 4.8 we show the reconstructions corresponding to the values of β that provide the maximum value of SSIM, the minimum value of the relative rms and the ones related to the lower and upper bounds of the interval of values of β including the crossing point of the discrepancy function with 1.

For the spacecraft, the values of $D_y(\beta)$ range between 0.9996 and 1.002 for values of β in the interval $[5.9 \times 10^{-4}, 10^{-3}]$, therefore the crossing point lies inside this interval. The corresponding rms errors range from 30.4% to 31%. The minimum value of $\rho(\beta)$ is obtained for $\beta = 3.3 \times 10^{-4}$ and the corresponding error is 30.3%. Again it is reached for a value of β smaller than that provided by the discrepancy principle. In the previous interval the values of KL(β) range in the interval $[9.68 \times 10^5, 9.70 \times 10^5]$; the minimum value is observed in the lower bound of this interval and therefore the error is 30.4%. Finally, in the same interval, the values of SSIM(β) range between 0.848 and 0.840, while the maximum value 0.863 is observed for $\beta = 6.0 \times 10^{-5}$ and corresponds to an error of 33.5%. In figure 4.10 we show the reconstructions corresponding to the values of β provided by the four criteria we are considering.

The results obtained with these numerical experiments indicate that the discrepancy principle can provide sensible results also in the case of real images if the

Figure 4.8. Reconstructions of the cameraman provided by the different criteria. Upper left panel: SSIM, $\beta = 5.8 \times 10^{-2}$. Upper right panel: $\rho(\beta)$, $\beta = 3.3 \times 10^{-3}$. Lower panels: the reconstructions correspond to the lower extreme 1.0×10^{-2} (left) and to the upper extreme 1.8×10^{-2} (right) of the interval of values of β including the crossing point corresponding to the discrepancy principle (see the text).

conditions indicated above are satisfied. We should point out that very frequently these conditions are not satisfied in the real world. In particular the pre-processing of the raw data may heavily perturb their statistics.

We conclude this section by discussing a criterion similar to equation (4.50) and based on a quadratic approximation of the data-fidelity function; it is derived in [2] and can be obtained in the following way.

We consider the function

$$\phi_y(x) = y \ln \frac{y}{x} + x - y, \quad x, y > 0, \tag{4.51}$$

and we represent the logarithm as a power series of $(x - y)/x$; then, after some manipulation, we obtain

$$\phi_y(x) = \frac{1}{2}\frac{(x-y)^2}{x}\left\{1 + \sum_{n=1}^{\infty} \frac{2}{(n+1)(n+2)}\left(\frac{x-y}{x}\right)^n\right\}, \tag{4.52}$$

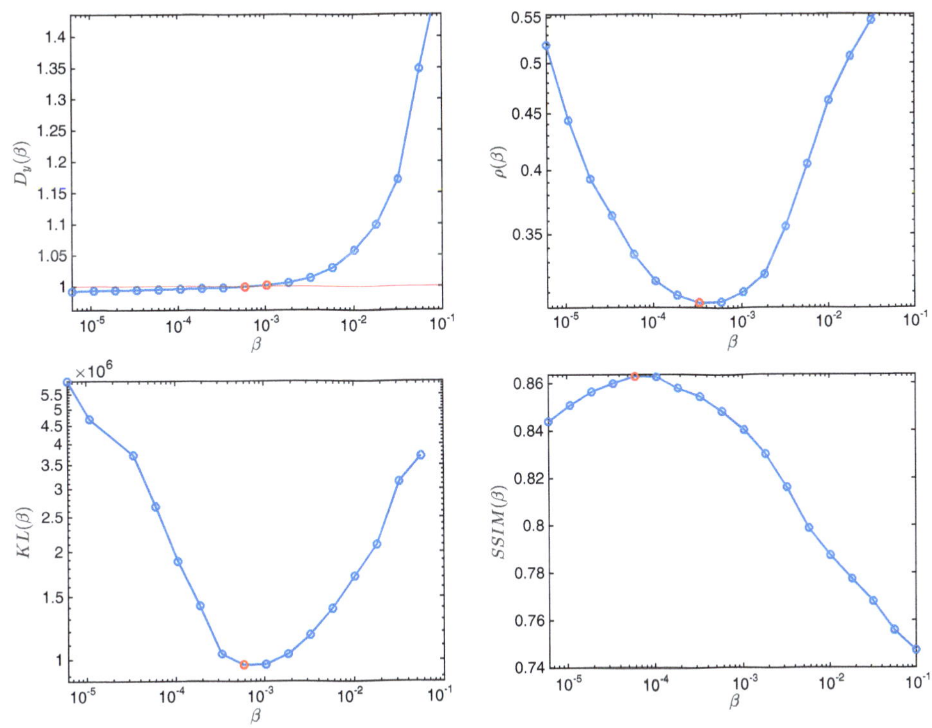

Figure 4.9. Plot of the FOMs as functions of β in the case of the spacecraft: discrepancy function (upper left); relative rms (upper right); KL (lower left); SSIM (lower right).

the series being convergent for $x > y/2$. It is easy to check that, for moderate and large values of y (let us say $y \geqslant 10$), the leading term of this series,

$$\psi_y(x) = \frac{1}{2}\frac{(x-y)^2}{x}, \tag{4.53}$$

provides a sufficiently accurate approximation of $\phi_y(x)$ for $|x - y| \leqslant \sqrt{y}$. If we apply this approximation to equation (4.7) we obtain the new data-fidelity function

$$\bar{f}_0(x; y) = \frac{1}{2}\sum_{i\in\mathcal{I}} \frac{(Hx + b - y)_i^2}{(Hx + b)_i}, \quad x \geqslant 0. \tag{4.54}$$

In [27], it is proved that all the results discussed in this chapter for $f_0(x; y)$ hold true also for $\bar{f}_0(x; y)$. In particular $\bar{f}_0(x; y)$ is convex and coercive and is strongly convex under the conditions which apply to $f_0(x; y)$. Therefore, it provides an approximation preserving all the good properties of $f_0(x; y)$. In addition, if Y_λ is the Poisson r.v. of lemma 4.1, we have the obvious result

$$E\left[\frac{(Y_\lambda - \lambda)^2}{\lambda}\right] = 1. \tag{4.55}$$

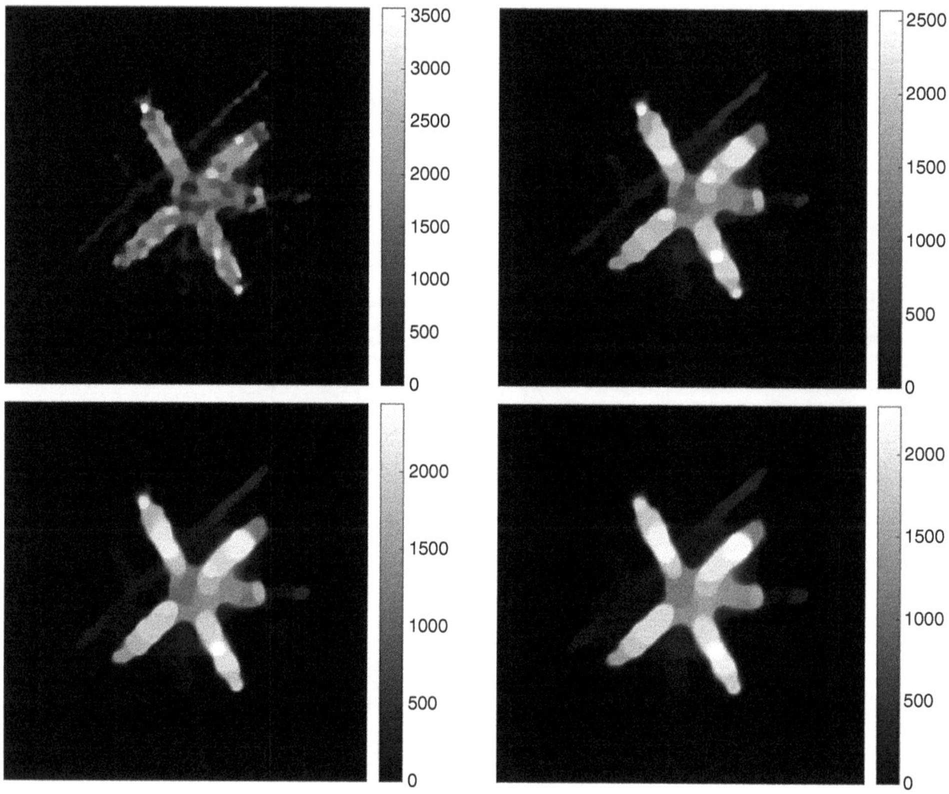

Figure 4.10. Reconstructions of the spacecraft provided by the different criteria. Upper left panel: SSIM, $\beta = 6.0 \times 10^{-5}$. Upper right panel: $\rho(\beta)$, $\beta = 3.3 \times 10^{-4}$. Lower panels: the reconstructions correspond to the lower extreme 5.9×10^{-4} (left) and to the upper extreme 10^{-3} (right) of the interval of values of β including the crossing point corresponding to the discrepancy principle (see the text). The two reconstructions are very similar, indicating a negligible variation of the reconstruction over this interval.

Therefore, by repeating the arguments leading to equation (4.50), we obtain the new discrepancy criterion

$$\bar{D}_y(\beta) = 1, \tag{4.56}$$

where

$$\bar{D}_y(\beta) = \frac{2}{\#(\mathcal{I})}\bar{f}_0(x; y) = \frac{1}{\#(\mathcal{I})}\sum_{i\in S}\frac{(Hx + b - y)_i^2}{(Hx + b)_i}. \tag{4.57}$$

This criterion is that proposed in [2]; we remark that here x_β^* can be the regularized solution defined in this chapter or that obtained by regularizing the data-fidelity function defined in equation (4.54) as discussed in [27]. In the same paper a comparison on simulated data of the different discrepancy criteria applied to the different regularization functions is performed, showing that essentially equivalent results are obtained in all cases.

4.5 The Bregman iteration

An approach which looks very promising for circumventing the problem of the choice of the regularization parameter β is based on Bregman iteration [8]. It is proposed by Osher *et al* [23] in the case of images corrupted by Gaussian noise, for reducing the loss of contrast produced by TV and other regularization methods. It was later extended by Brune *et al* [9, 10] to the case of Poisson data.

We recall that the Bregman distance of a convex function F between x and z is defined as

$$D^pF(x, z) = F(x) - F(z) - p^T(x - z), \tag{4.58}$$

where p is a subgradient of F at z (see definition A.15). Then, the Bregman iteration consists in solving a sequence of sub-problems, similar to the problem (4.29) with $f_1(x)$ replaced by its Bregman distance at the current iterate [8], as follows:

$$x^{(k+1)} = \arg\min_{x \geqslant 0} Q_k(x, p^{(k)}) \equiv f_0(x; y) + \beta D^{p^{(k)}} f_1(x, x^{(k)}), \qquad k = 0, 1, \ldots \tag{4.59}$$

with $p^{(k)} \in \partial f_1(x^{(k)})$ and $p^{(0)} \equiv 0$.

The Bregman iterative scheme can avoid the issue of the choice of β in equation (4.29) because it allows the use of an overestimated value of β; in other words the value of β should be larger than its 'optimal' value, however this can be defined. This point is suggested by the subsequent discussion.

In the following proposition, we summarize the convergence features of the method based on Bregman iteration, as described in [15, 23], assuming that f_0 and f_1 are general functions; in the proposition, we drop the dependence of f_0 on y.

Proposition 4.4. *Let $f_0(x)$ and $f_1(x)$ be non-negative, proper, lower semi-continuous and convex functions, with* $\mathrm{dom} f_0 \subset \mathrm{dom} f_1$; *moreover, assume that the relative interiors of f_0 and f_1 have a point in common. If, for any k, there exists a minimizer $x^{(k+1)}$ of the sub-problem* (4.59), *then the following conditions hold:*

1. *the sequence $f_0(x^{(k)})$ is monotonically non-increasing and we have*

$$f_0(x^{(k+1)}) \leqslant f_0(x^{(k+1)}) + \beta D^{p^{(k)}} f_1(x^{(k+1)}, x^{(k)}) \leqslant f_0(x^{(k)}), \tag{4.60}$$

with $p^{(k)} \in \partial f_1(x^{(k)})$;

2. *if there exists x such that $f_1(x) < \infty$, we have*

$$\begin{aligned} &\beta(D^{p^{(k+1)}} f_1(x, x^{(k+1)}) + D^{p^{(k)}} f_1(x^{(k+1)}, x^{(k)})) + f_0(x^{(k+1)}) \\ &\leqslant f_0(x) + \beta D^{p^{(k)}} f_1(x, x^{(k)}) \end{aligned} \tag{4.61}$$

3. *if x^* is a minimizer of $f_0(x)$ such that $f_1(x^*) < \infty$, we have that*

$$D^{p^{(k+1)}} f_1(x^*, x^{(k+1)}) \leqslant D^{p^{(k)}} f_1(x^*, x^{(k)}), \tag{4.62}$$

and

$$f_0(x^{(k+1)}) \leqslant f_0(x^*) + \beta\frac{f_1(x^*) - f_1(x^{(0)})}{k+1}. \tag{4.63}$$

Moreover, if the level subsets of f_0 are bounded, any limit point of the sequence $\{x^{(k)}\}$ is a minimizer of $f_0(x)$; if x^* is the unique minimizer of $f_0(x)$, then $x^{(k)} \to x^*$ as $k \to \infty$.

Therefore, for noise-free data $\bar{y}$, the sequence of the iterates obtained by equation (4.59) converges to a minimizer of the data-fidelity function $f_0(x; \bar{y})$, which in general coincides with the original object $\bar{x}$; on the other hand, for noisy data y, thanks to proposition 4.4, $\{x^{(k)}\}$ does not converge to $\bar{x}$, but to a minimizer x^* of $f_0(x; y)$. In this case, the Bregman iteration has the typical semi-convergence behavior of other iterative schemes described in sections 5.1 and 7.1, i.e. the sequence $\{x^{(k)}\}$ first approaches the required solution $\bar{x}$ and then it goes away, converging toward the minimum point of $f_0(x; y)$.

If an estimate γ of the noise level is known, such that $f_0(\bar{x}; y) \leqslant \gamma$, then we can observe from equation (4.61) with $x = \bar{x}$, that, as long as $f_0(x^{(k)}; y) \geqslant \gamma$, we have that the Bregman distance of the iterates from the object $\bar{x}$ decreases:

$$D^{p^{(k)}} f_1(\bar{x}, x^{(k)}) \leqslant D^{p^{(k-1)}} f_1(\bar{x}, x^{(k-1)}). \tag{4.64}$$

Therefore, thanks to equation (4.60), a stopping criterion for the iterative procedure is to terminate at the iteration K such that

$$K = \max\{k:\ f_0(x^{(k)}; y) \geqslant \gamma\}. \tag{4.65}$$

In the case of Gaussian noise, the Morozov discrepancy principle can be a reasonable stopping criterion. In the case of Poisson noise, it is reasonable to use the discrepancy criterion described in section 4.4 and to stop the Bregman iteration at the iteration K such that

$$K = \max\left\{k:\ \frac{2}{\#\mathcal{I}} f_0(x^{(k)}; y) \geqslant 1\right\}; \tag{4.66}$$

numerical experience shows that a practical criterion is a visual control of the approximations obtained in a set of iterations around K. In general, for a raw overestimation of β, few iterations are required to obtain a satisfactory solution and, consequently, few instances of the sub-problem (4.59) have to be solved.

Indeed, since the first step requires the solution of the original problem with a large influence of the regularization term, the first iterate $x^{(1)}$ is an over-regularized approximation of the solution. The additional information available in $x^{(1)}$ can be used at the second step, when we have to minimize the same data-fidelity function combined with the Bregman distance of f_1 at $x^{(1)}$. The latter can be interpreted as the residual between the regularization term and its approximation around $x^{(1)}$. In this way, we obtain an enhanced approximation $x^{(2)}$ and so on [29]. Looking back at the statistical model in [13], the Bregman distance of f_1 in the iterative procedure can be interpreted as a refinement of the prior, motivating the observed contrast enhancement.

The convergence and the properties of the Bregman iteration hold when the sub-problems (4.59) are exactly solved, while in the case of Poisson data a closed formula for their minimizers is unavailable. In [4], the authors describe a strategy based on the inexact solution of the sub-problems, by investigating how the convergence of the iterative procedure can be preserved by using a suitable stopping criterion for the inner solver of the sub-problems (4.59). The approach is similar to that proposed in [20] where the notion of ϵ-subgradient (see definition A.16) is introduced to deal with inexact Bregman schemes; a criterion for monitoring the inaccuracy of the current iterate is proposed and the subgradient or the ϵ-subgradient of f_1 (required in the subsequent sub-problem) is obtained by the inner iterative solver used for the current sub-problem. Specialized algorithms, able to provide also the sub-gradients or the ϵ-sub-gradients in the sequence of generated iterates, are described in chapter 6. We refer to [4] for theoretical and numerical details.

In order to highlight the effectiveness of the Bregman iteration, we report the numerical results obtained by applying this method to the reconstruction of the two test problems considered in section 4.4. As in that case, the HS regularization is used with $\delta = 10^{-4} \max_i(y_i)$ (the inner solver is the SGP method described in section 6.3.1).

For the cameraman, the value of β is set equal to 0.1 (we recall that, in the previous section, the value of β corresponding to the minimum relative rms is 3.3×10^{-3}); in figure 4.11 we show the plots of the discrepancy function, of the relative rms error and of SSIM as functions of the iterations of the Bregman method. The minimum relative rms is obtained at $k = 18$ and it is equal to 11.1%, while the maximum value for SSIM is 0.779, achieved at the iteration $k = 11$. The reconstructions corresponding to the best value obtained for the SSIM and for the relative rms are reported in figure 4.12; in this figure, we show also the reconstructions that approximately satisfy the rule (4.66). We observe that the quality of these reconstructions is comparable with that of the reconstructions shown in figure 4.8.

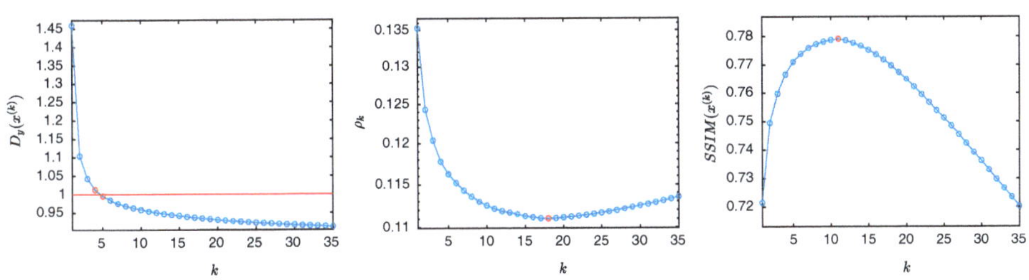

Figure 4.11. Numerical results of the Bregman iteration for the cameraman with $\beta = 0.1$. Left panel: plot of the discrepancy function as a function of the iterations. Middle panel: plot of the relative rms error. Right panel: plot of SSIM. The middle panel shows the semi-convergent behavior of the Bregman method.

Figure 4.12. Reconstructions of the cameraman provided by the Bregman methods at different iterations. Upper left panel: reconstruction at the iteration $k = 11$, corresponding to the maximum value of SSIM 0.779. Upper right panel: reconstruction at the iteration $k = 18$, corresponding to the minimum value of the relative rms 11.1%. Lower panels: the reconstructions correspond to the iterations 4 and 5, corresponding to the interval containing the crossing point of $D_y(x^{(k)})$ with the value 1.

For the spacecraft, we set $\beta = 0.005$, about an order of magnitude higher than the value corresponding to the minimum relative rms error in the experiments of the previous section; figure 4.13 show the plots of the discrepancy function, of the relative rms error and of SSIM as functions of the iterations. The minimum relative rms is obtained at $k = 16$ and it is equal to 31.1%; the maximum value for SSIM is 0.865, achieved at the iteration $k = 34$. Figure 4.14 shows the reconstructions corresponding to the best value obtained for SSIM and the relative rms error as well as those obtained at the iteration k and $k + 1$, where k is the iteration for which the rule (4.66) holds. These reconstructions are comparable with those shown in figure 4.10.

In figure 4.15 we show two video clips, one for the cameraman (left) and one for the spacecraft (right) showing the behavior of the reconstructions provided by Bregman iteration when the number of iterations increases.

In conclusion, we recognize a possible advantage of the Bregman method with respect to the criteria discussed in the previous section. First, in general it is not difficult to find an overestimate of the regularization parameter and, as

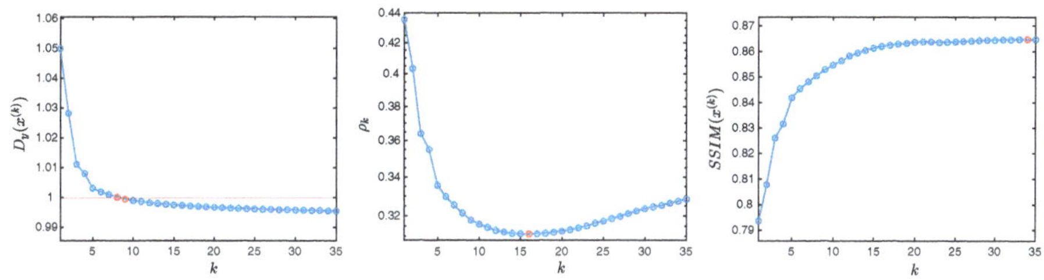

Figure 4.13. Numerical results of the Bregman iteration for the spacecraft with $\beta = 0.005$. Left panel: plot of the discrepancy function as a function of the number of Bregman iterations. Middle panel: plot of the relative rms error. Right panel: plot of SSIM. The middle panel shows the semi-convergent behavior of the Bregman method.

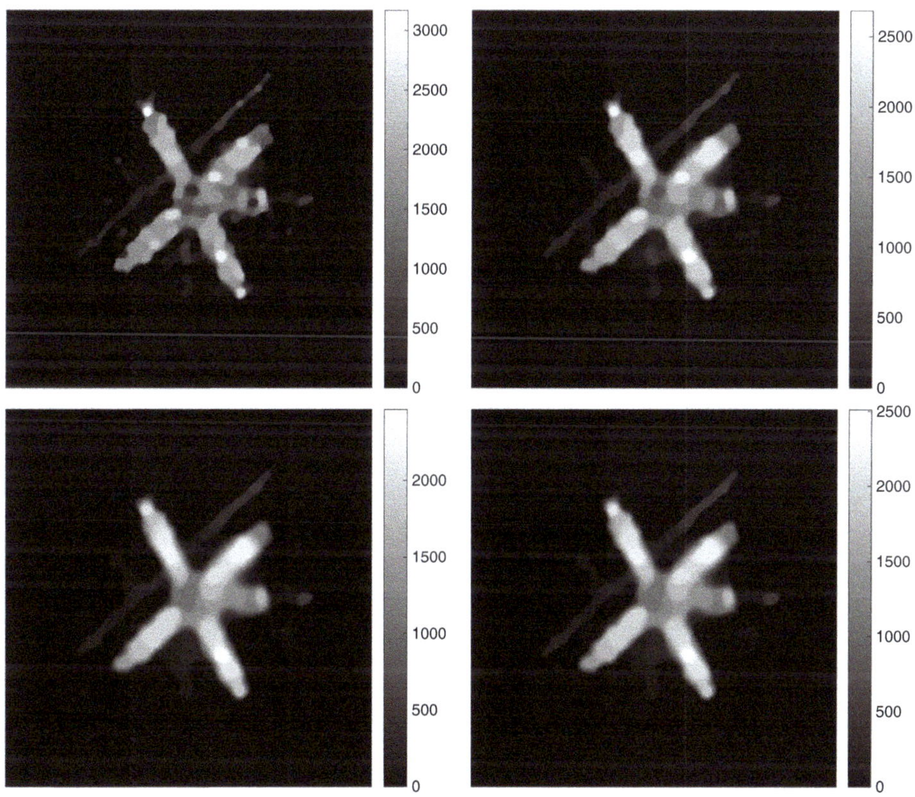

Figure 4.14. Reconstructions of the spacecraft provided by the Bregman method at different iterations. Upper left panel: reconstruction at the iteration $k = 34$, corresponding to the maximum value of SSIM 0.865. Upper right panel: reconstruction at the iteration $k = 16$, corresponding to the minimum value of the relative rms 31.1%. Lower panels: the reconstructions correspond to the iterations 8 and 9, corresponding to the interval containing the crossing point of $D_y(x^{(k)})$ with the value 1.

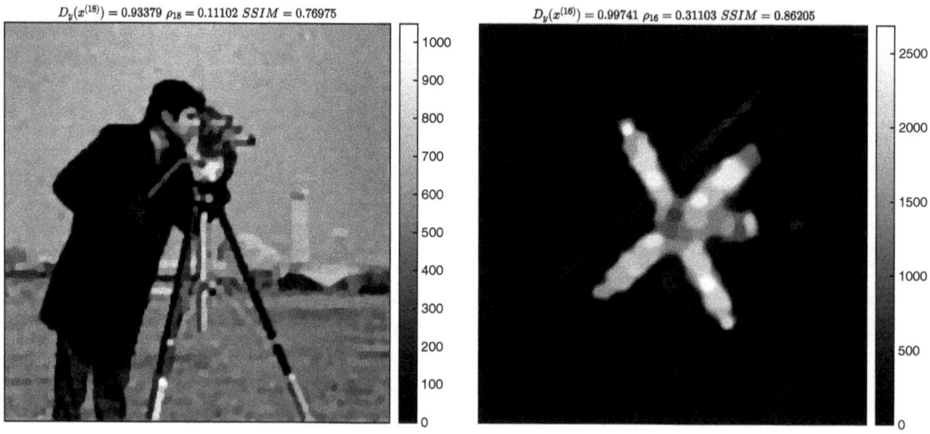

Figure 4.15. Video clips illustrating the reconstruction of cameraman and spacecraft as a function of Bregman iteration. For each frame the values of $D_y(x^{(k)})$, ρ_k and SSIM are indicated. Videos available at http://iopscience.iop.org/book/978-0-7503-1437-4.

demonstrated by the example of the cameraman, this estimate can be very raw. Indeed, in this case the value of β was two orders of magnitude greater than the value corresponding to minimum rms error. Secondly, since in general the number of iterations is not very large, one can easily choose by visual inspection the reconstruction that looks more sensible according to one's own criteria. In our opinion, image reconstruction software based on an automatic selection of the regularization parameter, even if very convenient for the non-expert user, may provide wrong results in some instances.

4.6 Supplementary material

4.6.1 Strong convexity of some regularized functions

In this section we prove that the null space of the Hessian of the objective function $f_\beta(x; y)$, corresponding to the regularization functions T1, T2 and HS, is trivial so that the function is strongly convex and the MAP estimate is unique. To this purpose we need the following lemma on the null space of the sum of two positive semi-definite matrices.

Lemma 4.3. *Let A and B be two symmetric positive semi-definite matrices. Then the following relationship holds true between their null spaces*

$$\mathcal{N}(A + B) = \mathcal{N}(A) \bigcap \mathcal{N}(B) \tag{4.67}$$

Proof. It is obvious that $\mathcal{N}(A) \cap \mathcal{N}(B) \subset \mathcal{N}(A + B)$. In order to prove the converse, let us assume that $v \in \mathcal{N}(A + B)$. Then we have $v^T(A + B)v = 0 = v^T Av + v^T Bv$. Since the two terms are non-negative, we must have $v^T Av = v^T Bv = 0$ and therefore v belongs to both $\mathcal{N}(A)$ and $\mathcal{N}(B)$. □

Another important point is that the null space of $\nabla^2_x f_0(x; y)$ is given by

$$\mathcal{N}(\nabla^2_x f_0(x; y)) = \{u : (Hu)_i = 0 \text{ if } i \in \mathcal{I}_1\} \tag{4.68}$$

and contains the null space of H. Under standard assumptions on H (see (3.31)), a constant vector does not belong to $\mathcal{N}(\nabla^2_x f_0(x; y))$. Therefore, the proof of the result consists in showing that the null space of the Hessian of the regularization functions T1, T2 and HS is trivial or contains only the constant vectors.

We indicate the entries of the vector $x \in \mathrm{R}^N$ as x_i, $i = 1, \ldots, N$ where, in the 2D and 3D cases, i corresponds to a multi-index. In particular, assuming a column-wise ordering, we define $x_i := x_{k,l}$, $i = (l-1)m + k$ for a 2D image $x \in \mathbf{R}^{m\times n}$ $(N = mn)$ and $x_i := x_{k,l,h}$, $i = (h-1)mn + (l-1)m + k$ for a 3D image $x \in \mathbf{R}^{m\times n\times r}$ $(N = mnr)$. In the following we use both notations, with a single index or a multi-index, taking into account the previous equivalence.

Moreover, we consider three kinds of boundary conditions [17]:

(A) **periodic boundary conditions**; for example, in the 2D case, we have: $x_{m+1,l} = x_{1,l}$, $x_{0,l} = x_{m,l}$, for $l = 1, \ldots, n$, and $x_{k,n+1} = x_{k,1}$, $x_{k,0} = x_{k,n}$, for $k = 1, \ldots, m$;

(B) **reflexive boundary conditions**; this means, in the 2D case, that we assume $x_{m+1,l} = x_{m,l}$, $x_{0,l} = x_{1,l}$, for $l = 1, \ldots, n$, and $x_{k,n+1} = x_{k,n}$, $x_{k,0} = x_{k,1}$, for $k = 1, \ldots, m$;

(C) **zero boundary conditions**; in the 2D case we have $x_{m+1,l} = 0$, $x_{0,l} = 0$, for $l = 1, \ldots, n$, and $x_{k,n+1} = 0$, $x_{k,0} = 0$, for $k = 1, \ldots, m$.

We first consider the case of HS regularization because the techniques used in this case can be easily applied to the case of T1 and T2. Then, using the above notation, the HS regularization function can be written as follows

$$f_1(x) = \sum_{i=1}^{N} (\sqrt{D_i} - \delta) \tag{4.69}$$

where we have $D_i = (\nabla x)_i^2 + \delta^2$ and

$$(\nabla x)_i^2 = \begin{cases} (x_{i+1} - x_i)^2 & \text{1Dcase,} \\ (x_{k+1,l} - x_{k,l})^2 + (x_{k,l+1} - x_{k,l})^2 & \text{2Dcase,} \\ (x_{k+1,l,h} - x_{k,l,h})^2 + (x_{k,l+1,h} - x_{k,l,h})^2 + (x_{k,l,h+1} - x_{k,l,h})^2 & \text{3Dcase.} \end{cases}$$

We can also write $(\nabla x)_i^2 = \|A_i x\|^2$ where A_i, $i = 1, \ldots, N$, is a matrix with N columns and 1, 2 or 3 rows, depending on the dimension of the object x. In particular, for periodic boundary conditions any A_i has exactly two non-zero elements for each row, which are equal to 1 and −1. For reflexive and zero boundary conditions, A_i has the same structure, unless when i is related to a boundary pixel: in this case, for zero boundary conditions, the row related to the discrete first order

$$\begin{pmatrix} -1 & 1 & & & \\ & -1 & 1 & & \\ & & -1 & 1 & \\ & & & -1 & 1 \\ 1 & & & & -1 \end{pmatrix} \begin{pmatrix} -1 & 1 & & & \\ & -1 & 1 & & \\ & & -1 & 1 & \\ & & & -1 & 1 \\ & & & & 0 \end{pmatrix} \begin{pmatrix} -1 & 1 & & & \\ & -1 & 1 & & \\ & & -1 & 1 & \\ & & & -1 & 1 \\ & & & & -1 \end{pmatrix}$$

Figure 4.16. Matrix A for 1D object of size $N = 5$ for **periodic** (left), **reflexive** (middle) and **zero** (right) boundary conditions.

difference with respect to the boundary direction has only one non-zero entry, equal to −1, while, for reflexive boundary conditions, this row has zero entries.

Now, in order to derive a matrix expression for the gradient and Hessian of (4.69), which is formally independent of the dimension of the object and of the boundary conditions, we define the block matrices

$$A = \begin{pmatrix} A_1 \\ A_2 \\ \cdots \\ A_N \end{pmatrix},$$

$$E(x) = \mathrm{diag}(D_i^{1/2} I_d), \quad F(x) = \mathrm{diag}\left(I_d - \frac{1}{D_i} A_i x x^T A_i^T \right); \; i = 1, \ldots, N. \tag{4.70}$$

The matrix A is a $q \times N$ matrix, with $q = dN$, where $d = 1, 2, 3$ is the dimension of the 1D, 2D or 3D object. The matrices $E(x), F(x) \in \mathbf{R}^{q \times q}$ are block diagonal matrices where the dimension d of the ith diagonal block is the number of rows of the difference matrix A_i. In figure 4.16 we show the matrix A for 1D object of size $N = 5$ with different boundary conditions.

Remark 4.2. *We observe that for a $m \times n$ 2D object, the $2mn \times mn$ matrix A is a permutation of a matrix obtained by suitable Kronecker products of the identity matrices I_n and I_m of order n and m, respectively, and the 1D matrices of the discrete first order differences along the vertical and horizontal directions, denoted by $A^{1D,\,m}$ and $A^{1D,\,n}$:*

$$A = P \begin{pmatrix} I_n \otimes A^{1D,\,m} \\ A^{1D,\,n} \otimes I_m \end{pmatrix} \tag{4.71}$$

where P is the square permutation matrix of order $2nm$ corresponding to the permutation $\{1, 2, 3, \ldots, 2nm\} \rightarrow \{1, mn + 1, 2, mn + 2, 3, mn + 3, \ldots, mn, 2mn\}$. The rank of A is equal to the rank of the matrix at the right-hand side of the previous equation.

In a similar way, for a $m \times n \times r$ 3D object, the matrix A can be obtained by a suitable permutation of a matrix similar to that of equation (4.71):

$$A = P \begin{pmatrix} I_{nr} \otimes A^{1D,\,m} \\ I_r \otimes A^{1D,\,n} \otimes I_m \\ A^{1D,\,r} \otimes I_{mn} \end{pmatrix}, \tag{4.72}$$

where P is a permutation matrix of order $3nmr$, and consequently A and the matrix at the right-hand side of (4.72) have the same rank.

The gradient and Hessian of $f_1(x)$ can be written in terms of the previously defined matrices as follows

$$\nabla f_1(x) = A^T E(x)^{-1} Ax, \qquad \nabla^2 f_1(x) = A^T E(x)^{-1} F(x) A.$$

In particular, $E(x)$ is diagonal with positive diagonal entries and the ith diagonal block of $F(x)$ is the difference between the identity matrix and a dyadic product; from the definition of D_i, we have that $\frac{1}{D_i} x^T A_i^T A_i x = \frac{\| A_i x \|^2}{D_i} < 1$. Consequently, from the Sherman–Morrison theorem (see proposition A.1), it follows that all the diagonal blocks of $F(x)$ and, thus, $F(x)$ itself, are non-singular. Taking into account the previous remarks, we can characterize the null space of $\nabla^2 f_1(x)$ in the following way.

Proposition 4.5. *The null space of the Hessian matrix $\nabla^2 f_1(x)$ is given by the set of the minimum points of $f_1(x)$, which is the subspace $\mathcal{N}(A)$.*

Proof. Since $E(x)^{-1}F(x)$ is non-singular for all $x \in \mathbf{R}^N$, we have that $\mathcal{N}(\nabla^2 f_1(x)) = \mathcal{N}(A)$. On the other hand, the minimum points of the convex function $f_1(x)$ satisfy the stationarity condition $\nabla f_1(x) = 0$, that is $A^T E(x)^{-1} Ax = 0$. Then, the set of the minimum points of $f_1(x)$ is the subspace $\mathcal{N}(A)$. □

Proposition 4.6. *For zero boundary conditions, $\mathcal{N}(A) = \{0\}$ while for periodic and reflexive boundary conditions $\mathcal{N}(A) = \{\alpha \mathbf{1}\}$.*

Proof. For 1D objects, the proof is immediate:

- for zero boundary condition, since A is a non-singular matrix, $\mathcal{N}(A) = \{0\}$;
- for periodic and reflexive boundary conditions, the rank of A is $N - 1$ (see the examples in figure 4.16): in the first case, where A is a circulant matrix, its eigenvalues are given by $\lambda_i^{(N)} = -1 + \cos \frac{2\pi(i-1)}{N} - \imath \sin \frac{2\pi(i-1)}{N}$, $i = 1, \ldots, N$ (here $\imath = \sqrt{-1}$); in the second case, where A is a triangular matrix, the eigenvalues of A are -1 with algebraic multiplicity $N - 1$ and 0 with algebraic multiplicity 1; thus, in both cases 0 is an eigenvalue of A with algebraic multiplicity equal to 1, A has rank $N - 1$ and $\mathcal{N}(A)$ is a one-dimensional subspace of $\mathbf{R}^N$, given by the eigenspace related to the eigenvalue 0 of A. Since $\mathbf{1}$ is an eigenvector related to the eigenvalue 0, we have $\mathcal{N}(A) = \{\alpha \mathbf{1}\}$.

For 2D objects, we consider the matrix $M = I_n \otimes A^{1D,\, m} + A^{1D,\, n} \otimes I_m$, that can be obtained as

$$M = I_n \otimes A^{1D,\, m} + A^{1D,\, n} \otimes I_m = (I_{nm} \quad I_{nm}) \begin{pmatrix} I_n \otimes A^{1D,\, m} \\ A^{1D,\, n} \otimes I_m \end{pmatrix} = (I_{nm} \quad I_{nm}) P^T A.$$

The rank of M is less than or equal to the minimum of the ranks of $(I_{nm} \quad I_{nm})$ and P^TA, which is equal to that of A. Thanks to the properties of the Kronecker product, the eigenvalues of M are $\lambda_k^{(m)} + \lambda_l^{(n)}$, $k = 1, \dots, m$, $l = 1, \dots, n$; then, recalling the 1D case, for zero boundary conditions, the eigenvalues of M are different from 0 and M is a non-singular matrix. It follows that the rank of A is mn and $\mathcal{N}(A) = \{0\}$. For periodic and reflexive boundary conditions, 0 is an eigenvalue of M with algebraic multiplicity equal to 1, so that the rank of M is $mn - 1$. It follows that the rank of A is greater than or equal to $mn - 1$. On the other hand, in these cases, $\mathcal{N}(A)$ is not trivial, since $A\mathbf{1} = 0$. Then the rank of A is $mn - 1$ and $\mathcal{N}(A) = \{\alpha\mathbf{1}\}$.

For 3D objects, the proof runs as in the 2D case, by considering that

$$M = I_{nr} \otimes A^{1D,\, m} + I_r \otimes A^{1D,\, n} \otimes I_m + A^{1D,\, r} \otimes I_{mn} = (I_{mnr}, I_{mnr}, I_{mnr})P^T A.$$

The eigenvalues of the previous matrix are given by $\lambda_k^{(m)} + \lambda_l^{(n)} + \lambda_h^{(r)}$, $k = 1, \dots, m$, $l = 1, \dots, n$, $h = 1, \dots, r$. Thus, for zero boundary conditions, the eigenvalues of M are different from 0, M and A have rank mnr and $\mathcal{N}(A)$ is trivial. For periodic and reflexive boundary conditions, 0 is an eigenvalue of M with algebraic multiplicity equal to 1, so that the rank of M is $mnr - 1$ and, consequently, the rank of A is greater than or equal to $mnr - 1$. Since $A\mathbf{1} = 0$, $\mathcal{N}(A) = \{\alpha\mathbf{1}\}$. □

As a consequence of the two previous propositions, in both cases of periodic and reflexive boundary conditions, the constant vectors are the minimum points of the function (4.69), while for zero boundary conditions $\mathcal{N}(A) = \{0\}$. Thus, for zero boundary conditions, it is immediate that $f_\beta(x; y)$ is strongly convex. For periodic or reflexive boundary conditions, the strong convexity of $f_\beta(x; y)$ follows from lemma 4.3 and the observation that constant vectors do not belong to $\mathcal{N}(\nabla_x^2 f_0(x; y))$.

Using similar arguments, we can prove the strong convexity of $f_\beta(x; y)$ when $f_1(x)$ is the T1 regularization (4.32); indeed T1 can be written as

$$f_1(x) = \frac{1}{2}\sum_{i=1}^{N} \|A_i x\|^2, \tag{4.73}$$

where A_i is defined as in the case of HS regularization. In the T1 case $f_1(x)$ is a convex quadratic function and its gradient and Hessian are given by

$$\nabla f_1(x) = A^T A x, \qquad \nabla^2 f_1(x) = A^T A,$$

where A is given in equation (4.70). Thus, following the proof of proposition 4.5 and taking into account proposition 4.6, we obtain that $\mathcal{N}(\nabla^2 f_1(x)) = \mathcal{N}(A)$; therefore, T1 has the same properties of HS, i.e. the minimum points of T1, in both cases of periodic and reflexive boundary conditions, are the constant vectors, while for zero boundary conditions $\mathcal{N}(A) = \{0\}$. Consequently, we obtain the strong convexity of $f_\beta(x; y)$.

Finally, we consider the case of T2 regularization (4.33), where we have

$$(\Delta x)_i^2 = \begin{cases} (x_{i+1} - 2x_i + x_{i-1})^2 & \text{1Dcase,} \\ (x_{k+1,\,l} - 2x_{k,\,l} + x_{k-1,\,l})^2 + (x_{k,\,l+1} - 2x_{k,\,l} + x_{k,\,l-1})^2 & \text{2Dcase,} \\ (x_{k+1,\,l,\,h} - 2x_{k,\,l,\,h} + x_{k-1,\,l,\,h})^2 + (x_{k,\,l+1,\,h} - 2x_{k,\,l,\,h} + x_{k,\,l-1,\,h})^2 + & \\ \quad + (x_{k,\,l,\,h+1} - 2x_{k,\,l,\,h} + x_{k,\,l,\,h-1})^2 & \text{3Dcase.} \end{cases}$$

We can also write $(\Delta x)_i^2 = \|B_i x\|^2$ where B_i, $i = 1, \dots, N$, is a matrix with N columns and 1, 2 or 3 rows, depending on the dimension of the object x. In particular, for periodic boundary conditions any B_i has exactly three non-zero elements for each row, two equal to 1 and one equal to −2. For reflexive and zero boundary conditions, B_i has the same structure, except when i is related to a boundary pixel: in this case, the row related to the discrete difference with respect to the boundary direction has only two non-zero entries, which are equal to −2 and 1 for zero boundary conditions and −1 and 1 for reflexive boundary conditions.

In order to derive a matrix expression for the gradient and Hessian of equation (4.33), which is formally independent on the dimension of the object and on the boundary conditions, we define the following $q \times N$ matrix

$$B = \begin{pmatrix} B_1 \\ B_2 \\ \cdots \\ B_N \end{pmatrix}, \tag{4.74}$$

with $q = dN$, where $d = 1, 2, 3$ is the dimension of the 1D, 2D or 3D object.

In figure 4.17 we show the matrix B for a 1D object of size $N = 5$ with different boundary conditions. As observed in remark 4.2 for matrix A, in the case of a 2D object, B is a permutation of a matrix obtained by suitable Kronecker products

$$B = P \begin{pmatrix} I_n \otimes B^{1D,\,m} \\ B^{1D,\,n} \otimes I_n \end{pmatrix} \tag{4.75}$$

where P is the square permutation matrix of order $2nm$ and $B^{1D,\,m}$, $B^{1D,\,n}$ are the 1D matrices of the discrete second order differences along the vertical and horizontal directions. In a similar way, for a $m \times n \times r$ 3D object, matrix B can be obtained by a suitable permutation of the following matrix:

$$\begin{pmatrix} -2 & 1 & & & 1 \\ 1 & -2 & 1 & & \\ & 1 & -2 & 1 & \\ & & 1 & -2 & 1 \\ 1 & & & 1 & -2 \end{pmatrix} \begin{pmatrix} -1 & 1 & & & \\ 1 & -2 & 1 & & \\ & 1 & -2 & 1 & \\ & & 1 & -2 & 1 \\ & & & 1 & -1 \end{pmatrix} \begin{pmatrix} -2 & 1 & & & \\ 1 & -2 & 1 & & \\ & 1 & -2 & 1 & \\ & & 1 & -2 & 1 \\ & & & 1 & -2 \end{pmatrix}$$

Figure 4.17. Matrix B for 1D object of size $N = 5$ for periodic (left), reflexive (middle) and zero (right) boundary conditions.

$$B = P\begin{pmatrix} I_{nr} \otimes B^{1D,\,m} \\ I_r \otimes B^{1D,\,n} \otimes I_n \\ B^{1D,\,r} \otimes I_{mn} \end{pmatrix}, \tag{4.76}$$

where P is a permutation matrix of order $3nmr$. Consequently, using the same arguments of the proof of proposition 4.6, we have that $\mathcal{N}(B) = \{0\}$ for zero boundary conditions and $\mathcal{N}(B) = \{\alpha\mathbf{1}\}$ for periodic and reflexive boundary conditions.

The gradient and Hessian of $f_1(x)$ can be written as

$$\nabla f_1(x) = B^T Bx, \qquad \nabla^2 f_1(x) = B^T B.$$

Therefore, using again the arguments of the proof of proposition 4.5, we can prove that $\mathcal{N}(\nabla^2 f_1(x)) = \mathcal{N}(B)$, obtaining that T2 has the same properties of T1 and HS, i.e. the minimum points of T2, in both cases of periodic or reflexive boundary conditions, are the constant vectors, while for zero boundary conditions $\mathcal{N}(B) = \{0\}$. Consequently, we obtain the strong convexity of $f_\beta(x; y)$ also in the case of T2 regularization.

4.6.2 Proof of lemma 4.1

Let us introduce a function $\phi(\xi)$ related to the values $F(n)$, $n = 0, 1, \ldots$ of the random variable $F(Y_\lambda)$ as follows

$$F(n) = 2\left\{n \ln\left(\frac{n}{\lambda}\right) + \lambda - n\right\} \doteq 2\lambda\, \phi(\xi) \tag{4.77}$$

where

$$\phi(\xi) = (1 + \xi)\ln(1 + \xi) - \xi, \qquad \xi = \frac{n - \lambda}{\lambda} \in [-1, +\infty), \tag{4.78}$$

with $z \ln(z) = 0$ if $z = 0$, so that the function is defined for $\xi \geqslant -1$. Then, from third order Taylor formula for $\phi(\xi)$ we obtain the expression

$$\phi(\xi) = \frac{1}{2}\xi^2 - \frac{1}{6}\xi^3 + R_3(\xi), \quad R_3(\xi) = -\frac{1}{3}\int_0^\xi \frac{(t - \xi)^3}{(1 + t)^3}dt, \tag{4.79}$$

which, inserted in equation (4.77), leads to

$$F(n) = \frac{(n - \lambda)^2}{\lambda} - \frac{1}{3}\frac{(n - \lambda)^3}{\lambda^2} + 2\lambda R_3\left(\frac{n - \lambda}{\lambda}\right). \tag{4.80}$$

By considering the central moments of the Poisson random variable Y_λ, defined by $\mu_k = E\{(Y_\lambda - \lambda)^k\}$, $k = 1, 2, \ldots$, and taking into account that $\mu_2 = \mu_3 = \lambda$, we get

$$E\{F(Y_\lambda)\} = \sum_{n=0}^{\infty} \frac{e^{-\lambda}\lambda^n}{n!}F(n) = 1 - \frac{1}{3\lambda} + \mathcal{R}(\lambda), \tag{4.81}$$

where

$$\mathcal{R}(\lambda) \doteq 2\lambda E\left\{R_3\left(\frac{Y_\lambda - \lambda}{\lambda}\right)\right\} = 2\lambda \sum_{n=0}^{\infty} \frac{e^{-\lambda}\lambda^n}{n!} R_3\left(\frac{n-\lambda}{\lambda}\right). \tag{4.82}$$

The proof of the lemma is based on a suitable estimate of $\mathcal{R}(\lambda)$ when $\lambda \to +\infty$.

To this purpose we split the series into two parts: a partial sum with n ranging from 0 to $\lfloor \lambda/2 \rfloor$ (corresponding to $\xi \in [-1, -1/2]$) and a series with n ranging from $\lfloor \lambda/2 \rfloor + 1$ to $+\infty$ (corresponding to $\xi \in (-1/2, +\infty)$).

As concerns the partial sum, we first observe that, since $0 \leqslant t - \xi < t + 1$, we obtain $|R_3(\xi)| \leqslant |\xi|/3 \leqslant 1/3$; next, we bound the sum by exploiting this inequality and remarking that $\lambda^m/m! < \lambda^n/n!$ when $m < n \leqslant \lfloor \lambda \rfloor$. Therefore, the sum can be bounded by the last term multiplied by the number of terms, so that, by assuming without loss of generality that $\lambda > 2$, we obtain

$$2\lambda \left| \sum_{n=0}^{\left\lfloor \frac{\lambda}{2} \right\rfloor} \frac{e^{-\lambda}\lambda^n}{n!} R_3\left(\frac{n-\lambda}{\lambda}\right) \right| \leqslant \frac{2}{3}\lambda e^{-\lambda} \left\lfloor \frac{\lambda}{2} \right\rfloor \frac{\lambda^{\left\lfloor \frac{\lambda}{2} \right\rfloor}}{\left\lfloor \frac{\lambda}{2} \right\rfloor!}. \tag{4.83}$$

If we use the Stirling lower bound of the factorial

$$n! \geqslant \sqrt{2\pi}\, n^{n+\frac{1}{2}} e^{-n}, \tag{4.84}$$

we have the following upper bound for the partial sum

$$2\lambda \left| \sum_{n=1}^{\left\lfloor \frac{\lambda}{2} \right\rfloor} \frac{e^{-\lambda}\lambda^n}{n!} n\, R_3\left(\frac{n-\lambda}{\lambda}\right) \right| \leqslant \frac{2\lambda}{\sqrt{2\pi}} e^{-\lambda + \left\lfloor \frac{\lambda}{2} \right\rfloor} \left\lfloor \frac{\lambda}{2} \right\rfloor^{\frac{1}{2}} \left(\frac{\lambda}{\left\lfloor \frac{\lambda}{2} \right\rfloor} \right)^{\left\lfloor \frac{\lambda}{2} \right\rfloor}. \tag{4.85}$$

Finally, if we remark that $\lfloor \lambda/2 \rfloor = O(\lambda/2)$, we obtain

$$2\lambda \left| \sum_{n=1}^{\left\lfloor \frac{\lambda}{2} \right\rfloor} \frac{e^{-\lambda}\lambda^n}{n!} n\, R_3\left(\frac{n-\lambda}{\lambda}\right) \right| = O\left(\lambda^{3/2} \left(\frac{e}{2}\right)^{-\lambda/2}\right), \tag{4.86}$$

and therefore the partial sum tends to zero exponentially fast.

As concerns the series with $n \geqslant \lfloor \lambda/2 \rfloor + 1$, observing that in the expression of $R_3(\xi)$, with $\xi > -1/2$, we have $1 + t > 1/2$, by considering separately the cases $\xi < 0$ and $\xi > 0$ we find that in both cases we have

$$|R_3(\xi)| \leqslant \frac{1}{3}\xi^4 \tag{4.87}$$

which implies

$$2\lambda\left|\sum_{n=\left[\frac{\lambda}{2}\right]+1}^{+\infty}\frac{e^{-\lambda}\lambda^n}{n!}R_3\left(\frac{n-\lambda}{\lambda}\right)\right| \leqslant \frac{2\lambda}{3}\sum_{n=0}^{\infty}\frac{e^{-\lambda}\lambda^n}{n!}\left(\frac{n-\lambda}{\lambda}\right)^4 = \tag{4.88}$$
$$\frac{2}{\lambda^3}\mu_4 = \frac{2}{\lambda^3}(3\lambda^2+\lambda) = O\left(\frac{1}{\lambda}\right),$$

where we have taken into account that the central moment μ_4 is given by $3\lambda^2 + \lambda$. Thus the proof of the lemma is complete.

4.6.3 Proof of lemma 4.2

Let x_β^* be the unique minimizer of $f_\beta(x; y)$ and let $0 \leqslant \beta_1 < \beta_2$; we need to prove the inequalities: (i) $f_0(x_{\beta_1}^*; y) < f_0(x_{\beta_2}^*; y)$ and (ii) $f_1(x_{\beta_1}^*) > f_1(x_{\beta_2}^*)$.

(i) The convexity of $f_0(x; y)$ gives

$$f_0(x_{\beta_2}^*; y) \geqslant f_0(x_{\beta_1}^*; y) + \nabla f_0\left(x_{\beta_1}^*; y\right)^T(x_{\beta_2}^* - x_{\beta_1}^*). \tag{4.89}$$

Since $x_{\beta_1}^*$ is the minimizer of $f_{\beta_1}(x; y)$, it satisfies the optimality condition

$$\nabla f_{\beta_1}\left(x_{\beta_1}^*; y\right)^T(x - x_{\beta_1}^*) \geqslant 0, \quad \forall x \in C; \tag{4.90}$$

then, letting $x = x_{\beta_2}^*$ we get

$$\left(\nabla f_0(x_{\beta_1}^*; y) + \beta_1\nabla f_1(x_{\beta_1}^*)\right)^T(x_{\beta_2}^* - x_{\beta_1}^*) \geqslant 0, \tag{4.91}$$

that is

$$\nabla f_0\left(x_{\beta_1}^*; y\right)^T(x_{\beta_2}^* - x_{\beta_1}^*) \geqslant -\beta_1\nabla f_1\left(x_{\beta_1}^*\right)^T(x_{\beta_2}^* - x_{\beta_1}^*). \tag{4.92}$$

The optimality of $x_{\beta_2}^*$ and the strict convexity of $f_\beta(x; y)$ yield

$$f_{\beta_2}(x_{\beta_1}^*; y) \geqslant f_{\beta_2}(x_{\beta_2}^*; y) > f_{\beta_2}(x_{\beta_1}^*; y) + \nabla f_{\beta_2}\left(x_{\beta_1}^*; y\right)^T(x_{\beta_2}^* - x_{\beta_1}^*) \tag{4.93}$$

so that

$$\nabla f_{\beta_2}\left(x_{\beta_1}^*; y\right)^T(x_{\beta_2}^* - x_{\beta_1}^*) < 0, \tag{4.94}$$

which implies

$$-\nabla f_0\left(x_{\beta_1}^*; y\right)^T(x_{\beta_2}^* - x_{\beta_1}^*) > \beta_2\nabla f_1\left(x_{\beta_1}^*\right)^T(x_{\beta_2}^* - x_{\beta_1}^*). \tag{4.95}$$

Adding inequalities (4.92) and (4.95) and using $(\beta_2 - \beta_1) > 0$ gives

$$\nabla f_1\left(x^*_{\beta_1}\right)^T (x^*_{\beta_2} - x^*_{\beta_1}) < 0; \tag{4.96}$$

thus, from equation (4.92) we obtain

$$\nabla f_0\left(x^*_{\beta_1}; y\right)^T (x^*_{\beta_2} - x^*_{\beta_1}) > 0 \tag{4.97}$$

and from inequality (4.89) we conclude $f_0(x^*_{\beta_1}; y) < f_0(x^*_{\beta_2}; y)$.

(ii) The convexity of $f_1(x)$ implies

$$f_1(x^*_{\beta_1}) \geqslant f_1(x^*_{\beta_2}) + \nabla f_1\left(x^*_{\beta_2}\right)^T (x^*_{\beta_1} - x^*_{\beta_2}). \tag{4.98}$$

From the optimality condition for $x^*_{\beta_2}$ we have

$$\left(\nabla f_0(x^*_{\beta_2}; y) + \beta_2 \nabla f_1(x^*_{\beta_2})\right)^T (x^*_{\beta_1} - x^*_{\beta_2}) \geqslant 0, \tag{4.99}$$

and consequently

$$\nabla f_0\left(x^*_{\beta_2}; y\right)^T (x^*_{\beta_1} - x^*_{\beta_2}) \geqslant -\beta_2 \nabla f_1\left(x^*_{\beta_2}\right)^T (x^*_{\beta_1} - x^*_{\beta_2}). \tag{4.100}$$

Moreover, the optimality of $x^*_{\beta_1}$ and the strict convexity of $f_\beta(x; y)$ imply

$$f_{\beta_1}(x^*_{\beta_2}; y) \geqslant f_{\beta_1}(x^*_{\beta_1}; y) > f_{\beta_1}(x^*_{\beta_2}; y) + \nabla f_{\beta_1}\left(x^*_{\beta_2}; y\right)^T (x^*_{\beta_1} - x^*_{\beta_2}), \tag{4.101}$$

so that

$$\nabla f_{\beta_1}\left(x^*_{\beta_2}; y\right)^T (x^*_{\beta_1} - x^*_{\beta_2}) < 0, \tag{4.102}$$

which implies

$$-\nabla f_0\left(x^*_{\beta_2}; y\right)^T (x^*_{\beta_1} - x^*_{\beta_2}) > \beta_1 \nabla f_1\left(x^*_{\beta_2}\right)^T (x^*_{\beta_1} - x^*_{\beta_2}). \tag{4.103}$$

By adding inequalities (4.100) and (4.103) and using $(\beta_1 - \beta_2) < 0$ we obtain

$$\nabla f_1\left(x^*_{\beta_2}\right)^T (x^*_{\beta_1} - x^*_{\beta_2}) > 0, \tag{4.104}$$

and from inequality (4.98) we conclude $f_1(x^*_{\beta_1}) > f_1(x^*_{\beta_2})$.

References

[1] Anconelli B, Bertero M, Boccacci P, Desiderá G, Carbillet M and Lantéri H 2006 Deconvolution of multiple images with high dynamic range and an application to LBT LINC-NIRVANA *Astron. Astrophys.* **460** 349–55

[2] Bardsley J M and Goldes J 2009 Regularization parameter selection methods for ill-posed Poisson maximum-likelihood estimation *Inverse Probl.* **25** 095005

[3] Barrett H H and Meyers K J 2003 *Foundations of Image Science* (New York: Wiley)

[4] Benfenati A and Ruggiero V 2013 Inexact Bregman iteration with an application to Poisson data reconstruction *Inverse Probl.* **29** 065016

[5] Bertero M and Boccacci P 1998 *Introduction to Inverse Problems in Imaging* (Bristol: Institute of Physics)

[6] Bertero M, Boccacci P, Talenti G, Zanella R and Zanni L 2010 A discrepancy principle for Poisson data *Inverse Probl.* **26** 105004

[7] Bonettini S and Ruggiero V 2010 On the uniqueness of the solution of image reconstruction problems with Poisson data *AIP Conf. Proc. ICNAAM 2010* vol **1281** 1803–6

[8] Bregman L M 1967 The relaxation method of finding the common points of convex sets and its applications to the solution of problems in convex optimization *USSR Comput. Math. Math. Phys.* **7** 200–17

[9] Brune C, Sawatzky A and Burger M 2009 Bregman-EM-TV methods with application to optical nanoscopy Scale Space and Variational Methods in Computer Vision. SSVM 2009 ed X C Tai, K Mørken, M Lysaker and K A Lie (Lecture Notes in Computer Science vol 5567) (Berlin, Heidelberg: Springer) pp 235–46

[10] Brune C, Sawatzky A and Burger M 2010 Primal and dual Bregman methods with application to optical nanoscopy *Int. J. Comput. Vis.* **92** 211–29

[11] Engl H W, Hanke M and Neubauer A 1996 *Regularization of Inverse Problems* (Dordrecht: Kluwer)

[12] Forthman P, Kohler T, Begemann P G C and Defrise M 2007 Penalized maximum-likelihood sinogram restoration for dual focal spot computed tomography *Phys. Med. Biol.* **52** 4513–23

[13] Geman S and Geman D 1984 Stochastic relaxation, Gibbs distributions, and the Bayesian restoration of images *IEEE Trans. Pattern Anal. Mach. Intell.* **6** 721–41

[14] Geman S and McClure D E 1985 Bayesian image analysis: An application to single photon emission tomography, *Proc. Statist. Comput. Sect., American Statist. Assoc.* pp 12–8 www.dam.brown.edu/people/geman/Homepage/Image%20processing,%20image%20analysis,%20Markov%20random%20fields,%20and%20MCMC/1985GemanMcClureASA.pdf

[15] Goldstein T and Osher S 2009 The split Bregman method for ℓ_1 regularized problems *SIAM J. Imaging Sci.* **2** 323–43

[16] Grenander U 1984 *Tutorial in Pattern Theory* (Providence, RI: Brown University)

[17] Hansen P C, Nagy J G and O'Leary D P 2006 *Deblurring Images: Matrices, Spectra and Filtering* (Philadelphia, PA: SIAM)

[18] Harmany Z T, Marcia R F and Willett R M 2012 This is SPIRAL-TAP: Sparse Poisson intensity reconstruction algorithms - Theory and practice *IEEE Trans. Image Process* **21** 1084–96

[19] Kaipio J and Somersalo E 2005 *Statistical and Computational Inverse Problems* (Berlin: Springer)

[20] Kiwiel K C 1997 Proximal minimization methods with generalized Bregman functions *SIAM J. Control Optim.* **35** 1142–68

[21] Kullback S and Leibler R A 1951 On information and sufficiency *Ann. Math. Stat.* **22** 79–86

[22] Natterer F and Wübbeling F 2001 *Mathematical Methods in Image Reconstruction* (Philadelphia, PA: SIAM)

[23] Osher S, Burger M, Goldfarb D, Xu J and Yin W 2005 An iterative regularization method for total variation-based image restoration *SIAM J. Multiscale Model. Simul.* **4** 460–89

[24] Papoulis A 1965 *Probability, Random Variables and Stochastic Processes* (New York: McGraw-Hill)

[25] Puetter R C, Gosnell T R and Yahil A 2005 Digital image reconstruction: Deblurring and denoising *Annu. Rev. Astron. Astrophys.* **43** 139–94

[26] Snyder D L and Miller M I 1985 The use of sieves to stabilize images produced with the EM algorithm for emission tomography *IEEE Trans. Nucl. Sci.* **32** 3864–72

[27] Staglianó A, Boccacci P and Bertero M 2011 Analysis of an approximate model for Poisson data reconstruction and a related discrepancy principle *Inverse Probl.* **27** 125003

[28] Wang Z, Bovik A C, Sheikh H R and Simoncelli E P 2004 Image quality assessment: From error visibility to structural similarity *IEEE Trans. Image Process* **13** 600–12

[29] Yin W, Osher S, Goldfarb D and Darbon J 2008 Bregman iterative algorithms for l_1-minimization with applications to compressed sensing *SIAM J. Imaging Sci.* **1** 143–68

[30] Zanella R, Boccacci P, Zanni L and Bertero M 2009 Efficient gradient projection methods for edge-preserving removal of Poisson noise *Inverse Probl.* **25** 045010

[31] Zanella R, Boccacci P, Zanni L and Bertero M 2013 Corrigendum: Efficient gradient projection methods for edge-preserving removal of Poisson noise *Inverse Probl.* **29** 119501

[32] Zanni L, Benfenati A, Bertero M and Ruggiero V 2015 Numerical methods for parameter estimation in Poisson data inversion *J. Math. Imaging Vis.* **52** 397–413

Chapter 5

Simple reconstruction methods

In this chapter we discuss the first iterative methods proposed for solving the minimization problems derived from the ML and Bayesian approaches and described in the previous chapter. Firstly, we focus on the renowned iterative method known as the *expectation maximization* method in emission tomography and as *Richardson–Lucy* method in microscopy and astronomy. Both convergence and semi-convergence properties of this method are discussed as well as its acceleration provided by the *ordered subset expectation maximization* method in the case of emission tomography. Next, we describe methods, obtained with simple modifications of the previous one, for the computation of MAP estimates. These methods are particular cases of scalar gradient methods and, consequently, they are only applicable to the case of differentiable regularization functions (smooth problems). In addition, they implement in a natural way the constraint of non-negativity. Therefore, they provide a useful introduction to the specific optimization methods discussed in the next chapter. The supplementary material contains convergence proofs and discussion of additional regularization functions.

5.1 Expectation maximization (EM) or Richardson–Lucy (RL) method

In the case of Tikhonov regularization of the least-squares problem, explicit solutions of the corresponding minimization problems are available in several instances, for instance in deconvolution problems. Examples can be found in [3]. No such example is known in the case of the data-fidelity function for Poisson data and of its regularized versions such as those introduced in the previous chapter. Therefore, the unique way for computing the minimizers is to use iterative methods, i.e. methods which, starting from an initial guess, step by step improve the reconstructed source object by comparing the corresponding computed data to the detected ones.

The first method we consider applies to the minimization of the data-fidelity function $f_0(x; y)$ and nowadays is a classic one, since it is still used in many applications and is considered a benchmark for comparison with other methods solving the same problem. It was introduced as a method for computing ML solutions in emission tomography by L A Shepp and Y Vardi [36] and in this context is known as *expectation maximization* (EM), since it is a particular case of a general method in statistics, with this denomination, introduced for computing ML solutions [6]. In a different context the same iterative method was also proposed, independently, by W H Richardson [32] and L B Lucy [26] and, for this reason, it is also known as Richardson–Lucy (RL) method for applications to astronomy and microscopy. In this book we combine the two denominations by calling the algorithm EM–RL, to specify that it is a particular application of EM to Poisson data inversion.

A first application to microscopy is investigated in [13–15] (see also [34] for an application to 4Pi microscopy), while application to astronomy was boosted by the Hubble space telescope (HST) optical aberration problem at the beginning of its mission, before the implementation of the corrective optics in late 1993. The four years from early 1990 to late 1993 were years of very active research [11, 41], mainly based on this algorithm which, in that context, was called the Lucy method.

We give the algorithm in the form proposed in [31, 38] for taking into account possible background emission. In fact, in the presence of a background, the non-negativity constraint (which is implicit in EM–RL) is active, in the reconstruction of the object above the background, only if this modified form of the iteration is used; otherwise, artifacts may appear. Usually they take the form of a sequence of alternating dark and bright rings around very bright sources and, for this reason, they are usually called *ringing effects* [25]. An example is shown in figure 5.1.

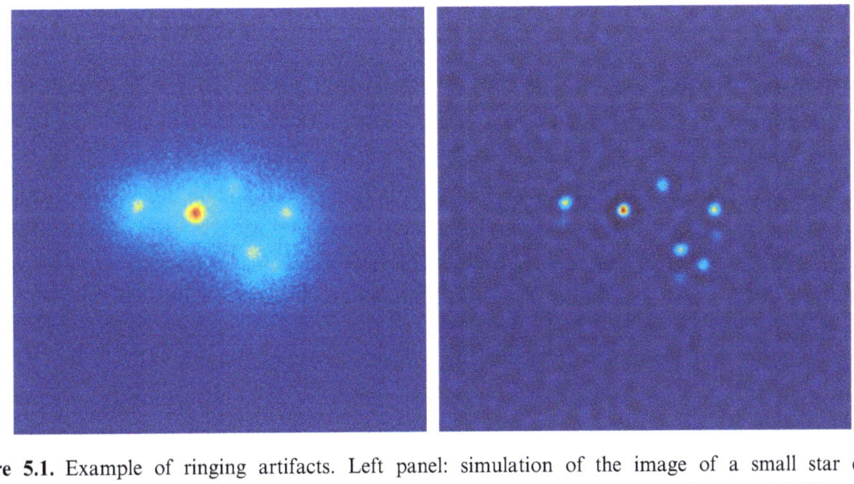

Figure 5.1. Example of ringing artifacts. Left panel: simulation of the image of a small star cluster superimposed to a constant background. Right panel: reconstruction obtained by the EM–RL method without background correction.

If $v > 0$ is the vector introduced in equation (4.11), then, for $k = 0, 1, \ldots$, the iteration, which takes into account background emission (see equation (3.30)), has the following form

$$x^{(k+1)} = \frac{x^{(k)}}{v} H^T \frac{y}{Hx^{(k)} + b}, \tag{5.1}$$

where, as in equation (4.15), quotients and products of vectors are component-wise. The iteration is, in general, initialized with a constant array/cube. If the imaging matrix satisfies the normalization condition of equation (3.32), then $v = 1$.

As already remarked, the algorithm was derived by Shepp and Vardi using the EM approach to the solution of maximum likelihood problems. Their derivation is not easy and therefore we omit it in this book. The interested reader can look at their paper or at our previous book ([3], appendix G). Here, for the convenience of the reader, we give a short heuristic derivation. Indeed, by taking into account that $v = H^T\mathbf{1}$, we can write the first KKT condition of equation (4.15) as the following fixed point equation

$$x^* = \frac{x^*}{v} H^T \frac{y}{Hx^* + b}. \tag{5.2}$$

By applying the fixed point method we just obtain equation (5.1).

We summarize the main properties of the EM–RL iteration.

- If $x^{(0)}$ is strictly positive, then, thanks to the conditions (3.31) satisfied by the imaging matrix, one can easily prove by induction that all the iterates are strictly positive, since $Hx^{(0)}$ is also strictly positive. Therefore, the algorithm is well defined.
- If $b = 0$ then, for any k, $x^{(k)}$ satisfies the flux condition already proved for the minimizers of $f_0(x; y)$ (see equation (4.20)); this condition implies (4.21) if the matrix H satisfies the normalization conditions (3.32) (see lemma 5.1).
- If $b = 0$, the iteration decreases the value of the objective function, i.e. $f_0(x^{(k)}; y) \geqslant f_0(x^{(k+1)}; y)$ (see lemma 5.2); therefore EM–RL is a descent method.
- If $b = 0$, the iteration converges to a minimizer x^* of the data-fidelity function. The flux condition (4.20) or (4.21) is a key point in all the convergence proofs we know [19, 20, 22, 29, 39]. A proof based on analytical techniques, derived from [29], is given in [30], chapter 5.3.2. For the convenience of the reader we give this proof in section 5.5.1. We remark that the convergence of the algorithm with $b \neq 0$ was proved only recently [33].
- The algorithm is easily implementable and the main computational cost per iteration is the computation of two matrix–vector multiplications. If H is a cyclic convolution this implies the computation of four FFT.

Remark 5.1. *As already remarked, an important feature of the EM–RL iteration in the case $b = 0$ is that, besides non-negativity, the iteration satisfies automatically the* *flux condition. Indeed, if we multiply by H both sides of equation (5.1) and we sum with respect to the index i, we obtain the following relationship*

$$\sum_{i\in\mathcal{I}}(Hx^{(k)})_i = \sum_{i\in\mathcal{I}} \frac{(Hx^{(k-1)})_i}{(Hx^{(k-1)} + b)_i} y_i, \tag{5.3}$$

whose interpretation is not easy if $b \neq 0$. But, if $b = 0$, then one obtains the condition (4.20) (with x^* replaced by $x^{(k)}$) or condition (4.21) if H satisfies the normalization condition. Therefore, in the case of $b \neq 0$, as well as in the case of Bayesian regularization, it could be important to consider a constrained minimization of the data-fidelity function, or in general of the Bayesian objective function, with a constraint set given by equation (4.22). Such a constrained minimization is possible with the optimization methods introduced in chapter 6.

Since the limit of the sequence generated by EM–RL iteration is a minimizer of the data-fidelity function $f_0(x; y)$, as follows from the discussion in chapter 4 this limit is not a sensible solution, except in particular cases (an example is shown in figure 4.3). However, numerical practice shows that the algorithm has a *semi-convergent* behavior: roughly speaking, this means that the iterations first approach a sensible solution and then go away. In other words, early stopping of the iterations has a sort of regularization effect. This property can be proved for some iterative methods applicable to the solution of the least-squares problem; one example is the well-known Landweber method (for a discussion see, for instance, [3]). As far as we know, no such proof is available in the case of EM–RL (however, see section 9.1), so that semi-convergence should be considered an empirical property of this algorithm. This property is very useful in practice because, in such a way, one can easily obtain a first reconstruction of the source object, even if this reconstruction is not very accurate.

Another important point is that, if we consider source objects with different numbers of emitted photons (i.e. different observation times, or different brightness in astronomy, or different doses in medical imaging), the quality of the reconstruction of the sources with higher numbers of photons is higher than that of the sources with lower number of photons; but the number of iterations required in the first case is larger than that of the second case. Since EM–RL is very slow, hundreds or thousands of iterations may be required to obtain a sensible solution. Therefore, the fast semi-convergent methods discussed in the next chapter are very useful in practice, especially in the case of very large images.

Note that a larger number of photons implies a smaller relative error on the data (see the discussion in section 3.4) and therefore it is quite reasonable to argue that the reconstruction improves if the relative error decreases: a goal that can be reached if the observation time or the dose administered to the patient increases. Again, in the case of least-squares problems it is possible to give quantitative arguments justifying this property (for the Landweber method see [3], chapter 6) while, in the case of EM–RL this is also an empirical fact.

We illustrate the semi-convergence of EM–RL by means of a numerical example. The synthetic source object is the one shown in figure 4.2 where we also show the reconstruction provided by a minimizer of the data-fidelity function, which is the limit of the sequence generated by EM–RL algorithm. We consider also a second brighter version which is obtained by multiplying by 15 the first object. In both cases we add a background equal to $b = 6.8 \times 10^3$. Next, both results are convolved with an ideal PSF and perturbed by Poisson noise. The value of b is inserted in the iteration (5.1), initialized with a constant array. Since the PSF is normalized to 1, in the same equation we have $\upsilon = 1$. Moreover, we assume that the best reconstruction is that provided by the value of k which minimizes the relative rms error defined by

$$\rho^{(k)} = \frac{\|x^{(k)} - \bar{x}\|}{\|\bar{x}\|}, \tag{5.4}$$

where $\|.\|$ denotes the ℓ_2 norm and $\bar{x}$ is the synthetic source object.

In figure 5.2 we show one still of a video where, for the two cases, we show in the upper panels the rms error as a function of the number of iterations (that with higher noise to the left and the other to the right); the running of the iterations is represented by a red circle moving along the curve. In the lower panels the corresponding reconstructions are given. Since the frames represented in the figure are obtained with 600 iterations, in the first case (left panels) the iteration is already beyond the minimum value of the rms error and the checkerboard effect (see figure 4.2) already

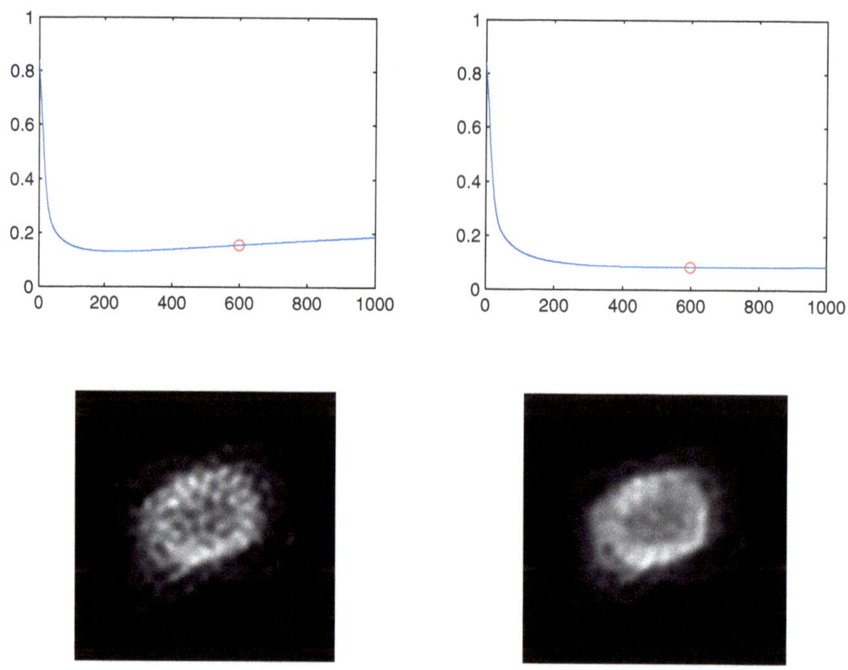

Figure 5.2. Video clip of the motion of the iterates along the error curve for the two examples described in the text, with a simultaneous representation of the corresponding reconstructions. The figure shows one still of this animation. Video available at http://iopscience.iop.org/book/978-0-7503-1437-4.

appears in the reconstruction, while in the second one the reconstruction is not yet the optimal one.

To show the best results achievable with this approach, in the first line of figure 5.3 we show the blurred and noisy images, while in the second line we show the reconstructions corresponding to the minimum of the rms error. In the first case (left panels) the minimum is reached after 230 iterations and its value is 13.4% while, in the second case, the minimum is reached after 340 iterations and its value is 6.3%. In these reconstructions the background should be zero; in practice it takes very small positive values since, as we know, the iterates are strictly positive. A comparison with figure 4.2, shows that early stopping of the EM–RL iteration definitely provides a much better result than the minimizer of the data-fidelity function, i.e. the limit of the sequence generated by EM–RL.

An important remark, which opens the way to the introduction of important accelerations of the EM–RL method is the following: if we take into account the expression of the gradient of the data-fidelity function (see equation (4.12)) and we remember that we set $H^T\mathbf{1} = v$, it is easy to recognize that the EM–RL iteration can be written in the following form

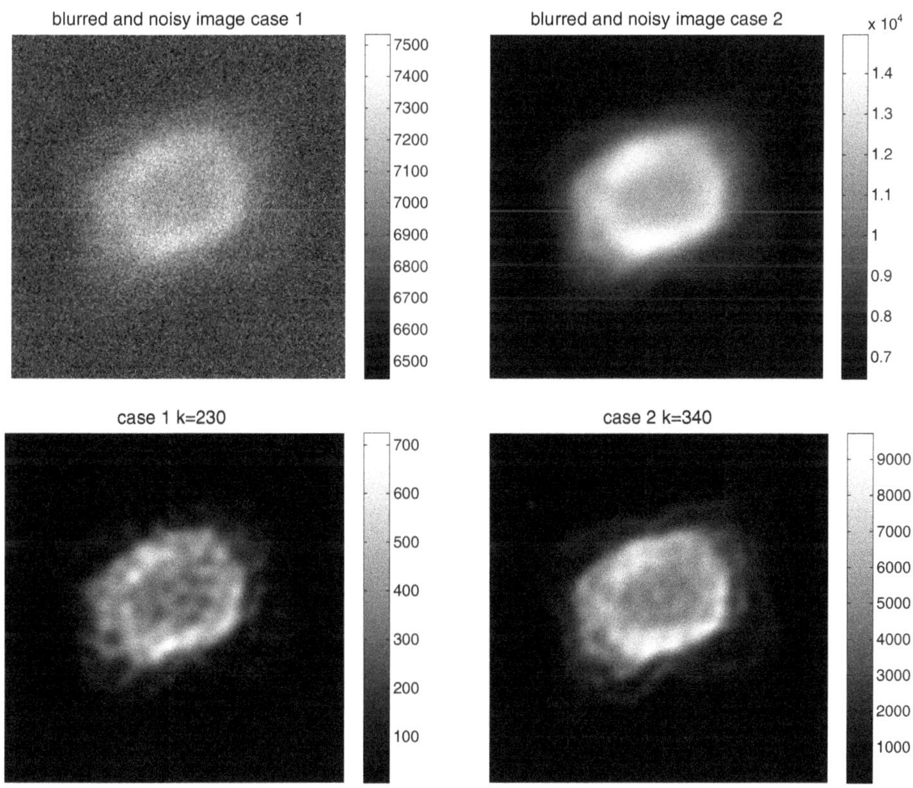

Figure 5.3. Examples of reconstructions provided by EM–RL with early stopping of the iterations. Upper panels: the two images with different brightness. Lower panels: the corresponding reconstructions.

$$x^{(k+1)} = x^{(k)} - \frac{x^{(k)}}{v}\nabla f_0(x^{(k)}; y); \tag{5.5}$$

therefore, the EM–RL method is a descent method with a descent direction given by the negative gradient multiplied by a diagonal and positive scaling factor given by the current iteration; the step-size in this direction is 1.

This remark suggests a first modification of EM–RL which was proposed for reducing the number of iterations and consists in introducing a suitable step-size α_k, depending on the iteration, so that the iteration takes the following form

$$x^{(k+1)} = x^{(k)} - \alpha_k\frac{x^{(k)}}{v}\left(v - H^T\frac{y}{Hx^{(k)} + b}\right). \tag{5.6}$$

The main issue is the choice of the step-size that, in the EM–RL literature is sometimes called *relaxation parameter* (see, for instance, [16] in the case of microscopy, [1] in the case of astronomy and, in a more general context, [24]). The approach is able to reduce the number of iterations but implies an increase of the computational cost per iteration due to the search of the relaxation parameter, so that the final gain may not be very relevant.

The approach proposed in [24] requires that, if the current iteration is non-negative, then also $x^{(k+1)}$ must be non-negative. From equation (5.6) it is easy to derive that the maximum value of α_k allowed by this condition is given by

$$\alpha_{\text{boundary}}(x^{(k)}) = \min_{i\in\mathcal{G}_k}\frac{v}{\left(v - H^T\frac{y}{Hx^{(k)} + b}\right)_i} \tag{5.7}$$

where $\mathcal{G}_k$ is the intersection of the indices of the positive components of $x^{(k)}$ and of the positive components of $\left(v - H^T\frac{y}{Hx^{(k)}+b}\right)$. We note that $\alpha_{\text{boundary}}(x^{(k)})$ is certainly greater than 1. Next an optimal step-size can be obtained by a line search procedure in an interval $(0, \alpha_{\text{lim}}(x^{(k)})$ in the direction $d^{(k)} = -\text{diag}(x^{(k)})\nabla f_0(x^{(k)}; y)$, with $\alpha_{\text{lim}}(x^{(k)}) \leqslant \alpha_{\text{boundary}}(x^{(k)})$.

In section 5.5.2 we give a proof of convergence of relaxation methods as applied to EM–RL as well as to other iterative methods considered in this chapter, thanks to their similar structure as descent methods with a descent direction given by a diagonal scaling of the negative gradient. The proof is derived from a proof given in [21] for the relaxed version of the method discussed in section 5.3. However, it is important to remark that a crucial point in the proof is the strict convexity of the objective function, i.e. the data-fidelity function $f_0(x; y)$ or the regularized function $f_\beta(x; y)$. Since this condition is not always satisfied by these functions, the result does not apply to all cases.

5.2 Ordered subset expectation maximization method

Since the convergence of EM–RL is very slow, a large number of iterations may be required even when early stopping is used to obtain a sensible solution. The number of iterations can be reasonable in cases where the number of detected photons is not too large. Such a situation may occur in some instances of microscopy and of medical imaging. However, in the latter case, the computational cost can still be quite high; indeed, the matrix describing the forward problem, even if sparse, is still very large without a structure allowing fast matrix–vector computation.

A considerable gain in computational time can be achieved by a method proposed by Husdson and Larkin in 1994 [18], with application to emission tomography. The idea consists in partitioning the data into (in general disjoint) ordered subsets and applying an EM–RL iteration to each subset; an iteration of the method, called by the authors *ordered subset expectation maximization* (OSEM), consists in a cycle over the selected subsets. The efficiency gained with this approach is such that OSEM can be a practical method in CT [2] and PET [5].

Subdivide the set $\mathcal{I}$ into (in general, non-overlapping) subsets $\mathcal{I}^{(l)}$, $l = 1, \dots, L$, such that $\mathcal{I} = \cup_{l=1}^{L}\mathcal{I}^{(l)}$. Denote as $y^{(l)}$ the data vector with components y_i, $i \in \mathcal{I}^{(l)}$ and as $H^{(l)}$ the block of the matrix H consisting of the rows with $i \in \mathcal{I}^{(l)}$. Moreover, for each l, introduce the vector $v^{(l)}$, defined by

$$v_j^{(l)} = \sum_{i\in\mathcal{I}^{(l)}} H_{ij},\ j \in \mathcal{J}. \tag{5.8}$$

We give the detailed scheme in algorithm 5.1.

Algorithm 5.1. OSEM method

Choose $x^{(0)} > 0$ and a subdivision of $\mathcal{I}$ into subsets $\mathcal{I}^{(l)}$, $l = 1, \dots, L$;
FOR $k = 0, 1, \dots$ UNTIL CONVERGENCE
STEP 1. set $x^{(k,0)} = x^{(k)}$;
STEP 2. FOR $l = 1, \dots, L$

$$x^{(k,l)} = \frac{1}{v^{(l)}} x^{(k,l-1)} (H^{(l)})^T \frac{y^{(l)}}{H^{(l)}x^{(k,l-1)} + b}$$

END
STEP 3. set $x^{(k+1)} = x^{(k,L)}$;
END

As a preliminary remark we observe that, when $b = 0$, from definition (5.8) and the algorithm, we obtain

$$\sum_{i\in\mathcal{I}^{(l)}} (Hx^{(k,l)})_i = \sum_{j\in\mathcal{J}} v_j^{(l)} x_j^{(k,l)} = \sum_{i\in\mathcal{I}^{(l)}} y_i; \tag{5.9}$$

this relationship indicates that, in order to avoid excessive oscillations of the ℓ_1 norm of the computed data inside one OSEM cycle, it is a good policy to select the subsets in such a way that they contain approximately the same number of photons.

As concerns the convergence of the algorithm, we remark that the lth step of the internal cycle is an EM–RL step for the minimization of the functional

$$f_0(x; y^{(l)}) = \sum_{i \in \mathcal{I}^{(l)}} \left\{ y_i \ln \frac{y_i}{(H^{(l)}x + b)_i} + (H^{(l)}x + b)_i - y_i \right\}; \tag{5.10}$$

therefore convergence can hold true only if all these functionals have the same non-negative minimizer x^*. Indeed, if $b = 0$, a proof of convergence exists in the so-called *consistent case*, i.e. there exists a non-negative solution x^* of the linear equation $Hx = y$, with the additional condition that the matrices $H^{(l)}$ are *balanced*, i.e. $v^{(l)}$ is independent of l. The proof, given in [18], is a straightforward extension of the proof of EM–RL (see section 5.5.1).

The proof of the convergence of OSEM in the non-consistent case may be a hopeless task; indeed, the best one can hope to prove is its cyclic convergence, i.e. each one of the sub-sequences $x^{(k, l)}$, with l fixed, is convergent to a minimizer of $f_0(x; y^{(l)})$. To our knowledge such a proof is not available.

The lack of a proof of convergence does not prevent the practical use of this algorithm in tomography where each subset consists of a suitable number of projections (see section 3.1.2). The important point is that OSEM, as EM–RL, has a semi-convergent behavior so that its acceleration effect means that we need a much smaller number of iterations for reaching a sensible solution similar to that provided by EM–RL. The reduction depends in a crucial way on the choice of the subsets and on their order in the internal cycle [18].

To this purpose a first important remark is that, by inspecting the two algorithms, EM–RL and OSEM, it is not difficult to realize that the cost per iteration in the two cases is approximately the same, at least in medical imaging which is the main application of OSEM. In particular, this is true in the specific case of SPECT (the application considered by Hudson and Larkin), where data can be structured into a sequence of projections. Therefore, a subset of OSEM can consist of a number of projections, possibly the same number for each subset. The number of subsets is called by Hudson and Larkin the OS *level*. Then, the main result, supported by theoretical arguments and numerical experiments is that, in the sense of semi-convergence, the number of iterations required by OSEM for reaching approximately the same result as EM–RL is about the number of EM–RL iterations divided by the OS level. Thanks to the previous remark this is also the gain in computational time.

We illustrate these features by means of a numerical simulation. We take as source object the well-known Shepp and Logan phantom [35], frequently used for testing numerical methods in tomography. We consider a 128×128 version with an average number of photons per pixel of 10, a maximum number of 34 and a total number of 3.7×10^5. We compute (see remark 3.2) 180 projections, from 0° to 179° with uniform spacing (1°). The corresponding sinogram is perturbed with Poisson noise. We compare FBP and EM–RL reconstructions with OSEM reconstructions corresponding to different OS levels. In figure 5.4 we plot the behavior of the relative rms error, defined in equation (5.4), as a function of the number of iterations. We consider a maximum of 200 iterations. The horizontal line corresponds to FBP, 28% error, while the curves represents the errors for different OS levels, this number being indicated above the curve. Therefore 1 corresponds to EM–RL which reaches the minimum after 179 iterations with a 16% error. The same error is obtained with an OS level of 18 after 9 iterations (which is approximately 179/18) while an increase of the OS order produces reconstructions with a higher error. As concerns the order of the subsets inside the OSEM cycle, for each OS level we take an angle of 90° between adjacent subsets, so that they contain significantly different information about the source object.

In figure 5.5 we compare the reconstruction provided by FBP with that provided by OSEM with 18 subsets (that provided by EM–RL is very similar).

We conclude this section by indicating an application of OSEM to astronomy [4], even if it refers to a very specific problem, namely Fizeau interferometry, which, as

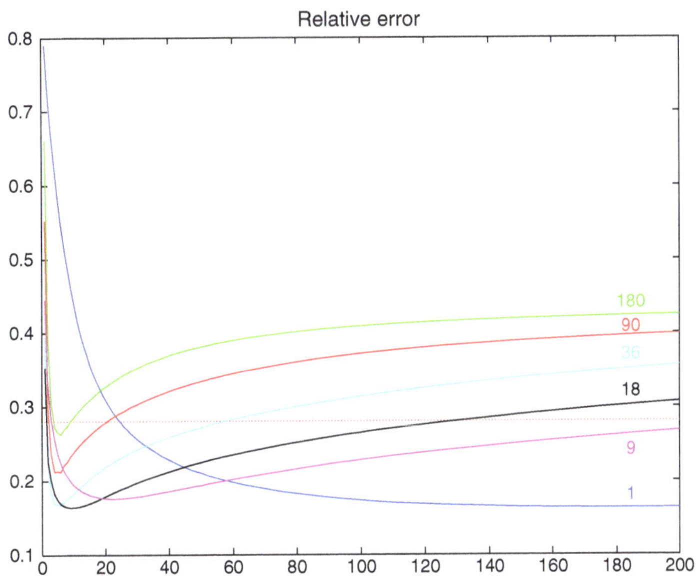

Figure 5.4. Reconstruction error $\rho^{(k)}$, equation (5.4), as a function of the number of iterations for EM–RL (blue line) and OSEM with different OS level (indicated above the curve). The horizontal line corresponds to the FBP error.

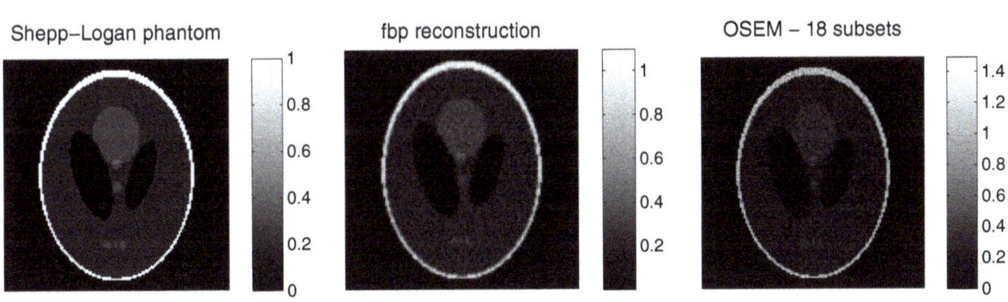

Figure 5.5. Shepp and Logan phantom (left panel), FBP reconstruction (central panel) and OSEM reconstruction with 18 subsets (right panel).

described in section 2.3.4.1, is a particular feature of only one telescope, the large binocular telescope (LBT), operating in the Mount Graham International Observatory, Arizona.

A Fizeau interferometer acquires L different images of the same astronomical target, corresponding to L different orientations of the baseline, $y^{(l)}, l = 1, 2, \dots, L$. Each image is associated with a specific PSF $h^{(l)}$, so that, thanks to the statistical independence of the different images, the data-fidelity function is given by

$$f_0(x; y) = \sum_{l=1}^{L} \sum_{i \in \mathcal{I}} \left\{ y_i^{(l)} \ln \frac{y_i^{(l)}}{(h^{(l)} * x + b^{(l)})_i} + (h^{(l)} * x + b^{(l)})_i - y_i^{(l)} \right\}. \tag{5.11}$$

Since all PSFs are normalized to unit ℓ_1 norm, it is easy to recognize that the EM–RL method for the minimization of this function is given by

$$x^{(k+1)} = x^{(k)} \frac{1}{L} \sum_{l=1}^{L} (h^{(l)})^T * \frac{y^{(l)}}{h^{(l)} * x^{(k)} + b^{(l)}}. \tag{5.12}$$

If now we select the different subsets of the given data as coinciding with the different acquired images, then it is obvious that algorithm 5.1 applies immediately to this situation. However, it is important to remark that the cost per iteration of EM–RL, as given by equation (5.12), and that of OSEM, is not the same. Indeed, for each iteration of EM–RL we must compute $3L + 1$ FFTs (only the FFT of $x^{(k)}$ must be computed) while we must compute $4L$ FFT for one OSEM iteration (the FFTs of all $x^{(k,l)}$ must be computed), with an increase in the computational cost given approximately by $4L/(3L + 1)$. Since, as it has been verified by numerical experiments, also in this case the reduction in the number of iterations is given by L, we obtain a gain in computational time approximately given by $4/(3L + 1)$. For instance, in the case of 9 images, the gain is a factor of 7, not very large but significant in the case of very large images.

We conclude by remarking that OSEM is also proposed as an acceleration of a regularized EM–RL in terms of Gibbs priors [18]. We discuss this point in the next sections.

5.3 One-step late (OSL) method

The checkerboard effect, characterizing the minimizers of the data-fidelity function and discussed in section 3.1, affects also the EM–RL iterates since they converge to one of these minimizers. In general, the effect appears after a large number of iterations since the convergence is slow, so that one can profit by the semi-convergent behavior of the algorithm discussed in section 5.1. In the case of OSEM the effect already appears after a small number of iterations.

First attempts to overcome this difficulty are based on the Bayesian approach proposed by Geman and Geman [7] and they are discussed in section 4.2. For this reason the focus is on a regularization function suggested in that paper: given a non-negative, symmetric and differentiable function ϕ, the following regularization term is introduced

$$f_1(x) = \frac{1}{2} \sum_{j \in \mathcal{I}} \sum_{j' \in \mathcal{N}_j} \phi\left(\frac{x_j - x_{j'}}{\epsilon_{j,j'}}\right), \tag{5.13}$$

where $\mathcal{N}_j$ is a suitable neighborhood of the pixel j, $\epsilon_{j,j'}$ is 1 for interactions between horizontal and orthogonal nearest neighbors and $\sqrt{2}$ for interactions between diagonal nearest neighbors. As first papers in this direction we can mention [8, 9, 12] and [10]. The interesting feature of the last paper is that a simple algorithm is proposed for the minimization of the regularized objective function $f_\beta(x; y)$, defined in equation (4.29). The algorithm, which consists in a simple modification of EM–RL, is easily implementable and is called by Green *one-step late* (OSL). According to his approach, it can be introduced as follows.

As in the case of EM–RL, let us write the first KKT condition for the minimization of $f_\beta(x; y)$

$$x_\beta^* \nabla f_\beta(x_\beta^*; y) = 0, \tag{5.14}$$

x_β^* being a minimizer of the objective function. Then, by taking into account the expression (4.12) of the gradient of the data-fidelity function, from the previous equation, writing, as usual, $v = H^T\mathbf{1}$, we obtain

$$x_\beta^*\left(v - H^T \frac{y}{Hx_\beta^* + b} + \beta \, \nabla f_1(x_\beta^*)\right) = 0; \tag{5.15}$$

by simple algebraic manipulations, we obtain the fixed point equation

$$x_\beta^* = \frac{x_\beta^*}{v + \beta \, \nabla f_1(x_\beta^*)} H^T \frac{y}{Hx_\beta^* + b}. \tag{5.16}$$

Finally, the OSL method is obtained by applying the fixed point method to this fixed point equation

$$x^{(k+1)} = \frac{x^{(k)}}{v + \beta \, \nabla f_1(x^{(k)})} H^T \frac{y}{Hx^{(k)} + b}. \tag{5.17}$$

If $v + \beta \;\; \nabla f_1(x) > 0$ on the non-negative orthant and the initial guess $x^{(0)}$ is positive, then, as in the case of EM–RL, a nice property is that all the iterates are positive if the initial guess $x^{(0)}$ is positive.

The main difficulties with OSL are

- the condition on the gradient of $f_1(x)$ indicated above is not satisfied by several regularization functions used in practice for arbitrary non-negative values of β;
- no convergence result is available.

The first problem is already considered in [10] where it is remarked that, if the modulus of the derivative of the function ϕ is bounded, then, for a sufficiently small value of the regularization parameter β, the denominator in equation (5.17) is positive. As a particular example, the function $\phi(t) = \ln \cosh(t)$ is considered; it has all the good properties since its derivative is bounded by 1. Moreover, it is easy to remark that the function behaves as t^2 for small t and as $|t|$ for large t so that the regularization function (5.13) has a behavior similar to that of the HS regularization introduced in section 4.2.

Both problems mentioned above were investigated in a paper by Lange [21]. As concerns the first one, he characterizes a class of regularization functions with a bounded gradient. As concerns convergence, he starts from the remark that the OSL iteration can be written in the following form

$$x^{(k+1)} = x^{(k)} - \frac{x^{(k)}}{v + \beta \nabla f_1(x^{(k)})} \nabla f_\beta(x^{(k)}; y), \tag{5.18}$$

so that, if the positivity of the denominator is assured, OSL appears as a descent method with a descent direction given by a diagonal and positive scaling of the negative gradient (remember that the iterates are positive). As for EM–RL algorithm, this remark suggests to look for a relaxed version, by introducing a line search along the descent direction for computing a suitable step-size α_k

$$x^{(k+1)} = x^{(k)} - \alpha_k \frac{x^{(k)}}{v + \beta \nabla f_1(x^{(k)})} \nabla f_\beta(x^{(k)}; y). \tag{5.19}$$

In the paper of Lange [21] an approximate line search is proposed and the convergence of the iterates to a solution is proved for strictly convex $f_\beta(x; y)$. As already remarked, in section 5.5.2 we report his proof in a form which can be also applied to the other relaxed methods considered in this chapter.

We conclude by remarking that OSEM method can also be applied to the OSL method if, in algorithm 5.1, one EM–RL iteration is replaced by one OSL iteration. Also, in this case no convergence proof is available.

5.4 Split gradient method (SGM)

The main drawback of OSL for an application to arbitrary differentiable regularization functions is the restriction to regularization functions f_1 with a bounded gradient. This drawback does not affect a similar method proposed by Lantéri *et al* [23]. Even if the method was proposed having in mind application to astronomical imaging, application to any kind of image affected by Poisson noise is possible (see, for instance, [40] for an application to 3D fluorescence microscopy).

The basic point is to consider a decomposition of the gradient of the following form

$$-\nabla f_1(x) = U_1(x) - V_1(x), \tag{5.20}$$

where U_1, V_1 are non-negative arrays (remark that, in other applications, one should require $V_1 > 0$). Such a decomposition always exists even if it is not unique. One choice can be given by the positive and negative part of the gradient and, given one choice, others can be obtained by adding an arbitrary non-negative array to both U_1, V_1. The interesting fact is that, in the case of the differentiable functions introduced in section 4.2 a 'natural' choice is suggested by the computation of the gradient. The method based on this decomposition is called by the authors *split gradient method* (SGM).

If we insert equation (5.20) into equation (5.15) we obtain

$$x_\beta^* \left(v - H^T \frac{y}{Hx_\beta^* + b} - \beta\, U_1(x_\beta^*) + \beta\, V_1(x_\beta^*) \right) = 0, \tag{5.21}$$

and from this condition, by simple algebraic manipulations, we derive the fixed point equation

$$x_\beta^* = \frac{x^*}{v + \beta\, V_1(x^*)} \left\{ H^T \frac{y}{Hx_\beta^* + b} + \beta\, U_1(x^*) \right\}. \tag{5.22}$$

With a procedure which is now standard in this chapter, we obtain the iterative method

$$x^{(k+1)} = \frac{x^{(k)}}{v + \beta\, V_1(x^{(k)})} \left\{ H^T \frac{y}{Hx^{(k)} + b} + \beta\, U_1(x^{(k)}) \right\}. \tag{5.23}$$

The iteration is slightly more complex that that of OSL but it is obvious that it can be used for any differentiable regularization function and that, if the iteration is initialized with a positive $x^{(0)}$, then all the iterates are strictly positive.

No convergence proof is available for this iterative method. However, it can be transformed into a scaled gradient method, as in the case of the previous algorithms, if we remark that it can be written in the following form

$$x^{(k+1)} = x^{(k)} - \frac{x^{(k)}}{v + \beta\, V_1(x^{(k)})} \nabla f_\beta(x^{(k)}; y). \tag{5.24}$$

The algorithm was presented by the authors in a relaxed form with the addition of a line search along the descent direction

$$x^{(k+1)} = x^{(k)} - \alpha_k \frac{x^{(k)}}{v + \beta\, V_1(x^{(k)})} \nabla f_\beta(x^{(k)}; y). \tag{5.25}$$

If $f_\beta(x; y)$ is strictly convex, the convergence proof given in section 5.5.2 applies also to this case. Also, in this case OSEM method can be applied if, in algorithm 5.1, one EM–RL iteration is replaced by one SGM iteration. Again, no convergence proof is available.

We give a few examples of possible choices of the arrays U_1, V_1 with reference to the regularization functions given in section 4.2. As already remarked, these choices are not unique but they have been already used in the applications of the method and they provided satisfactory results. They can be derived by computing the gradient of the regularization function (this is an easy but sometimes tedious exercise) and taking into account that the variable x belongs to the non-negative orthant since we are considering constrained minimizations. As concerns the discrete gradient of x we use the notations introduced in section 4.6.1.

- **Tikhonov regularizer T0**

$$U_1(x) = 0, \quad V_1(x) = x. \tag{5.26}$$

- **Tikhonov regularizer T1**—in the 2D case we have

$$[U_1(x)]_{k,l} = x_{k-1,l} + x_{k,l-1} + x_{k+1,l} + x_{k,l+1}, \; [V_1(x)]_{k,l} = 4x_{k,l}, \tag{5.27}$$

  and, in the 3D case

$$\begin{aligned}[U_1(x)]_{k,l,h} = {} & x_{k-1,l,h} + x_{k,l-1,h} + x_{k,l,h-1} + x_{k+1,l,h} \\ & + x_{k,l+1,h} + x_{k,l,h+1}, \; [V_1(x)]_{k,l,h} = 6x_{k,l,h}.\end{aligned} \tag{5.28}$$

- **Tikhonov regularizer T2**—in the 2D case we have

$$\begin{aligned}[U_1(x)]_{k,l} &= 4(x_{k,l-1} + x_{k,l+1} + x_{k-1,l} + x_{k+1,l}), \\ [V_1(x)]_{k,l} &= 12x_{k,l} + x_{k-2,l} + x_{k+2,l} + x_{k,l-2} + x_{k,l+2}\end{aligned} \tag{5.29}$$

  and, in the 3D case

$$\begin{aligned}[U_1(x)]_{k,l} = {} & 4\,(x_{k,l-1,h} + x_{k,l+1,h} + x_{k,1,h-1} + x_{k,l,h+1} \\ & + x_{k-1,l,h} + x_{k+1,l,h}) \\ [V_1(x)]_{k,l} = {} & 18x_{k,l,h} + x_{k-2,l,h} + x_{k+2,l,h} + x_{k,l-2,h} \\ & + x_{k,l+2,h} + x_{k,l,h-2} + x_{k,l,h+2}.\end{aligned} \tag{5.30}$$

- **Sparsity regularization in pixel space**

$$U_1(x) = 0, \quad V_1(x) = 1. \tag{5.31}$$

- **HS regularization**—in the 2D case we have

$$\begin{aligned}[U_1(x)]_{k,l} &= \frac{x_{k+1,l} + x_{k,l+1}}{\sqrt{\delta^2 + (\Delta x)^2_{k,l}}} + \frac{x_{k-1,l}}{\sqrt{\delta^2 + (\Delta x)^2_{k-1,l}}} + \frac{x_{k,l-1}}{\sqrt{\delta^2 + (\Delta x)^2_{k,l-1}}} \\ [V_1(x)]_{k,l} &= \frac{2x_{k,l}}{\sqrt{\delta^2 + (\Delta x)^2_{k,l}}} + \frac{x_{k,l}}{\sqrt{\delta^2 + (\Delta x)^2_{k-1,l}}} + \frac{x_{k,l}}{\sqrt{\delta^2 + (\Delta x)^2_{k,l-1}}}\end{aligned} \tag{5.32}$$

  and, in the 3D case

$$
\begin{aligned}
U_1(x)]_{k,l,h} &= \frac{x_{k+1,l,h} + x_{k,l+1,h} + x_{k,l,h+1}}{\sqrt{\delta^2 + (\Delta x)^2_{k,l,h}}} + \frac{x_{k-1,l,h}}{\sqrt{\delta^2 + (\Delta x)^2_{k-1,l,h}}} \\
&+ \frac{x_{k,l-1,h}}{\sqrt{\delta^2 + (\Delta x)^2_{k,l-1,h}}} + \frac{x_{k,l,h-1}}{\sqrt{\delta^2 + (\Delta x)^2_{k,l,h-1}}} \\
V_1(x)]_{k,l,h} &= \frac{3x_{k,l,h}}{\sqrt{\delta^2 + (\Delta x)^2_{k,l,h}}} + \frac{x_{k,l,h}}{\sqrt{\delta^2 + (\Delta x)^2_{k-1,l,h}}} \\
&+ \frac{x_{k,l,h}}{\sqrt{\delta^2 + (\Delta x)^2_{k,l-1,h}}} + \frac{x_{k,l,h}}{\sqrt{\delta^2 + (\Delta x)^2_{k,l,h-1}}}.
\end{aligned}
\tag{5.33}
$$

We leave as an exercise to the reader to compute the functions U_1, V_1 at the boundaries according to the selected boundary conditions. Such a computation is obviously required for the implementation of the method. Other examples of regularization functions and of corresponding U, V decomposition are given in section 5.5.3.

5.5 Supplementary material

5.5.1 Convergence of the EM–RL algorithm

For the convenience of the reader in this subsection we give the proof of the convergence of the EM–RL algorithm. The proof is an adaptation to our notations of the proof given in [29, 30]. As already remarked, the convergence of the iterates is obtained only in the case $b = 0$, so that the algorithm takes the following form

$$x^{(k+1)} = x^{(k)} H^T \frac{y}{Hx^{(k)}}. \tag{5.34}$$

In order to avoid purely technical difficulties, we assume $y > 0$; moreover we assume that the matrix H satisfies condition (3.32) and that all its entries are positive, so that we have $Hx > 0$ for any x with at least one positive component. We remark that this condition on H is, in general, satisfied in microscopy and astronomy, but not in tomography where the matrices are sparse.

Lemma 5.1. *If* $x^{(0)} > 0$, *then, for any* k, *we have* $x^{(k)} > 0$. *Moreover, the following identities hold true*

$$\sum_{i \in \mathcal{I}} (Hx^{(k)})_i = \sum_{j \in \mathcal{J}} x_j^{(k)} = \sum_{i \in \mathcal{I}} y_i. \tag{5.35}$$

Proof. The first statement can be proved by induction, taking into account the assumptions on y and the matrix H.

As concerns the identities, the first one can be derived from remark 5.1. The second derives from the iteration (5.34). Indeed, we have

$$\sum_{j \in \mathcal{J}} x_j^{(k)} = \sum_{j \in \mathcal{J}} x_j^{(k-1)} \sum_{i \in \mathcal{I}} H_{i,j} \frac{y_i}{(Hx^{(k-1)})_i} = \sum_{i \in \mathcal{I}} y_i \sum_{j \in \mathcal{J}} \frac{H_{i,j} x_j^{(k-1)}}{(Hx^{(k-1)})_i}, \tag{5.36}$$

and the last sum over j is 1. □

Lemma 5.2. *The following inequalities hold true*

$$f_0(x^{(k)}; y) - f_0(x^{(k+1)}; y) \geqslant \mathrm{KL}(x^{(k+1)}; x^{(k)}) \geqslant 0 \tag{5.37}$$

$$\mathrm{KL}(x^*; x^{(k)}) - \mathrm{KL}(x^*; x^{(k+1)}) \geqslant f_0(x^{(k)}; y) - f_0(x^*; y) \geqslant 0, \tag{5.38}$$

where x^* is a generic limit point of the sequence $x^{(k)}$, $k = 0, 1, \ldots$.

Proof. Consider the first inequality. Thanks to the identities (5.35) we can write

$$f_0(x^{(k)}; y) - f_0(x^{(k+1)}; y) = \sum_{i \in \mathcal{I}} y_i \ln (Hx^{(k+1)})_i - \sum_{i \in \mathcal{I}} y_i \ln (Hx^{(k)})_i. \tag{5.39}$$

Consider now the following identity, which holds true for any $h, \; x > 0$

$$\sum_{i \in \mathcal{I}} y_i \ln (Hx)_i = \sum_{i \in \mathcal{I}} y_i \sum_{j \in \mathcal{J}} \frac{H_{i,j} h_j}{(Hh)_i} \left[\ln(H_{i,j} x_j) - \ln \left(\frac{H_{i,j} x_j}{(Hx)_i} \right) \right]. \tag{5.40}$$

If we insert this identity in the previous equation with $h = x^{(k)}$ and x respectively $x^{(k+1)}$ and $x^{(k)}$, by collecting terms we obtain

$$f_0(x^{(k)}; y) - f_0(x^{(k+1)}; y) = \sum_{i \in \mathcal{I}} y_i \sum_{j \in \mathcal{J}} \frac{H_{i,j} x_j^{(k)}}{(Hx^{(k)})_i} \left[\ln \left(\frac{x_j^{k+1}}{x_j^{(k)}} \right) - \ln \left(\frac{x_j^{k+1} (Hx^{(k)})_i}{x_j^{(k)} (Hx^{(k+1)})_i} \right) \right].$$

If in the first term we exchange the two sums and we take into account the iteration (5.34) we obtain

$$\begin{aligned} & f_0(x^{(k)}; y) - f_0(x^{(k+1)}; y) \\ & = \sum_{j \in \mathcal{J}} x_j^{(k+1)} \ln \left(\frac{x_j^{(k+1)}}{x_j^{(k)}} \right) - \sum_{i \in \mathcal{I}} y_i \sum_{j \in \mathcal{J}} \frac{H_{i,j} x_j^{(k)}}{(Hx^{(k)})_i} \ln \left(\frac{x_j^{k+1} (Hx^{(k)})_i}{x_j^{(k)} (Hx^{(k+1)})_i} \right). \end{aligned} \tag{5.41}$$

Since the negative logarithm is a convex function, by Jensen's inequality the second term is bounded by

$$
\begin{aligned}
&-\sum_{i\in\mathcal{I}} y_i \sum_{j\in\mathcal{J}} \frac{H_{i,j}x_j^{(k)}}{(Hx^{(k)})_i} \ln\left(\frac{x_j^{k+1}(Hx^{(k)})_i}{x_j^{(k)}(Hx^{(k+1)})_i}\right) \geqslant \\
&-\sum_{i\in\mathcal{I}} y_i \ln\left(\sum_{j\in\mathcal{J}} \frac{H_{i,j}x_j^{(k)}}{(Hx^{(k)})_i} \frac{x_j^{k+1}(Hx^{(k)})_i}{x_j^{(k)}(Hx^{(k+1)})_i}\right) = 0;
\end{aligned}
\tag{5.42}
$$

indeed, in the second term the sum between bracket is 1. Therefore, $f_0(x^{(k)}; y) - f_0(x^{(k+1)}; y)$ is bounded from below by the first term in equation (5.41) and, by taking into account equation (5.35) this is inequality (5.37).

As concerns inequality (5.38), let x^* be a limit point of the sequence $x^{(k)}$, $k = 0, 1, \ldots$. Some of its components (in general, many of them) can be zero but, if at least one is not zero, thanks to the restrictive assumptions on H, we have $Hx^*>0$. Let us denote as $\mathcal{J}_1$ the set of indices such that $x_j^* > 0$, and remark that x^* has the following properties

$$
x^* = x^* H^T \frac{y}{Hx^*} \quad \rightarrow \quad \left(H^T \frac{y}{Hx^*}\right)_j = 1, \ j \in \mathcal{J}_1. \tag{5.43}
$$

Then, for $j \in \mathcal{J}_1$ introduce the quantities

$$
\xi_{i,j}^{(k)} = \frac{x_j^{(k)}}{x_j^{(k+1)}} H_{i,j} \frac{y_i}{(Hx^{(k)})_i}, \quad \xi_{i,j}^* = H_{i,j} \frac{y_i}{(Hx^*)_i}; \ i \in \mathcal{I}, j \in \mathcal{J}_1. \tag{5.44}
$$

Using the iteration (5.34) and property (5.43), we obtain

$$
\sum_{i\in\mathcal{I}} \xi_{i,j}^{(k)} = 1 = \sum_{i\in\mathcal{I}} \xi_{i,j}^*. \tag{5.45}
$$

By taking into account these properties in the computation of the KL divergence, we obtain

$$
\begin{aligned}
&0 \leqslant \sum_{j\in\mathcal{J}_1} x_j^* \mathrm{KL}\left(\xi_{\cdot,j}^*; \xi_{\cdot,j}^{(k)}\right) = \sum_{j\in\mathcal{J}_1} x_j^* \sum_{i\in\mathcal{I}} \xi_{i,j}^* \ln \frac{\xi_{i,j}^*}{\xi_{i,j}^{(k)}} = \\
&\sum_{j\in\mathcal{J}_1} x_j^* \sum_{i\in\mathcal{I}} H_{i,j} \frac{y_i}{(Hx^*)_i} \ln \frac{(Hx^{(k)})_i x_j^{(k+1)}}{(Hx^*)_i x_j^{(k)}} = \\
&\sum_{j\in\mathcal{J}_1} x_j^* \sum_{i\in\mathcal{I}} H_{i,j} \frac{y_i}{(Hx^*)_i} \left[\ln \frac{(Hx^{(k)})_i}{(Hx^*)_i} + \ln \frac{x_j^{(k+1)}}{x_j^{(k)}}\right] = \\
&\sum_{i\in\mathcal{I}} y_i \ln \frac{(Hx^{(k)})_i}{(Hx^*)_i} + \sum_{j\in\mathcal{J}_1} x_j^* \ln \frac{x_j^{(k+1)}}{x_j^{(k)}},
\end{aligned}
\tag{5.46}
$$

where, in the last step, we used equation (5.43). If we write the result in terms of data-fidelity functions and KL divergences, by taking into account that also x^* satisfies equation (4.20), we obtain

$$0 \leqslant f_0(x^*;y) - f_0(x^{(k)}; y) + \mathrm{KL}(x^*;x^{(k)}) - \mathrm{KL}(x^*;x^{(k+1)}), \tag{5.47}$$

and this is the first part of inequality (5.38); we obtain the second part if we also observe that, thanks to inequality (5.37), $f_0(x^{(k)}; y) \geqslant f_0(x^*;y)$ for any limit point of the sequence. □

Inequality (5.37) states that the data-fidelity function decreases if $x^{(k+1)} \neq x^{(k)}$ while inequality (5.38) states that the iterates become closer to any limit point, in the sense of the KL divergence, when k increases.

Theorem 5.1. *Let $x^{(0)} > 0$; then, the sequence $x^{(k)}$, $k = 1, 2, \ldots$, generated by the iteration (5.34) converges to a minimizer of the data-fidelity function $f_0(x; y)$.*

Proof. Let us first remark that the sequence has certainly limit points because it is bounded, thanks to equation (5.35). If x^* is one of the limit points, there exists a sub-sequence $x^{(k_l)}$ converging to x^* and, for this sub-sequence, $\mathrm{KL}(x^*; x^{(k_l)})$ tends to zero. Since lemma 5.2 implies that $\mathrm{KL}(x^*; x^{(k)})$ is a non-increasing function of k, it follows that also $\mathrm{KL}(x^*; x^{(k)})$ tends to zero. But, thanks to proposition 4.1 the unique vector annihilating $\mathrm{KL}(x^*; x)$ is $x = x^*$, so that $x^{(k)}$ must converge to x^*.

It is still necessary to prove that x^* is a minimizer of $f_0(x; y)$, i.e. that it satisfies the KKT conditions. For the positive components of x^* this is obvious since we know that they satisfy equation (5.43). As concerns the null components of x^*, let us remark that, for these components we have

$$x_j^{(k)} = x_j^{(0)} \prod_{l=1}^{k-1} \left(H^T \frac{y}{Hx^{(l)}}\right)_j \to 0, \quad \left(H^T \frac{y}{Hx^{(k)}}\right)_j \to \left(H^T \frac{y}{Hx^*}\right)_j \tag{5.48}$$

but this is possible only if

$$\left(H^T \frac{y}{Hx^*}\right)_j \leqslant 1, \tag{5.49}$$

and therefore also the null components satisfy KKT conditions. □

5.5.2 Convergence of relaxed methods

In this subsection, following the arguments developed in [21], we give the proof of the convergence of methods which can be viewed as relaxed schemes of the simple iterative methods introduced in this chapter. Indeed, as follows from the previous sections, in order to solve the problem

$$\min_{x \geqslant 0} f_\beta(x; y) \tag{5.50}$$

with $\beta \geqslant 0$, we can use relaxed methods whose basic iteration, has the following form

$$x^{(k+1)} = x^{(k)} - \alpha \frac{x^{(k)}}{R(x^{(k)})} \nabla f_\beta(x^{(k)}; y), \tag{5.51}$$

where $\alpha > 0$ is the relaxation parameter, $x^{(k)}$ has non-negative entries and R: $\mathbf{R}^n \to \mathbf{R}^n$ is a continuous vector function with entries strictly positive on the non-negative orthant, i. e. $R(x) > 0$ for any $x \geqslant 0$. The diagonal matrix $\mathrm{diag}\left(\frac{x}{R(x)}\right)$ has the role of a scaling matrix in the gradient iteration. As a consequence of the assumption on R, each diagonal entry of the scaling matrix is 0 or positive depending on whether x_j is 0 or positive.

In particular, if $\alpha = 1$, for $\beta = 0$ and $R(x) = \upsilon$ equation (5.51) is the EM–RL iteration (5.1) (see also equation (5.5)); for $\beta > 0$ and $R(x) = 1 + \beta \nabla f_1(x)$, we obtain the OSL iteration (5.17) or (5.19), while for $R(x) = 1 + \beta V_1(x)$, we have the SGM iteration (5.23) or (5.24).

A crucial assumption for obtaining convergence results is to require *coercivity and strong convexity of the objective function*, so that the solution of the problem (5.50) exists and is unique. According to the notation used in [21], we introduce the definition of a weak stationary point for this problem.

Definition 5.1. *A point x^* is a weak stationary point for f_β if $x_j^* \frac{\partial f_\beta}{\partial x_j}(x^*) = 0$ with $x^* \geqslant 0$, $j = 1, \dots, N$.*

From this definition, we have that any jth component of a weak stationary point x^* is such that $x_j^* = 0$ or $\frac{\partial f_\beta}{\partial x_j}(x^*) = 0$. We observe that a weak stationary point may not satisfy the first-order conditions of a minimum point for the problem (5.50), since it is allowed the existence of an index j such that $x_j^* = 0$ and $\frac{\partial f_\beta}{\partial x_j}(x^*) < 0$.

In order to simplify the notation, in the following we omit the dependence on y of $f_\beta(x; y)$ and we denote by $d(x^{(k)})$ the following vector

$$d(x^{(k)}) = -\frac{x^{(k)}}{R(x^{(k)})} \nabla f_\beta(x^{(k)}), \tag{5.52}$$

which is zero if $x^{(k)}$ is a weak stationary point.

If $d(x^{(k)}) \neq 0$, we remark that the single variable function $\alpha \to f_\beta(x^{(k)} + \alpha d(x^{(k)}))$ has first-order Taylor expansion

$$\begin{aligned} f_\beta(x^{(k)} + \alpha d(x^{(k)})) &= f_\beta(x^{(k)}) + \alpha \nabla f_\beta(x^{(k)})^T d(x^{(k)}) + O(\alpha^2) \\ &= f_\beta(x^{(k)}) - \alpha \nabla f_\beta(x^{(k)})^T \frac{x^{(k)}}{R(x^{(k)})} \nabla f_\beta(x^{(k)}) + O(\alpha^2), \end{aligned} \tag{5.53}$$

so that, if $\alpha > 0$ is sufficiently small, the following inequality holds true

$$f_\beta(x^{(k)} + \alpha d(x^{(k)})) < f_\beta(x^{(k)}), \tag{5.54}$$

and we conclude that an iteration as equation (5.51) can be viewed as a descent method.

Next, we must introduce conditions on α assuring strict positivity of $x^{(k+1)}$ for a given strictly positive $x^{(k)}$. To this purpose, we set $\epsilon_1 \in (0, 1)$; for any given $x > 0$, we define

$$\alpha_{\lim}(x) = \max\{\alpha \geqslant 0\colon \nabla f_\beta(x + \alpha d(x))^T d(x) \leqslant 0,\ x + \alpha d(x) \geqslant \epsilon_1 x\}. \tag{5.55}$$

and we require that, if x is the current iterate, the next iterate has the form

$$x + \alpha d(x)$$

with $0 \leqslant \alpha \leqslant \alpha_{\lim}$. Note that $\nabla f_\beta(x + \alpha d(x))^T d(x)$ is the derivative of the function $\alpha \to f_\beta(x + \alpha d(x))$. Because of the coercivity and strict convexity of $f_\beta(x)$, $\alpha_{\lim}(x) < \infty$ unless x is a weak stationary point (indeed, in this case $d(x) = 0$).

In equation (5.55) the condition $\nabla f_\beta(x + \alpha d(x))^T d(x) \leqslant 0$ is required for assuring that the next iterate is on the descent slope of the curve $\alpha \to f_\beta(x + \alpha d(x))$ and never beyond the minimum point on the curve. Moreover, the conditions $x + \alpha d(x) \geqslant \epsilon_1 x$ are introduced to prevent violations of strict positivity and to keep the algorithm in the interior of the non-negative orthant.

The following lemma states a crucial feature of $\alpha_{\lim}(x)$.

Lemma 5.3. *$\alpha_{\lim}(x)$ is continuous in x on the set of non-weak stationary points.*

Proof. α satisfies the condition $x + \alpha d(x) \geqslant \epsilon_1 x$ if and only if

$$\alpha \leqslant \alpha_{\text{boundary}}(x) = \min_j \begin{cases} \infty, & d_j(x) \geqslant 0 \\ \dfrac{(\epsilon_1 - 1)x_j}{d_j(x)}, & d_j(x) < 0. \end{cases} \tag{5.56}$$

If we let $\alpha_{\text{root}}(x)$ be the unique scalar satisfying $\nabla f_\beta(x + \alpha d(x))^T d(x) = 0$, then

$$\alpha_{\lim}(x) = \min\{\alpha_{\text{boundary}}(x), \alpha_{\text{root}}(x)\}. \tag{5.57}$$

Thus, it is sufficient to prove that $\alpha_{\text{root}}(x)$ is continuous in x. To this purpose, let us consider a sequence $x^{(k)}$ with limit x. Let $x^{(k_i)}$ be any sub-sequence such that $\alpha_{k_i} = \alpha_{\text{root}}(x^{(k_i)})$ has a limit finite or infinite. If the limit is finite, by continuity we have

$$\nabla f_\beta\left(x + \lim_{i\to\infty} \alpha_{k_i} d(x)\right)^T d(x) = 0$$

and by uniqueness

$$\lim_{i \to \infty} \alpha_{k_i} = \alpha_{\text{root}}(x).$$

We can rule out the case $\lim_{i \to \infty} \alpha_{k_i} = \infty$ as follows. On one hand we have

$$\limsup_{i \to \infty} f_\beta(x^{(k_i)} + \alpha_{k_i} d(x^{(k_i)})) \leqslant \lim_{i \to \infty} f_\beta(x^{(k_i)}) = f_\beta(x);$$

on the other hand, since $d(x^{(k_i)}) \to d(x) \neq 0$ (x is a non-weak stationary point),

$$\|x^{(k_i)} + \alpha_{k_i} d(x^{(k_i)})\| \to \infty,$$

so that, thanks to the coercivity of f_β, we have $f_\beta(x^{(k_i)} + \alpha_{k_i} d(x^{(k_i)})) \to \infty$, obtaining a contradiction. □

In order to determine a suitable value of α, we select a value such that

(A1) $(1 - \epsilon_2)\alpha_{\text{lim}}(x) \leqslant \alpha \leqslant \alpha_{\text{lim}}(x)$, with $\epsilon_2 \in (0, 1)$.

Since α is not uniquely determined, the rule $z = x + \alpha d(x)$, with all possible values of α satisfying condition (A1), generates the set of all possible new points z from the current point x. We denote this set as $T(x)$ and the correspondence $x \to T(x)$ is called *iteration map* since it maps a point x into a set. For a weak stationary point x^*, we have $d(x^*) = 0$ and $T(x^*)$ reduces to the singleton $\{x^*\}$. For a non-weak stationary point x, the condition (A1) excludes that $x \in T(x)$.

The convergence theory for algorithms dealing with a point to set mapping as $T(x)$ requires the map to be closed in the following sense: $T(x)$ is closed at x if, whenever $\lim_{k \to \infty} x^{(k)} = x$, $z^{(k)} \in T(x^{(k)})$ and $\lim_{k \to \infty} z^{(k)} = z$, then $z \in T(x)$.

Lemma 5.4. *The iteration map $T(x)$ satisfying the condition (A1) is closed at every non-weak stationary point x.*

Proof. Suppose that the non-weak stationary point x is the limit of the sequence $x^{(k)}$, that $z^{(k)} \in T(x^{(k)})$ and $z = \lim_{k \to \infty} z^{(k)}$ where

$$z^{(k)} = x^{(k)} + \alpha_k d(x^{(k)})$$

for some α_k satisfying $(1 - \epsilon_2)\alpha_{\text{lim}}(x^{(k)}) \leqslant \alpha \leqslant \alpha_{\text{lim}}(x^{(k)})$. Since $d(x)$ is continuous and x is a non-weak stationary point, $\lim_{k \to \infty} \|d(x^{(k)})\|$ exists and is positive. Thus, $\lim_{k \to \infty} \alpha_k = \lim_{k \to \infty} \frac{\| z^{(k)} - x^{(k)} \|}{\| d(x^{(k)}) \|} = \alpha$ exists, and

$$z = x + \alpha d(x).$$

The continuity of $\alpha_{\text{lim}}(x)$ now implies $0 < (1 - \epsilon_2)\alpha_{\text{lim}}(x) \leqslant \alpha \leqslant \alpha_{\text{lim}}(x)$. Hence $z \in T(x)$. □

We prove now the convergence of the relaxed algorithm (5.51) where, at each iteration k, we choose for α a value α_k satisfying condition (A1) with $x = x^{(k)}$; we assume also that $x^{(0)}$ is chosen in the interior of the non-negative orthant; then, condition (A1) assures that any subsequent iterate is also an interior point. Due to the monotone behavior of the modified algorithm, all the iterates belong to the initial level set $\mathcal{L}_0 = \{x: f_\beta(x) \leqslant f_\beta(x^{(0)})\}$. This set is compact because of the coercivity of f_β.

Lemma 5.5. *Let* $\{x^{(k)}\}$ *be the sequence generated by the relaxed algorithm (5.51). We have* $\|x^{(k+1)} - x^{(k)}\| \to 0$ *as* $k \to \infty$.

Proof. By construction, $x^{(k)}$ is the maximum point and $x^{(k+1)}$ is the minimum point of the function $\alpha \to f_\beta((1 - \alpha)x^{(k)} + \alpha x^{(k+1)})$ for $\alpha \in [0, 1]$; indeed, $\alpha_{\lim}(x^{(k)})$ is defined in such a way to keep $x^{(k+1)}$ on the downward slope of the function $f_\beta(x^{(k)} + \alpha d(x^{(k)}))$. Now, assume that $x^{(k_i+1)} - x^{(k_i)} \to x^a \neq 0$ for some sub-sequence $x^{(k_i)}$. Since all the terms of the sub-sequence belong to the same compact set, we may also assume that $x^{(k_i)} \to x^b$ and $x^{(k_i+1)} \to x^c$. Letting $i \to \infty$ in

$$f_\beta(x^{(k_i+1)}) \leqslant f_\beta((1 - \alpha)x^{(k_i)} + \alpha x^{(k_i+1)}) \leqslant f_\beta(x^{(k_i)}) \tag{5.58}$$

we obtain

$$f_\beta(x^c) \leqslant f_\beta((1 - \alpha)x^b + \alpha x^c) \leqslant f_\beta(x^b) \tag{5.59}$$

for $\alpha \in [0, 1]$. Since the sequence $f_\beta(x^{(k)})$ is decreasing and bounded, $f_\beta(x^b) = f_\beta(x^c) = \lim_{k\to\infty} f_\beta(x^{(k)})$, so that $f_\beta(x)$ is constant on the line segment connecting x^b and x^c, with $x^b \neq x^c$. This contradicts the strong convexity of f_β. □

Lemma 5.6. *The limit points of the sequence generated by the relaxed method (5.51) are weak stationary points of* f_β *in the initial level set; their number is finite.*

Proof. The first part of the lemma follows immediately from lemma 5.4, the compactness of the initial level set $\mathcal{L}_0$ and the Zangwill's global convergence theorem (see proposition A.4).

As concerns the second part, let us recall that each component of a weak stationary point x^* is zero or satisfies the condition $\left(\nabla f_\beta(x^*)\right)_j = 0$. If we consider all possible subsets of the set $\mathcal{J}$ of the indices j (including the empty subset), the number of these subsets is obviously finite, even if it can be very large. Then, to each one of these subsets let us associate the subspace of the vectors x whose components with index in the subset are zero. Since $f_\beta(x)$ is strongly convex, it has a unique minimum point in this subspace, which can be a weak stationary point. It follows that the number of these points does not exceed the number of subsets and is finite. □

Lemma 5.7. *The set of the limit points of the iterates* $x^{(k)}$ *is compact and connected. Therefore, it consists of a single weak stationary point.*

Proof. According to a theorem of Ostrowski (see proposition A.5), the set of limit points of any sequence $x^{(k)}$ with $\|x^{(k+1)} - x^{(k)}\| \to 0$ is compact and connected. Since the weak stationary points are finite in number, they are disconnected and so the set of limit points of $x^{(k)}$ must consist of a single point. □

Theorem 5.2. *The limit point of the iterates* $x^{(k)}$ *exists and is the unique minimum point of* $f_\beta(x)$ *in the non-negative orthant.*

Proof. The existence and uniqueness of the limit is established from the previous lemma. Suppose that the weak stationary point x^* is this limit, but fails to represent the minimum of f_β on the non-negative orthant. According to the first-order optimality conditions, one has $\frac{\partial f_\beta}{\partial x_j}(x^*) < 0$ for some index j with $x_j^* = 0$. For $x^{(k)}$ close to x^*, by continuity we must have $d_j(x^{(k)}) > 0$. Then, since $x^{(k)}$ is an interior point of the non-negative orthant, we have

$$x_j^{(k+1)} = x_j^{(k)} + \alpha_k d_j(x^{(k)}) > x_j^{(k)},$$

and, consequently, $x_j^{(k)}$ does not tend to zero. This contradiction proves the theorem.

□

The previous arguments provide a proof of the convergence of the relaxed algorithm, provided that, at each iteration, the relaxation parameter α satisfies condition (A1).

From the practical point of view, a crucial step of equation (5.51) is the determination of $\alpha_{\mathrm{lim}}(x)$. When the basic iteration provides a decrease of the objective function, as for instance in the case of EM–RL, we can set $\alpha_{\mathrm{lim}}(x) = 1$. In other cases, an approximate value of $\alpha_{\mathrm{lim}}(x)$ can be determined by different techniques, aimed at estimating a suitable minimum point of $f_\beta(x + \alpha d(x))$ with respect to α (see for example the one-dimensional minimization schemes in [27]). Then, the selection of a value of α satisfying the condition (A1) can be obtained by a simple procedure of step-halving or step-doubling. For example, we can set $\epsilon_2 = 1/2$ and define the next iterate by $x + 2^m d(x)$ where m is the largest integer satisfying $2^m \leqslant \alpha_{\mathrm{lim}}(x)$. Indeed, starting with the guess $m = 0$, one can check that the condition $1/2\alpha_{\mathrm{lim}}(x) \leqslant 2^m \leqslant \alpha_{\mathrm{lim}}(x)$ is satisfied; otherwise, one can find the correct value of m by successive decrements of m by 1 when $\alpha_{\mathrm{lim}}(x) < 1$ and increments of m by 1 when $\alpha_{\mathrm{lim}}(x) > 1$.

5.5.3 Additional regularization functions

In this subsection we give other examples of regularization functions, sometimes used in the practice of image reconstruction with Poisson data. We comment on each of them and we also provide a non-negative decomposition $U_1, \quad V_1$ of their gradients for possible applications of the SGM method.

- **Cross-entropy (CE) regularization**—this kind of regularization is used in the so-called *maximum entropy* approach to the regularization of the least-squares problem and, in our notation, is given by

$$f_1(x) = \mathrm{KL}(x, \bar{x}) = \sum_{j \in \mathcal{J}} \left\{ x_j \ln \frac{x_j}{\bar{x}_j} + \bar{x}_j - x_j \right\} \tag{5.60}$$

  where $\bar{x}$ is a reference image. When $\bar{x}$ is a constant, then the cross-entropy becomes the negative Shannon entropy considered for instance in [37]. If both x and $\bar{x}$ have the same ℓ_1 norm, let us say c, then a possible choice for the functions $U_1, \quad V_1$ is

$$[U_1(x)]_i = -\ln \frac{x_i}{c}, \; [V_1(x)]_i = -\ln \frac{\bar{x}_i}{c}. \tag{5.61}$$

  We remark that, since the background is taken into account by the algorithms, some of the components of x can be zero; for this reason, in the computation of the gradient, it is convenient to add a small positive quantity to x. We also remark that, if $\bar{x}$ is a constant, i.e. $\bar{x}_i = c/\sharp\{\mathcal{J}\}$, then $V_1(x) = \ln(\sharp\{\mathcal{J}\})$.
- **MISTRAL (MIS) regularization**—this kind of regularization is used in the software packages MISTRAL [28] and AIDA [17]. It has properties similar to those of HS regularization and is given by

$$f_1(x) = \sum_{j \in \mathcal{J}} \left\{ |(\nabla x)_j| - \delta \ln \left(1 + \frac{|(\nabla x)_j|}{\delta} \right) \right\}, \tag{5.62}$$

  where δ is a thresholding parameter. If, in the 2D case, we set

$$\begin{aligned} s = |(\nabla x)_{k,l}| &= \sqrt{(x_{k+1,l} - x_{k,l})^2 + (x_{k,l+1} - x_{k,l})^2} \\ \phi(s) = s - \delta \ln\left(1 + \frac{s}{\delta}\right), &\quad \phi'(s) = \frac{s}{s + \delta} \end{aligned} \tag{5.63}$$

  by simple computations we obtain

$$\begin{aligned} [U_1(x)]_{k,l} &= \frac{x_{k+1,l} + x_{k,l+1}}{\delta + |(\nabla x)_{k,l}|} + \frac{x_{k-1,l}}{\delta + |(\nabla x)_{k-1,l}|} + \frac{x_{k,l-1}}{\delta + |(\nabla x)_{k,l-1}|} \\ [V_1(x)]_{k,l} &= \frac{2x_{k,l}}{\delta + |(\nabla x)_{k,l}|} + \frac{x_{k,l}}{\delta + |(\nabla x)_{k-1,l}|} + \frac{x_{k,l}}{\delta + |(\nabla x)_{k,l-1}|}. \end{aligned} \tag{5.64}$$

Similarly, in the 3D case

$$
\begin{aligned}
[U_1(x)]_{k,l,h} &= \frac{x_{k+1,l,h} + x_{k,l+1,h} + x_{k,l,h+1}}{\delta + |(\nabla x)_{k,l,h}|} + \frac{x_{k-1,l,h}}{\delta + |(\nabla x)_{k-1,l,h}|} \\
&+ \frac{x_{k,l-1,h}}{\delta + |(\nabla x)_{k,l-1,h}|} + \frac{x_{k,l,h-1}}{\delta + |(\nabla x)_{k,l,h-1}|} \\
[V_1(x)]_{k,l,h} &= \frac{3x_{k,l,h}}{\delta + |(\nabla x)_{k,l,h}|} + \frac{x_{k,l,h}}{\delta + |(\nabla x)_{k-1,l,h}|} \\
&+ \frac{x_{k,l,h}}{\delta + |(\nabla x)_{k,l-1,h}|} + \frac{x_{k,l,h}}{\delta + |(\nabla x)_{k,l,h-1}|}.
\end{aligned}
\tag{5.65}
$$

- **Markov random field (MRF) regularization**—this is the regularization given in equation (5.13). If we take into account the symmetry of the function ϕ and if we assume that the neighborhood of the pixel j is symmetric, then the gradient is given by

$$
\left(\nabla f_1\right)_j(x) = \sum_{j' \in \mathcal{N}_j} \frac{1}{\epsilon_{j,j'}} \phi'\left(\frac{x_j - x_{j'}}{\epsilon_{j,j'}}\right), \tag{5.66}
$$

where $\phi'(s)$ is the derivative of the function $\phi(s)$, which is assumed non-negative and differentiable.

For obtaining a decomposition of the gradient in the vein of the previous computations, it is necessary to specify the function ϕ. We give the result for two functions. The first is $\phi(s) = \sqrt{\delta^2 + s^2}$ and an expression valid both in 2D and 3D case is the following

$$
\begin{aligned}
[U_1(x)]_j &= \sum_{j' \in \mathcal{N}_j} \frac{x_{j'}}{\epsilon_{j,j'}\sqrt{\delta^2 + \left(\frac{x_j - x_{j'}}{\epsilon_{j,j'}}\right)^2}}, \\
[V_1(x)]_j &= \sum_{j' \in \mathcal{N}_j} \frac{x_j}{\epsilon_{j,j'}\sqrt{\delta^2 + \left(\frac{x_j - x_{j'}}{\epsilon_{j,j'}}\right)^2}}.
\end{aligned}
\tag{5.67}
$$

The second is the function considered in [10], which is given by $\phi(s) = \ln(\cosh \frac{s}{\delta})$ and has a behavior similar to that of the previous one for small and large values of s. In such a case we have

$$
\begin{aligned}
[U_1(x)]_j &= \sum_{j' \in \mathcal{N}_j} \frac{e^{-\frac{x_j - x_{j'}}{\delta \epsilon_{j,j'}}}}{2\delta\epsilon_{j,j'} \cosh\left(\frac{x_j - x_{j'}}{\delta\epsilon_{j,j'}}\right)}, \\
[V_1(x)]_j &= \sum_{j' \in \mathcal{N}_j} \frac{e^{\frac{x_j - x_{j'}}{\delta \epsilon_{j,j'}}}}{2\delta\epsilon_{j,j'} \cosh\left(\frac{x_j - x_{j'}}{\delta\epsilon_{j,j'}}\right)}.
\end{aligned}
\tag{5.68}
$$

References

[1] Adorf H M, Hook R N, Lucy L B and Murtagh F D 1992 *Accelerating the Richardson-Lucy restoration algorithm Proc. of* 4th *ESO/ST-ECF Data Analysis Workshop* (Baltimore, MD: Space Telescope Science Institute) pp 99–103

[2] Beister M, Kolditz D and Kalender W A 2012 Iterative reconstruction methods in X-ray CT *Phys. Med.* **28** 94–108

[3] Bertero M and Boccacci P 1998 *Introduction to Inverse Problems in Imaging* (Bristol: Institute of Physics)

[4] Bertero M and Boccacci P 2000 Application of the OS-EM method to the restoration of LBT images *Astron. Astrophys. Suppl. Ser.* **144** 181–6

[5] Defrise M, Kinahan P E and Michel C J 2004 *Image Reconstruction Algorithms in PET* (Berlin: Springer) pp 63–91

[6] Dempster A P, Laird N M and Rubin D B 1997 Maximum likelihood from incomplete data via the EM algorithm *J. R. Stat. Soc.* B **39** 1–38

[7] Geman S and Geman D 1984 Stochastic relaxation, Gibbs distributions, and the Bayesian restoration of images *IEEE Trans. Pattern Anal. Mach. Intell.* **6** 721–41

[8] Geman S and McClure D E 1985 Bayesian image analysis: an application to single photon emission tomography, *Proc. Statist. Comput. Sect., American Statist. Assoc.* pp 12–8

[9] Geman S and McClure D E 1987 Statistical methods for tomographic image reconstruction *Bull. Int. Stat. Inst.* **LII–4** 5–21

[10] Green P J 1990 Bayesian reconstructions from emission tomography data using a modified EM algorithm *IEEE Trans. Med. Imaging* **9** 84–93

[11] Hanisch R J and White R L 1994 The restoration of HST images and spectra—II *Proc. of a Workshop held at the Space Telescope Science Institute (Baltimore, 18–19 November 1993)* (Baltimore: The Space Telescope Science Institute)

[12] Hebert T and Leahy R 1989 A generalized EM algorithm for 3-D Bayesian reconstruction from Poisson data using Gibbs priors *IEEE Trans. Med. Imaging* **8** 194–202

[13] Holmes T J 1988 Maximum-likelihood image restoration adapted for noncoherent optical imaging *J. Opt. Soc. Am.* A **5** 666–73

[14] Holmes T J 1989 Expectation-maximization restoration of band-limited, truncated point-process intensities with application in microscopy *J. Opt. Soc. Am.* A **6** 1006–14

[15] Holmes T J and Liu Y H 1989 Richardson-Lucy/maximum likelihood image restoration algorithm for fluorescence microscopy: further testing *Appl. Opt.* **28** 4930–8

[16] Holmes T J and Liu Y H 1991 Acceleration of maximum-likelihood image restoration for fluorescence microscopy *J. Opt. Soc. Am.* A **8** 893–907

[17] Horn E F Y, Marchis F, Lee T K, Haase S, Agard D A and Sedat J W 2007 AIDA: an adaptive image deconvolution algorithm with application to multi-frame and three dimensional data *J. Opt. Soc. Am.* A **24** 1580–600

[18] Hudson H M and Larkin R S 1994 Accelerated image reconstruction using ordered subsets of projection data *IEEE Trans. Med. Imaging* **13** 601–9

[19] Iusem A N 1991 Convergence analysis for a multiplicatively relaxed EM algorithm *Math. Methods Appl. Sci.* **14** 573–93

[20] Iusem A N 1992 A short convergence proof of the EM algorithm for a specific Poisson model *REBRAPE* **6** 57–67

[21] Lange K 1990 Convergence of EM image reconstruction algorithms with Gibbs smoothing *IEEE Trans. Med. Imaging* **9** 439–46

[22] Lange K and Carson R 1984 EM reconstruction algorithms for emission and transmission tomography *J. Comput. Assist. Tomogr.* **8** 306–16
[23] Lantéri H, Roche M and Aime C 2002 Penalized maximum likelihood image restoration with positivity constraints: multiplicative algorithms *Inverse Probl.* **18** 1397–419
[24] Lantéri H, Roche M, Cuevas O and Aime C 2001 A general method to devise maximum-likelihood signal restoration multiplicative algorithms with nonnegativity constraints *Signal Process.* **81** 945–74
[25] Lantéri H, Roche M, Gaucherel P and Aime C 2002 Ringing reduction in image restoration algorithms using a constraint on the inferior bound of the solution *Signal Process.* **82** 1481–504
[26] Lucy L B 1974 An iterative technique for the rectification of observed distributions *Astronom. J.* **79** 745–54
[27] Luenberger D G 1989 *Linear and Nonlinear Programming* 2nd edn (Reading, MA: Addison-Wesley)
[28] Mugnier L M, Fusco T and Conan J M 2004 MISTRAL: a myopic edge-preserving image restoration method, with application to astronomical adaptive-optics-corrected long-exposure images *J. Opt. Soc. Am.* A **21** 1841–54
[29] Mülthei H N and Schorr B 1989 On properties of the iterative maximum likelihood reconstruction method *Math. Methods Appl. Sci.* **11** 331–42
[30] Natterer F and Wübbeling F 2001 *Mathematical Methods in Image Reconstruction* (Philadelphia, PA: SIAM)
[31] Politte D G and Snyder D L 1991 Correction for accidental coincidences and attenuation in maximum-likelihood image reconstruction for positron-emission tomography *IEEE Trans. Med. Imaging* **10** 82–9
[32] Richardson W H 1972 Bayesian-based iterative method of image restoration *J. Opt. Soc. Am.* A **62** 55–9
[33] Salvo K and Defrise M 2018 A convergence proof of MLEM with fixed background *IEEE Trans. Med. Imaging* submitted *IEEE Trans Med Imaging* September 2018 https://ieeexplore.ieee.org/document/8467338
[34] Schrader M, Hell S W and van der Voort H T M 1998 Three-dimensional super-resolution with a 4Pi-confocal microscope using image restoration *J. Appl. Phys.* **84** 4033–42
[35] Shepp L A and Logan B F 1974 The Fourier reconstruction of a head section *IEEE Trans. Nucl. Sci.* **21** 21–43
[36] Shepp L A and Vardi Y 1982 Maximum likelihood reconstruction for emission tomography *IEEE Trans. Med. Imaging* **1** 113–22
[37] Skilling J and Bryan R K 1984 Maximum entropy image reconstruction: general algorithm *Mon. Not. R. Astron. Soc.* **211** 111–24
[38] Snyder D L 1990 Modifications of the Lucy-Richardson iteration for restoring Hubble Space-Telescope imagery *The Restoration of HST Images and Spectra* ed R L White and R J Allen (Baltimore, MD: The Space Telescope Science Institute) pp 56–61
[39] Vardi Y, Shepp L A and Kaufman L 1985 A statistical model for positron emission tomography *J. Am. Stat. Assoc.* **80** 8–37
[40] Vicidomini G, Boccacci P, Diaspro A and Bertero M 2009 Application of the split-gradient method to 3D image deconvolution in fluorescence microscopy *J. Microsc.* **234** 47–61
[41] White R L and Allen R J 1990 The restoration of HST images and spectra *Proc. of a Workshop held at the Space Telescope Science Institute (Baltimore, 20–21 August 1990)* (Baltimore, MD: The Space Telescope Science Institute)

IOP Publishing

Inverse Imaging with Poisson Data
From cells to galaxies
Mario Bertero, Patrizia Boccacci and Valeria Ruggiero

Chapter 6

Optimization methods

In recent years several optimization methods have been proposed and investigated for the solution of the minimization problems introduced in chapter 4. Taking into account the limitations of the methods discussed in chapter 5, the main reasons for proposing new optimization methods are the following:

- propose more efficient approaches than those described in chapter 5 for solving smooth problems with non-negativity constraint;
- solve non-smooth problems with particular reference to TV regularization;
- solve smooth and non-smooth problems with additional constraints besides non-negativity, such as box constraints (upper and/or lower bounds) and, possibly, a single linear equality constraint.

Research is still in progress even if the literature is already very broad. In this chapter we attempt to organize the many proposed methods into a few large classes of optimization approaches, focusing on convex problems.

6.1 Some basic tools: proximity operators and conjugate functions

As shown in the previous chapters, image reconstruction problems with Poisson data typically require the numerical solution of large-scale optimization problems with simple constraints, as box constraints or their combination with a single linear constraint. The size of the problems is very large, frequently the involved matrices are not available in memory and only the operators of the matrix–vector products can be computed. The constrained minimization problems introduced in chapter 4, i.e. problems (4.13) and (4.30), can be viewed as special instances of the following general (unconstrained) optimization problem

$$\min_x F(x) := \Phi(x) + \Psi(x) \tag{6.1}$$

where $\Phi(x)$ is a continuously differentiable function in its effective domain (denoted by $\mathrm{dom}\,\Phi$ in the following) and $\Psi(x)$ is a non-smooth proper, convex and lower

doi:10.1088/2053-2563/aae109ch6

semi-continuous (l.s.c.) function (see definitions A.6 and A.12 in appendix A), with dom $\Psi \subseteq$ dom Φ.

We give an explicit relationship between the problems of chapter 4 and the general optimization problem (6.1).

- **Smooth problems**—in the case of the ML estimates (4.13), $\Phi(x)$ is the data-fidelity function $f_0(x; y)$ while, in the case of MAP estimates (4.30) with a continuously differentiable regularization term (as, for example, T0, T1, T2 or HS), $\Phi(x)$ is the objective function $f_\beta(x; y)$; in both cases, $\Psi(x)$ is the indicator function of a convex and closed set $\iota_\Omega(x)$, where Ω can be the non-negative orthant or the convex and compact set defined in equation (4.22). We remark that in these problems the function $\Phi(x)$ is also coercive and convex. However, for the smooth problems treated in section 6.3 the convexity condition for $\Phi(x)$ is not always required; we will point out the theoretical results for which the assumption is essential.
- **Non-smooth problems**—in the case of MAP estimates with a non-smooth regularization, such as TV or frame analysis regularization, $\Phi(x)$ is the data-fidelity function $f_0(x; y)$, while $\Psi(x)$ includes the non-smooth part of the objective function, i.e. $\Psi(x) = \beta f_1(x) + \iota_\Omega(x)$, the set Ω being defined as in the previous case. The methods for solving these problems are treated in sections 6.4, 6.5 and 6.6 and, in this case, the convexity of the function $F(x)$ is required; this condition, as we know, is satisfied by $f_0(x; y)$.

Most of the methods devised to address the problem (6.1) can be brought back to the general *proximal splitting* methods for the minimization of convex functions or in general for the determination of zeroes of maximal monotone operators [10, 60, 92]. They are called *proximal* because each non-smooth function is involved via its proximity operator, which introduces an ℓ_2 penalization term. The key role covered by scaling techniques in the simple methods discussed in chapter 5, leads to generalizing the usual metric tools and to consider the proximity operator with respect to more general metrics, induced by the scaling used in the reconstruction method. We give in the following some basic definitions and properties, useful in this chapter.

Given a vector $x \in \mathbf{R}^N$ and a symmetric positive definite (s.p.d.) matrix D of order N, the vector norm of x with respect to the matrix D can be defined as:

$$\|x\|_D = \sqrt{x^T D x} \tag{6.2}$$

and it can be called the D-norm of x. The standard ℓ_2 norm $\|x\|$ is recovered for $D = I$. In order to introduce the definition of proximity operator with respect to the D-norm of a generic proper, convex and l.s.c. function F: $\mathbf{R}^N \to \bar{\mathbf{R}}$, $\bar{\mathbf{R}} = \mathbf{R} \cup \{\pm\infty\}$, we observe that, for any $z \in \mathbf{R}^N$ the problem

$$\min_{x \in \mathbf{R}^N} F(x) + \frac{1}{2}\|x - z\|_D^2 \tag{6.3}$$

admits a unique solution $\bar{x}$; indeed the objective function is coercive and strongly convex.

Definition 6.1. *Given a proper, convex and l.s.c. function F, for any* $z \in \mathbf{R}^N$ *the unique solution* $\bar{x}$ *of the problem* (6.3) *is defined as the proximal point of F at z with respect to the D-norm and it is denoted by*

$$\bar{x} = \mathrm{prox}_{F,\,D}(z) := \arg\min_{x \in \mathbf{R}^N} F(x) + \frac{1}{2}\|x - z\|_D^2; \tag{6.4}$$

the operator $z \to \mathrm{prox}_{F,\,D}(z)$ *is defined as the proximity operator of F with respect to the D-norm.*

The classic proximity operator is recovered for $D = I$; this is denoted in the following by $\bar{x} := \mathrm{prox}_F(z)$, dropping the metric symbol in this special case. As a consequence of the first order optimality conditions for the problem (6.3), the proximity operator $\mathrm{prox}_{F,\,D}(z)$ is characterized by the following inclusion

$$\forall (z, \bar{x}) \in \mathbf{R}^N \times \mathbf{R}^N, \quad \bar{x} = \mathrm{prox}_{F,\,D}(z) \Leftrightarrow D(z - \bar{x}) \in \partial F(\bar{x}) \tag{6.5}$$

where $\partial F(\bar{x})$ denotes the subdifferential of F at $\bar{x}$ (see section A.6). Thus, the proximity operator provides a practical way to individuate an element of the set $\partial F(\bar{x})$. A survey of the properties of proximity operators and of the closed-form expression of the proximity operators of various functions is given in [48].

An important proximity operator is the projection on a set. Consider a non-empty, convex and closed subset Ω of $\mathbf{R}^N$; in such a case, the indicator function $\iota_\Omega(x)$ of Ω is a proper, convex, l.s.c. function, whose effective domain is equal to Ω. The proximity operator of ι_Ω with respect to a D-norm is the projection operator $z \to \mathrm{P}_{\Omega,\,D}(z)$ with respect to the D-norm, defined as

$$\bar{x} := \mathrm{P}_{\Omega,\,D}(z) = \arg\min_{x \in \Omega} \frac{1}{2}\|x - z\|_D^2. \tag{6.6}$$

When $D = I$, the standard ℓ_2 projection of z on Ω is recovered, denoted by $\bar{x} := \mathrm{P}_\Omega(z)$, dropping the metric symbol.

An important tool in convex optimization is provided by the conjugate function F^*, which is defined as follows.

Definition 6.2. *Let* $F\colon \mathbf{R}^N \to \bar{\mathbf{R}}$ *be a convex function. The conjugate function of F is the function* $F^*\colon \mathbf{R}^N \to \bar{\mathbf{R}}$ *defined by*

$$F^*(z) = \sup_{x \in \mathbf{R}^N} (x^T z - F(x)) = -\inf_{x \in \mathbf{R}^N} (F(x) - x^T z). \tag{6.7}$$

Figure 6.1 shows the geometrical interpretation of the conjugate function for a function of a real variable. To clarify the previous definition, we give some simple examples:

- given a non-empty convex set Ω, the conjugate function of the indicator function $\iota_\Omega(x)$ is the support function $\sigma_\Omega(z) = \sup_{x\in\Omega} x^T z$;
- for $F(x) = \frac{1}{2}\|x\|^2$, we have $F^*(z) = \sup_{x\in\mathbf{R}^N}\left(x^T z - \frac{1}{2}\|x\|^2\right) = \frac{1}{2}\|z\|^2$;
- for the ℓ_2 vector norm $\|x\|$, since $\|x\| = \sup_{\|z\|\leqslant 1} z^T x$, the conjugate function is the indicator function $\iota_{B(0,1)}(z)$ of the closed unit ball with respect to this norm, i.e. $B(0, 1) = \{z: \|z\| \leqslant 1\}$; indeed, $\iota_{B(0,1)}(z) = \sup_{x\in\mathbf{R}^N}(x^T z - \|x\|)$; more in general, given a vector norm $\|\cdot\|_+$ and its dual $\|\cdot\|_*$ (for example ℓ_1 and ℓ_∞), we have $x^T z \leqslant \|x\|_+ \|z\|_*$ so that $\|x\|_+ = \sup_{\|z\|_*\leqslant 1} z^T x$ and $\|z\|_* = \sup_{\|x\|_+\leqslant 1} x^T z$; consequently, the conjugate function of $\|x\|_+$ ($\|x\|_*$) is the indicator function $\iota_{B^*(0,1)}(z)$ ($\iota_{B^+(0,1)}(z)$), where $B^*(0, 1)$ ($B^+(0, 1)$) is the closed unit ball with respect to the $\|\cdot\|_*$ norm ($\|\cdot\|_+$ norm, respectively);
- for $F(x)$ equal to the Kullback–Leibler divergence $\mathrm{KL}(y; x)$, with $y > 0$, (see equation (4.4)), the conjugate function is given by

$$\mathrm{KL}^*(y; z) = \begin{cases} -y^T \ln(\mathbf{1} - z) & \text{for } z < 1 \\ \infty & \text{otherwise,} \end{cases} \tag{6.8}$$

where the operator ln applied to a vector is intended component-wise.

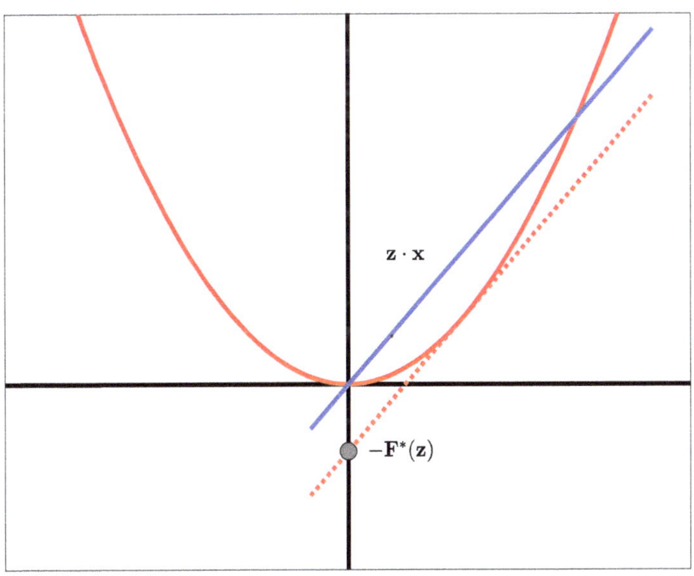

Figure 6.1. Geometrical interpretation of the conjugate function of $F(x) = x^2$: $-F^*(z)$ is the infimum of the difference between F and the continuous linear function zx.

For a l.s.c. function F, we have that the conjugate function of F^* is F, so that F can be written in terms of its conjugate as

$$F(x) = \sup_{z \in \mathbf{R}^N} (z^T x - F^*(z)). \tag{6.9}$$

In some applications, the proximal operator of the conjugate function F^* of F can be computed more easily than that of F; in these cases, we can use the following relation (Moreau identity) to obtain the proximal operator of F:

$$\text{prox}_{\gamma F,\, D}(z) = z - \gamma D^{-1} \text{prox}_{\frac{1}{\gamma} F^*,\, D^{-1}}\left(\frac{1}{\gamma} Dz\right) \tag{6.10}$$

where γ is a positive scalar and D is a s.p.d. matrix (see for instance [10, 107]).

Many convergence results about variable metric methods require that the sequences of matrices inducing the metric belong to a compact set of s.p.d. matrices and satisfy suitable assumptions on a partial ordering on this set. In the following, we denote by $\mathcal{D}_L$ the compact set of s.p.d. matrices such that all their eigenvalues are in the interval $[\frac{1}{L}, L]$, $L \geqslant 1$. As concerns the partial ordering the following definition is appropriate.

Definition 6.3. *Let Q, R be s.p.d. matrices of order N. The notation $Q \succcurlyeq R$ indicates that $Q - R$ is a symmetric positive semidefinite matrix or, equivalently, $x^T Q x \geqslant x^T R x$ for $x \in \mathbf{R}^N$ (Loewner partial ordering).*

6.2 The family of forward–backward (FB) splitting methods

In the framework of signal processing, one of the most popular approaches is the family of *forward–backward* (FB) *splitting* methods [45] which are a generalization of the projected gradient method [87] and were initially proposed in decomposition methods for solving variational inequalities (see for instance [93]).

With reference to the general problem (6.1) and in the setting of *variable metric* methods, namely different D_k norms are used at different iterations k, any iteration of the forward–backward iterative scheme can be broken up into a forward (explicit) step using the function $\Phi(x)$, and a backward (implicit) step using the function $\Psi(x)$:

$$\begin{aligned} g^{(k)} &= x^{(k)} - \alpha_k D_k^{-1} \nabla \Phi(x^{(k)}), && \text{forward step} \\ x^{(k+1)} &= \text{prox}_{\alpha_k \Psi,\, D_k}(g^{(k)}), && \text{backward step} \end{aligned} \tag{6.11}$$

where $\{D_k\}$ is a suitable sequence of s.p.d. matrices. When $D_k = I$ for all $k \geq 0$, the standard FB iteration is recovered. The forward step in (6.11) is known also as scaled gradient step of Φ with step-length α_k, while the backward step is the application of the proximity operator of Ψ with respect to the D_k-metric with parameter α_k. Furthermore, a relaxation parameter λ_k can be included:

$$x^{(k+1)} = x^{(k)} + \lambda_k(\text{prox}_{\alpha_k \Psi,\, D_k}(x^{(k)} - \alpha_k D_k^{-1} \nabla \Phi(x^{(k)})) - x^{(k)}). \tag{6.12}$$

With the purpose of providing a simple example, let us remark that, when $\Phi = f_0(x; y)$ and $\Psi(x)$ is the indicator function of the non-negative orthant $C = \{x \geqslant 0\}$, if we set $D_k^{-1} = \text{diag}(x^{(k)})$ as suggested by the EM–RL method, then we obtain

$$\begin{aligned} g^{(k)} &= x^{(k)} - \alpha_k \text{diag}(x^{(k)}) \nabla f_0(x^{(k)}; y) \\ x^{(k+1)} &= P_C(g^{(k)}); \end{aligned} \tag{6.13}$$

indeed $P_{C, D_k}(\cdot) \equiv P_C(\cdot)$, since the projection operator on the non-negative orthant with respect to any D-norm coincides with the ℓ_2 projection.

The features and convergence properties of the classic FB algorithm are deeply investigated (see [10, 34, 42, 45, 48, 88, 116] and references therein), while variable metric FB methods are analyzed in several papers (see for example [20, 24, 26, 31, 46, 47]).

It should be noted that the convergence of FB method (6.11) to a critical point of F can be proved even when Φ and Ψ are non-convex functions [4, 6, 98], provided that the function F satisfies the Kurdyka–Lojasiewicz inequality; the same result is proved also in the context of variable metric, even when a closed formula for the proximity operator of Ψ is not available [25, 43].

In the next sections we limit ourselves to the case where Φ is differentiable and Ψ is a convex function or both functions are convex. In particular, in section 6.3 we focus on variable metric FB algorithms where the proximal step is a projection while the case where both functions are convex and one is non-differentiable, is addressed in section 6.4. We report some specific variants of this general scheme, tailored to the features of specific problems, detailing the assumptions required for obtaining convergence results and useful properties. In particular, we focus on methods that are the basis of available software packages.

6.3 FB methods for smooth problems of image reconstruction

In this section we discuss methods for solving the problems (4.13) and (4.30) with a differentiable regularization term $f_1(x)$, by considering the general constrained optimization problem

$$\min_{x \in \Omega} \Phi(x), \tag{6.14}$$

where $\Phi(x)$ is a continuously differentiable function (possibly non-convex) and Ω is a non-empty convex and closed subset of $\mathbf{R}^N$; this problem is equivalent to the unconstrained problem (6.1) with $\Psi(x) = \iota_\Omega(x)$.

We restrict the discussion to two approaches which look particularly suitable for image reconstruction problems with Poisson data: the scaled gradient projection method [27, 31, 104] and the projected Newton-like method [8, 15, 83].

6.3.1 Scaled gradient projection (SGP) method

When the projection on the set Ω is easily computable by a closed formula or by a cheap algorithm, a simple and effective scheme for computing a numerical solution of the problem (6.14) is a gradient projection method. We stress that this method does not require that $\Phi(x)$ is a convex function, but only that it is continuously differentiable on dom Φ, with $\Omega \subseteq$ dom Φ.

Given the step-length parameters λ_k, $\alpha_k > 0$, a sequence of s.p.d. matrix $\{D_k\}$ and a starting point $x^{(0)} \in \Omega$, the scaled gradient projection (SGP) method generates a sequence of iterates as follows:

$$\begin{aligned} w^{(k)} &= \mathrm{P}_{\Omega,\, D_k}(x^{(k)} - \alpha_k D_k^{-1} \nabla \Phi(x^{(k)})), \\ x^{(k+1)} &= x^{(k)} + \lambda_k (w^{(k)} - x^{(k)}), \end{aligned} \tag{6.15}$$

which is precisely the expression assumed by the FB splitting method (6.12) in the considered case.

Different SGP methods correspond to different strategies for selecting the parameters α_k, λ_k, D_k. The version of the SGP method in [27, 31] is based on the idea that the vector $d^{(k)} = w^{(k)} - x^{(k)}$ computed by equation (6.15) is a descent direction for the function Φ at the current iterate $x^{(k)}$. The parameter λ_k is used to force a sufficient decrease of the objective function and is computed by a backtracking procedure that allows one to prove the convergence of the scheme. The other two ingredients of the method are a positive step-length parameter α_k which must belong to a closed interval $[\alpha_{\min}, \alpha_{\max}]$, with $\alpha_{\min} > 0$, and a sequence of s.p.d. matrices $\{D_k\}$, chosen in the compact set $\mathcal{D}_L$ of the s.p.d. matrices of order N, with $L \geqslant 1$ (see the definition at the end of section 6.1). It is worth stressing that any choice of the step-length α_k in the closed interval $[\alpha_{\min}, \alpha_{\max}]$ and of the scaling matrices $\{D_k\}$ in the set $\mathcal{D}_L$ is permitted. This is very important from a practical point of view, since it allows one to make the updating rules of α_k and D_k oriented at optimizing the performance of the method. The main steps of SGP are detailed in algorithm 6.1.

Algorithm 6.1. Scaled gradient projection (SGP)

Choose the starting point $x^{(0)} \in \Omega$, set the parameters c, $\rho \in (0, 1)$, $0 < \alpha_{\min} < \alpha_{\max}$, $L \geqslant 1$ and fix a positive integer M.

FOR $k = 0, 1, 2, \ldots$ DO THE FOLLOWING STEPS:

STEP 1. choose the parameter $\alpha_k \in [\alpha_{\min}, \alpha_{\max}]$ and the scaling matrix $D_k \in \mathcal{D}_L$;

STEP 2. compute $w^{(k)} = \mathrm{P}_{\Omega,\, D_k}(x^{(k)} - \alpha_k D_k^{-1} \nabla \Phi(x^{(k)}))$;

if $w^{(k)} = x^{(k)}$, then stop, declaring $x^{(k)}$ is a stationary point;

STEP 3. set $d^{(k)} = w^{(k)} - x^{(k)}$;

STEP 4. set $\lambda_k = 1$ and $\Phi_{max} = \max_{0 \leqslant j \leqslant \min\{k,\, M-1\}} \Phi(x^{(k-j)})$;

STEP 5. backtracking loop:

  IF $\Phi(x^{(k)} + \lambda_k d^{(k)}) \leqslant \Phi_{max} + c\lambda_k \nabla \Phi(x^{(k)})^T d^{(k)}$

   go to STEP 6;

  ELSE

   set $\lambda_k = \rho\lambda_k$ and go to STEP 5;

  END

STEP 6. set $x^{(k+1)} = x^{(k)} + \lambda_k d^{(k)}$;

If the projection performed in step 2 returns a vector $w^{(k)}$ equal to $x^{(k)}$, then $x^{(k)}$ is a stationary point and the iterate terminates (see [31], lemma 2.2). When $w^{(k)} \neq x^{(k)}$, it is possible to prove that $d^{(k)}$ is a descent direction for Φ at $x^{(k)}$ ([31], lemma 2.3) and the backtracking loop in step 5 terminates with a finite number of runs; thus the algorithm is well-defined. In step 5, a non-monotone line-search strategy is implemented: this procedure ensures that $\Phi(x^{(k+1)})$ is lower than the maximum of the objective function on the last M iterations [74]; of course, if $M = 1$, then the strategy reduces to the standard monotone Armijo line-search procedure. In figure 6.2 we show one step of SGP when Ω is a box region, detailing the forward and the backward steps which make SGP a special FB method where the step involving the proximity operator is a projection.

6.3.1.1 Convergence

When Φ is a differentiable function in dom $\Phi \supseteq \Omega$, and $D_k = I$ for all $k \geqslant 0$, any limit point of the sequence generated by the algorithm 6.1 is a stationary point of problem (6.14) [19]. A similar result is obtained in [20] under the assumption that the sequence of matrices $\{D_k\}$ are in $\mathcal{D}_L$, $L \geqslant 1$; nevertheless, in the version of SGP proposed in [20], the definition of D_k includes also α_k (i.e. $D_k := \frac{D_k}{\alpha_k}$), while in [31] the step-length α_k and the scaling matrix are managed separately. In particular, in [31] and [27] the following convergence results are proved.

First, the limit of any convergent subsequence of $\{x^{(k)}\}$, generated by the algorithm 6.1, is a stationary point of Φ. Next, under the additional assumption that the objective function Φ of (6.14) is convex and that the set of solutions X^* is

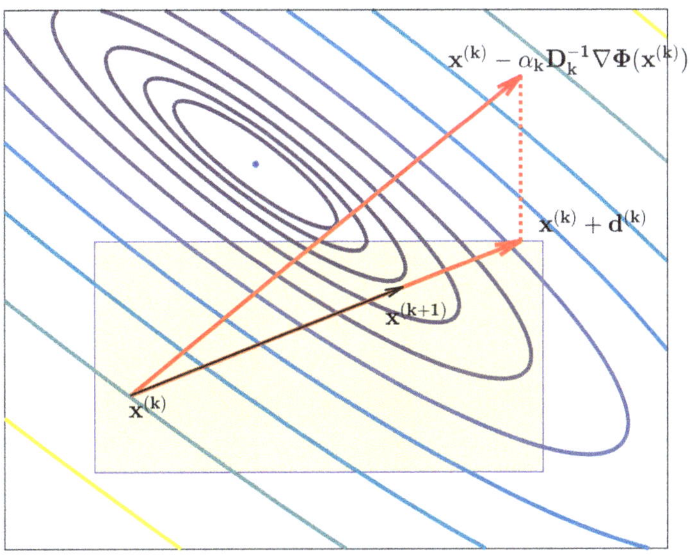

Figure 6.2. An iteration of SGP method for a minimization problem on a box constraint (indicated by the yellow region). Starting from $x^{(k)}$ along a descent direction $-D_k^{-1}\nabla\Phi(x^{(k)})$, a point $g^{(k)} = x^{(k)} - \alpha_k D_k^{-1}\nabla\Phi(x^{(k)})$ is computed (forward step); then, a projection on Ω, i.e. a backward step, provides the point $w^{(k)} = x^{(k)} + d^{(k)}$; finally by a backtracking procedure, the new iterate $x^{(k+1)}$ is obtained.

non-empty, the whole sequence $\{x^{(k)}\}$ converges to a solution of the problem (6.14), provided that the sequence of s.p.d. scaling matrices $\{D_k\}$, $D_k \in \mathcal{D}_L$, $L \geqslant 1$, satisfies the additional condition:

$$D_{k+1} \preccurlyeq (1 + \xi_k)D_k \qquad \xi_k \geqslant 0 \quad \sum_{k=0}^{\infty} \xi_k < \infty. \tag{6.16}$$

In such a case, the sequence $\{D_k\}$ asymptotically approaches a constant matrix [46].

Remark 6.1. *A practical rule for choosing the sequence* $\{D_k\}$ *in such a way that the condition* (6.16) *holds, is as follows:*

$$\begin{aligned} &D_k \in \mathcal{D}_{L_k} \text{ with } L_k \leqslant L \\ &L_k^2 = 1 + \zeta_k, \text{ where } \zeta_k \geqslant 0, \ \sum_{k=0}^{\infty} \zeta_k < \infty. \end{aligned} \tag{6.17}$$

Indeed, we observe that in the case of a s.p.d. matrix U of order N for any $x \in \mathbf{R}^N$, *the following inequalities hold:*

$$\mu_{\min}(U)\|x\|^2 \leqslant x^T U x \leqslant \mu_{\max}(U)\|x\|^2,$$

where $\mu_{\min}(U)$ *and* $\mu_{\max}(U)$ *are, respectively, the minimum and the maximum eigenvalue of* U. *By taking into accounts the bounds on the eigenvalues of* D_k, D_{k+1}, *for any* $x \in \mathbf{R}^N$, *we have*

$$x^T D_{k+1} x \leqslant \frac{L_{k+1}L_k}{L_k}\|x\|^2 \leqslant L_{k+1}L_k \ x^T D_k x.$$

Let us define $\xi_k = L_{k+1}L_k - 1 = \sqrt{(1 + \zeta_{k+1})(1 + \zeta_k)} - 1$ *and observe that the series* $\sum \xi_k$ *and* $\sum \zeta_k$ *have the same behavior, since* $\lim_{z\to 0}(\sqrt{1 + z} - 1)/z = 1/2$. *Therefore, for any* $x \in \mathbf{R}^N$, *we can write for* $k \geqslant 0$

$$x^T D_{k+1} x \leqslant (1 + \xi_k) x^T D_k x,$$

with $\xi_k \geqslant 0$ *and* $\sum \xi_k < \infty$, *so that a sequence of matrices* $\{D_k\}$ *chosen according to the rule* (6.17) *satisfies the assumption* (6.16). *Furthermore, it is immediate that the sequence* $\{D_k\}$ *asymptotically approaches the identity matrix.*

The practical rule (6.17) does not limit the positive impact that a suitable choice of scaling matrices has on the performance of SGP. On the contrary, the convergence result is coherent with the numerical experience which shows that the advantages of using a particular scaling matrix are observed mainly at the initial iterations.

Finally, if Φ is convex and $\nabla\Phi$ is Lipschitz-continuous on Ω, it can be proved that the theoretical rate of convergence of the objective function is $\mathcal{O}\left(\frac{1}{k}\right)$ [27]; furthermore, by the argument of [95], theorem 2.1.6, it is possible to state that, for any choice of the diagonal scaling matrices, there exists at least a problem for which the rate of convergence cannot be $\mathcal{O}\left(\frac{1}{k^2}\right)$. However, suitable choices of the step-lengths α_k and of the scaling matrices D_k provide, in the first iterations, a practical behavior very similar to a super-linear rate of convergence.

6.3.1.2 Step-length selection

The step-length α_k has a huge impact on the convergence rate. Starting from the seminal work of Barzilai and Borwein (BB) [9], many step-length updating strategies have been devised to accelerate the slow convergence exhibited in most cases by standard gradient methods (see for example [49, 50, 51, 52, 58, 68, 69, 120]). Taking into account the recent advances on the BB-like updating rules, SGP exploits an updating strategy based on an alternating rule for the step-length which is very efficient in the applications [68].

The basic idea of this strategy comes from the observation that the forward step of SGP at the current iterate $x^{(k)}$ is the solution of the auxiliary problem

$$\min_x \left(\Phi(x^{(k)}) + \nabla\Phi(x^{(k)})^T(x - x^{(k)}) + \frac{1}{2\alpha_k}(x - x^{(k)})^T D_k (x - x^{(k)}) \right).$$

If the matrix $U(\alpha) = \frac{1}{\alpha} D_k$ is regarded as an approximation of the Hessian $\nabla^2\Phi(x^{(k)})$, it is possible to derive two updating rules for α_k by forcing quasi-Newton properties on $U(\alpha)$:

$$\alpha_k^{\mathrm{BB1}} = \arg\min_{\alpha\in\mathbf{R}} \|U(\alpha)s^{(k-1)} - z^{(k-1)}\|,$$
$$\alpha_k^{\mathrm{BB2}} = \arg\min_{\alpha\in\mathbf{R}} \|s^{(k-1)} - U(\alpha)^{-1}z^{(k-1)}\|,$$

where $s^{(k-1)} = x^{(k)} - x^{(k-1)}$ and $z^{(k-1)} = \nabla\Phi(x^{(k)}) - \nabla\Phi(x^{(k-1)})$. In this way, we obtain

$$\alpha_k^{\mathrm{BB1}} = \frac{s^{(k-1)T} D_k D_k s^{(k-1)}}{s^{(k-1)T} D_k z^{(k-1)}}, \qquad \alpha_k^{\mathrm{BB2}} = \frac{s^{(k-1)T} D_k^{-1} z^{(k-1)}}{z^{(k-1)T} D_k^{-1} D_k^{-1} z^{(k-1)}}, \tag{6.18}$$

and the standard BB rules are recovered when $D_k = I$. Inspired by the step-length alternation successfully implemented in the framework of non-scaled gradient methods [68, 120], the step-length selection proposed for SGP follows a very effective adaptive switching rule, named $\mathrm{ABB}_{\mathrm{min}}$ rule and detailed in algorithm 6.2 [31].

Another choice for the step-length updating is the strategy recently proposed in [104] (see [67] for unconstrained optimization). The principal aim of this strategy is to acquire second-order information by considering a small number (namely $m \ll N$) of scaled gradients $D_i^{-1/2}\nabla\Phi(x^{(i)})$, $i = k, k-1, \dots, k-m+1$, computed at the previous iterations; in any gradient, the elements corresponding to the entries of the current iterate $x^{(i)}$ satisfying the box constraints are set to 0. These scaled and reduced gradients enable to determine approximations of the eigenvalues of the Hessian matrix, named Ritz values. The inverse of these values are used as step-lengths. We refer to [67, 103, 104] for further details. In [103, 104] it is shown that the SGP version exploiting this Ritz-like step-length selection strategy outperforms the one based on the alternating Barzilai–Borwein rules.

Algorithm 6.2. ABB$_{\min}$ step-length selection.

IF $k = 0$

set $\alpha_0 \in [\alpha_{\min}, \alpha_{\max}]$, $\tau_1 \in (0, 1)$ and a non-negative integer M_α;

ELSE

IF $s^{(k-1)T} D_k z^{(k-1)} \leqslant 0$

$\alpha_k^{\mathrm{BB1}} = \alpha_{\max}$;

ELSE

$$\alpha_k^{\mathrm{BB1}} = \max\left\{\alpha_{\min}, \min\left\{\frac{s^{(k-1)^T} D_k D_k s^{(k-1)}}{s^{(k-1)^T} D_k z^{(k-1)}}, \alpha_{\max}\right\}\right\};$$

END

IF $s^{(k-1)T} D_k^{-1} z^{(k-1)} \leqslant 0$

$\alpha_k^{\mathrm{BB2}} = \alpha_{\max}$;

ELSE

$$\alpha_k^{\mathrm{BB2}} = \max\left\{\alpha_{\min}, \min\left\{\frac{s^{(k-1)^T} D_k^{-1} z^{(k-1)}}{z^{(k-1)^T} D_k^{-1} D_k^{-1} z^{(k-1)}}, \alpha_{\max}\right\}\right\};$$

END

IF $\frac{\alpha_k^{\mathrm{BB2}}}{\alpha_k^{\mathrm{BB1}}} \leqslant \tau_k$

$\alpha_k = \min_{j=\max\{1,\, k-M_\alpha\},\, \dots,\, k}\left\{\alpha_j^{\mathrm{BB2}}\right\}$;

$\tau_{k+1} = 0.9\tau_k$;

ELSE

$\alpha_k = \alpha_k^{\mathrm{BB1}}$;

$\tau_{k+1} = 1.1\tau_k$;

END

END

6.3.1.3 Variable metric selection

The introduction of a variable scaling matrix D_k, that is of a variable metric depending on the iteration k, has two main goals: improve the rate of convergence without increasing the computational complexity. A classic way for reaching these objectives is to choose a matrix with a *diagonal* structure, such that D_k approximates the Hessian of Φ:

$$(D_k)_{i,\,i} = \left(\frac{\partial^2 \Phi(x^{(k)})}{(\partial x_i)^2}\right), \qquad i = 1, \dots, N.$$

Since the computation of the Hessian could be very expensive, a practical choice should be to take a suitable approximation of it. On the other hand, the motivation underlying the choice of a variable metric is to be able to capture in the metric the local features of the problem. From this point of view, the simple methods discussed in chapter 5 are scaled gradient methods and therefore they can suggest very effective choices of $\{D_k\}$ both for ML problem (4.13) and for MAP problem (4.30) with a differentiable regularization term.

For ML problem, the remark that the EM–RL method can be viewed as a scaled gradient method for the minimization of the data-fidelity function $f_0(x; y)$, equation (4.7), with a scaling given by the current iterate, suggests to select a diagonal matrix D_k as follows:

$$(D_k)_{i,\,i} = \max\left\{\frac{1}{L_k},\ \min\left\{L_k,\ \frac{(H^T 1)_i}{x_i^{(k)}}\right\}\right\},\quad i = 1, \ldots, N \tag{6.19}$$

with $0 < L_k \leqslant L$, $L_k \geqslant 1$ (and the convention that $1/0 = \infty$); this definition takes into account the conditions of section 6.3.1.1 for the convergence of the iteration. Moreover, the theoretical assumptions for the convergence and the rate of convergence are satisfied when we adopt the practical rule $L_k^2 = 1 + \zeta_k$, with $\zeta_k = \mathcal{O}\left(\frac{1}{k^h}\right)$, $h > 1$.

For MAP problems with a differentiable regularization term, following the SGM approach [86], a strategy for a convenient choice of D_k is based on a decomposition of the gradient of $\Phi(x) := f_\beta(x; y)$. From the decomposition of ∇f_1 in equation (5.20)

$$-\nabla f_1(x) = U_1(x) - V_1(x),$$

with $U_1(x) \geqslant 0,\ V_1(x) > 0$ for any $x \geqslant 0$, following the suggestions of section 5.4, we can obtain a decomposition of $\nabla\Phi(x) := \nabla f_\beta(x; y)$:

$$-\nabla f_\beta(x; y) = \left(H^T \frac{y}{Hx + b} + \beta U_1(x)\right) - (H^T 1 + \beta V_1(x)).$$

As a consequence, an effective choice of D_k for the MAP problem is a diagonal matrix with diagonal entries defined as follows:

$$(D_k)_{i,\,i} = \max\left\{\frac{1}{L_k},\ \min\left\{L_k,\ \frac{(H^T 1 + \beta V_1(x^{(k)}))_i}{x_i^{(k)}}\right\}\right\},\quad i = 1, \ldots, N. \tag{6.20}$$

A Matlab® version of SGP for 2D and 3D images corrupted by Poisson noise is downloadable from http://www.unife.it/prin/software.

6.3.2 Projected Newton-like method

The projected Newton-like methods, proposed by Bertsekas in [15] for simple constrained problems, have been recently used for the differentiable regularization of image reconstruction problems with Poisson data when only the non-negativity constraint is imposed on the solution [8, 23, 83]. For the sake of simplicity, we present the basic idea of this class of methods when the feasible region Ω coincides with the non-negative orthant C, i.e., $\Omega = C$. However, these algorithms can be applied with minor modifications also to problems with box or linear inequality constraints.

The basic iteration of the method is given by

$$x^{(k+1)} = P_\Omega(x^{(k)} - \alpha_k D_k^{-1} \nabla\Phi(x^k)), \tag{6.21}$$

where D_k^{-1} is a suitable matrix and α_k is selected in such a way that the decrease of the objective function is assured. We remark that the iteration (6.21) is very similar to the first step of the SGP method, but there are two key differences. First, the iteration (6.21) embodies two scaling matrices, justifying the denomination of *two-metric projection method*: the matrix D_k^{-1}, which scales the gradient, and the identity matrix, which is used in the projection with respect to the ℓ_2 norm. This provides the possibility of including second-order derivative information within D_k as in the Newton's method, while maintaining the simplicity of the Euclidean projection on Ω. Next, unlike SGP, the parameter λ_k is fixed at unity and α_k has to be selected to assure a sufficient decrease of the objective function. This is obtained by a variant of the line-search strategy on the projection arc, defined as $\{x^{(k)}(\alpha): \alpha > 0\}$, with $x^{(k)}(\alpha) := P_\Omega(x^{(k)} - \alpha d^{(k)})$ and $d^{(k)} = D_k^{-1}\nabla\Phi(x^{(k)})$. This procedure consists in choosing $\alpha_k := \rho^{m_k}$, with $\rho \in (0, 1)$ and m_k the first non-negative integer m for which

$$\Phi(x^{(k)}) - \Phi(x^{(k)}(\rho^m)) \geqslant c\nabla\Phi(x^{(k)})^T(x^{(k)} - x^{(k)}(\rho^m)), \qquad c \in (0, 1). \tag{6.22}$$

Figure 6.3 illustrates the Armijo line-search rule on a projection arc for a box-constrained problem.

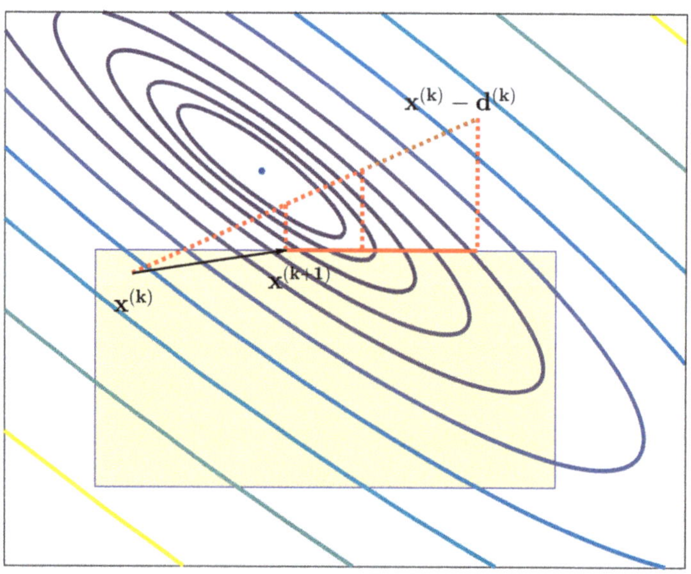

Figure 6.3. The Armijo line-search rule on a projection arc for a box-constrained problem; starting from $x^{(k)}$ along the direction $-d^{(k)}$, three successive points are tested by the Armijo rule along the projection arc, given by a portion of the boundary of the feasible region.

As concerns the choice of the matrices $\{D_k^{-1}\}$, the direction $-d^{(k)} = -D_k^{-1}\nabla\Phi(x^{(k)})$ is not in general a descent direction. A class of non-diagonal matrices for which the decrease of the objective function is guaranteed is the set of s.p.d. matrices which are diagonal with respect to the set of indices $\mathcal{I}^{+}(x^{(k)}) = \{i\colon x_i^{(k)} = 0, \frac{\partial\Phi(x^{(k)})}{\partial x_i} > 0\}$, i.e. $(D_k^{-1})_{i,j} = 0$ for any $i \in \mathcal{I}^{+}(x^{(k)}), j = 1, \dots, N, i \neq j$. Indeed, when D_k^{-1} is a s.p.d. with this particular structure, it is possible to prove ([15], proposition 1) that

- $x^{(k)}(\alpha) = x^{(k)}$ for any $\alpha \geqslant 0$ if and only if $x^{(k)}$ is a stationary point, i.e. it satisfies the necessary first order optimality conditions;
- if $x^{(k)}$ is not a stationary point, then there exists $\bar{\alpha} > 0$ such that $\Phi(x^{(k)}(\alpha)) < \Phi(x^{(k)})$ for any $\alpha \in (0, \bar{\alpha}]$.

Nevertheless, we remark that, given a sequence $\{x^{(k)}\}$ of points converging to a boundary point $\bar{x}$, the set $\mathcal{I}^{+}(x^{(k)})$ may be strictly contained in the set $\mathcal{I}^{+}(\bar{x})$. To avoid these theoretical difficulties and, at the same time, to ensure convergence of the iteration, we can add to the set $\mathcal{I}^{+}(x^{(k)})$ the indices of the entries of $x^{(k)}$ *close* to zero, satisfying $\frac{\partial\Phi(x^{(k)})}{\partial x_i} > 0$. Thus, the sequence $\{D_k^{-1}\}$ can be chosen as a set of matrices which are diagonal with respect to the sets of indices

$$\mathcal{I}_k^{+} = \left\{i\colon 0 \leqslant x_i^{(k)} \leqslant \epsilon_k, \frac{\partial\Phi(x^{(k)})}{\partial x_i} > 0\right\}, \tag{6.23}$$

where $\epsilon_k = \min\{\epsilon, \|x^{(k)} - P_\Omega(x^{(k)} - \nabla\Phi(x^k))\|\}$ and ϵ is a small positive parameter; as a consequence of this definition, we have $\mathcal{I}_k^{+} \supseteq \mathcal{I}^{+}(x^{(k)})$. Thus, following [15], in order to assure the decrease of the objective function, one has to set

$$x^{(k+1)} = x^{(k)}(\alpha_k) = P_\Omega(x^{(k)} - \alpha_k d^{(k)}), \tag{6.24}$$

where $d^{(k)} = D_k^{-1}\nabla\Phi(x^k)$ and $\alpha_k := \rho^{m_k}$, with $\rho \in (0, 1)$ and m_k the first non-negative integer m such that

$$\begin{aligned}\Phi(x^{(k)}) - \Phi(x^{(k)}(\rho^m)) \geqslant c\Bigg(&\rho^m \sum_{i \notin \mathcal{I}_k^{+}} \frac{\partial\Phi(x^{(k)})}{\partial x_i} d_i^{(k)} \\ &+ \sum_{i \in \mathcal{I}_k^{+}} \frac{\partial\Phi(x^{(k)})}{\partial x_i}(x_i^{(k)} - x_i^{(k)}(\rho^m))\Bigg),\end{aligned} \tag{6.25}$$

with $c \in (0, \frac{1}{2})$. This condition can be viewed as a combination of the Armijo rule (6.22) and the one usually employed in unconstrained minimization or along a feasible direction (as in SGP method with $M = 1$).

A crucial point in proving the convergence of equation (6.21) combined with the line-search strategy (6.25) is to devise a rule generating a sequence $\{D_k\}$ of matrices such that, for all $m \geqslant 0$, the two terms of the right-hand side of equation (6.25) are non-negative and their sum is positive if and only if $x^{(k)}$ is not a stationary point. For example, this condition is satisfied when we generate a sequence of s.p.d. matrices which are diagonal with respect to $\mathcal{I}_k^+$; indeed, in this case, by denoting as $g^{(k)}$ the vector such that $g_i^{(k)} = 0$ for $i \in \mathcal{I}_k^+$ and $g_i^{(k)} = \frac{\partial\Phi(x^{(k)})}{\partial x_i}$ for $i \notin \mathcal{I}_k^+$, we have

$$\sum_{i \notin \mathcal{I}_k^+} \frac{\partial\Phi(x^{(k)})}{\partial x_i} d_i^{(k)} = \sum_{i \notin \mathcal{I}_k^+} \frac{\partial\Phi(x^{(k)})}{\partial x_i} (D_k^{-1}\nabla\Phi(x^{(k)}))_i = (g^{(k)})^T D_k^{-1} g^{(k)} \geqslant 0; \qquad (6.26)$$

furthermore, since $\frac{\partial\Phi(x^{(k)})}{\partial x_i} > 0$ and $d_i^k > 0$ for $i \in \mathcal{I}_k^+$ (as follows from the definition of $\mathcal{I}_k^+$), we have

$$\begin{gathered} (x_i^{(k)} - x_i^{(k)}(\alpha)) \geqslant 0, \quad \forall \alpha \geqslant 0, \\ \sum_{i \in \mathcal{I}_k^+} \frac{\partial\Phi(x^{(k)})}{\partial x_i} (x_i^{(k)} - x_i^{(k)}(\alpha)) \geqslant 0, \quad \forall \alpha \geqslant 0, \end{gathered} \qquad (6.27)$$

with $k \geqslant 0$, $i \in \mathcal{I}_k^+$.

More in general, provided that the sequence $\{D_k\}$ satisfies the following assumptions for all $k \geqslant 0$

(a1) $(D_k^{-1})_{i,i} > 0$ for $i \in \mathcal{I}_k^+$,

(a2) $\sum_{i \notin \mathcal{I}_k^+} \frac{\partial\Phi(x^{(k)})}{\partial x_i} d_i^{(k)} \geqslant 0$ and the equality holds if and only if $d_i^{(k)} = \frac{\partial\Phi(x^{(k)})}{\partial x_i} = 0$ for all $i \notin \mathcal{I}_k^+$,

it is possible to generalize the statements proved in [15], proposition 2:

- if $x^{(k)}$ is not a stationary point, one of the inequalities (6.26) or (6.27) is strict for all $\alpha > 0$ so that the right-hand side of equation (6.25) is positive for all $m \geqslant 0$, in particular for all m sufficiently large, allowing to determine α_k via a finite number of steps;
- when $x^{(k)}$ is a stationary point, the right-hand side of equation (6.25) is zero, so that $x^{(k)} = x^{(k)}(\alpha)$ for all $\alpha \geqslant 0$; it follows that the inequality (6.25) is satisfied for $m = 0$ and, in this case, $x^{(k+1)} = x^{(k)}$.

In conclusion, for a suitable set of matrices, which are diagonal with respect to the sets $\mathcal{I}_k^+$, the algorithm (6.24)–(6.25) is well-defined and decreases the value of the objective function at each iteration k, except when $x^{(k)}$ is a stationary point and the algorithm terminates.

The following proposition which is a generalization of [15], proposition 2 (see also [83]), provides the assumptions required for stating that the limit point of the sequence $\{x^{(k)}\}$ generated by the iteration (6.24) combined with equation (6.25) is a stationary point for the constrained problem (6.14).

Proposition 6.1. *Assume that $\nabla\Phi$ is Lipschitz-continuous on each bounded set of $\mathbf{R}^N$. Assume also that $\{D_k\}$ satisfies the assumptions (a1)–(a2) and that there exist positive scalars μ_1, μ_2, μ such that, for $k \geqslant 0$, the following conditions hold true*

$$\mu_1 \leqslant (D_k^{-1})_{i,\,i} \leqslant \mu_2 \quad \forall\, i \in \mathcal{I}_k^+,$$
$$\mu_1 \sum_{i \notin \mathcal{I}_k^+} \left(\frac{\partial \Phi(x^{(k)})}{\partial x_i} \right)^2 \leqslant \sum_{i \notin \mathcal{I}_k^+} \frac{\partial \Phi(x^{(k)})}{\partial x_i} d_i^{(k)} \leqslant \mu_2 \sum_{i \notin \mathcal{I}_k^+} \left(\frac{\partial \Phi(x^{(k)})}{\partial x_i} \right)^2, \tag{6.28}$$
$$\sum_{i \notin \mathcal{I}_k^+} (d_i^{(k)})^2 \leqslant \mu \sum_{i \notin \mathcal{I}_k^+} \frac{\partial \Phi(x^{(k)})}{\partial x_i} d_i^{(k)}.$$

Then every limit point of the sequence generated by iteration (6.24) *combined with condition* (6.25) *is a stationary point.*

We can immediately verify that, when D_k are s.p.d. matrices with eigenvalues uniformly bounded above and away from zero and diagonal structure with respect to the set $\mathcal{I}_k^+$, the assumptions of proposition 6.1 are satisfied [15]. Nevertheless, other strategies to select a set of matrices diagonal with respect to $\mathcal{I}_k^+$ allow one to obtain the same theoretical convergence result [8, 83].

6.3.2.1 Selection of $\{D_k^{-1}\}$

As in SGP method, the matrices $\{D_k^{-1}\}$ have the role of scaling matrices which approximate the inverse of the Hessian of Φ at $x^{(k)}$ and, at the same time, take into account the constraints. Therefore, when $\Phi(x)$ is a strongly convex function in the level set $\{x\colon \Phi(x) \leqslant \Phi(x^{(0)})\}$, since $\nabla^2\Phi(x)$ is positive definite, a natural choice for the sequence of D_k^{-1} is to set

$$(D_k)_{i,j} = \begin{cases} 0 & \text{if } i \neq j \text{ and either } i \in \mathcal{I}_k^+ \text{ or } j \in \mathcal{I}_k^+ \\ \dfrac{\partial^2 \Phi(x^{(k)})}{\partial x_i \partial x_j} & \text{otherwise,} \end{cases}$$

i.e. the diagonal blocks of D_k^{-1} which are not related to the indices of $\mathcal{I}_k^+$, are the inverse of the corresponding blocks of the Hessian of $\Phi(x)$. In this setting, Bertsekas proved that the sequence $\{x^{(k)}\}$ generated by (6.24)–(6.25) converges to the unique solution x^* of the problem (6.14) and the rate of convergence is super-linear and at least quadratic if $\nabla^2\Phi$ is Lipschitz-continuous in a neighborhood of x^* ([15], proposition 4). With this approach, the most expensive step of the method is the computation of the Hessian of Φ at $x^{(k)}$ and the solution of the related block diagonal system for computing the descent direction $-d^{(k)}$:

$$D_k d = \nabla\Phi(x^{(k)}).$$

A different approach is proposed in [83]. Here, the following diagonal matrices E_k and F_k are introduced

$$(E_k)_{i,\,i} = \begin{cases} 1 & i \notin \mathcal{I}_k^+ \\ 0 & i \in \mathcal{I}_k^+ \end{cases} \qquad F_k = I - E_k;$$

then, the scaling matrix D_k^{-1}, diagonal with respect to the set $\mathcal{I}_k^+$, is defined by

$$D_k^{-1} = E_k U_k^{-1} E_k + F_k,$$

where U_k^{-1} is a symmetric square matrix. As a consequence, the search direction $d^{(k)}$ can be written as

$$d^{(k)} = D_k^{-1}\nabla\Phi(x^{(k)}) = E_k U_k^{-1} E_k \nabla\Phi(x^{(k)}) + F_k\nabla\Phi(x^{(k)}) = E_k v^{(k)} + F_k\nabla\Phi(x^{(k)}),$$

with $v^{(k)} = U_k^{-1}E_k\nabla\Phi(x^{(k)})$; $v^{(k)}$ can be computed as solution of the following linear system

$$U_k v = E_k\nabla\Phi(x^{(k)}). \tag{6.29}$$

By setting $U_k = \nabla^2\Phi(x^{(k)})$, it is possible to obtain an approximate solution of the system $\nabla^2\Phi(x^{(k)})v = E_k\nabla\Phi(x^{(k)})$ via the truncated conjugate gradient (CG) method. More precisely, $v^{(k)}$ can be set equal to the approximate solution of system (6.29), obtained after a number $\ell^{(k)} + 1$ of iterations of CG. Then, using the properties of the CG method ([89], p 246), the vector $v^{(k)}$ can be expressed as the following matrix–vector product

$$v^{(k)} = \pi_{\ell^{(k)}}(\nabla^2\Phi(x^{(k)}))(E_k\nabla\Phi(x^{(k)})),$$

where $\pi_{\ell^{(k)}}(\nabla^2\Phi(x^{(k)}))$ is a suitable polynomial of degree $\ell^{(k)}$ computed at $\nabla^2\Phi(x^{(k)})$. Consequently, the matrix U_k^{-1} can be set equal to $\pi_{\ell^{(k)}}(\nabla^2\Phi(x^{(k)}))$, the search direction $d^{(k)}$ is given by

$$d^{(k)} = \begin{cases} v_i^{(k)} & i \notin \mathcal{I}_k^+, \\ \dfrac{\partial\Phi(x^{(k)})}{\partial x_i} & i \in \mathcal{I}_k^+, \end{cases} \tag{6.30}$$

and, in view of the rule $d^{(k)} = D_k^{-1}\nabla\Phi(x^{(k)})$, D_k^{-1} is defined as follows

$$\left(D_k^{-1}\right)_{i,j} = \begin{cases} \delta_{i,j} & \text{if either } i \in \mathcal{I}_k^+ \text{ or } j \in \mathcal{I}_k^+ \\ \left(\pi_{\ell^{(k)}}\left(\nabla^2\Phi\left(x^{(k)}\right)\right)\right)_{i,j} & \text{otherwise.} \end{cases} \tag{6.31}$$

Under the assumption that $\nabla^2\Phi(x^{(k)})$ has positive uniformly bounded eigenvalues for $k \geqslant 0$, the above sequence $\{D_k^{-1}\}$ satisfies the assumptions (a1)–(a2) and the conditions (6.28), so that the sequence $\{x^{(k)}\}$ converges to a solution of the problem (6.14). This approach, named projected-Newton-CG method, requires at any iteration only matrix–vector products, in addition to the computation of the Hessian matrix of Φ.

Another strategy for obtaining an easily invertible approximation of $\nabla^2\Phi(x^{(k)})$ can be adopted for image deconvolution with T0 regularization [23]. In this case, the objective function is $\Phi(x) := f_0(x; y) + \frac{\beta}{2}\|x\|^2$ and the imaging matrix H has a Toeplitz structure with spectral decomposition $H = \mathcal{F}^*\Xi\mathcal{F}$. Given a threshold parameter $\sigma > 0$, the Hessian matrix of the objective function $\nabla^2\Phi(x^{(k)}) = H^T\mathrm{diag}\left(\frac{y}{\left(Hx^{(k)}+b\right)}\right)H + \beta I$ can be approximated by

$$U_k = \gamma_k H_\sigma^T H_\sigma + \beta I$$

where $H_\sigma = \mathcal{F}^* \Xi_\sigma \mathcal{F}$, $(\Xi_\sigma)_{i,\,i} = (\Xi)_{i,\,i}$ for $|(\Xi)_{i,\,i}| > \sigma$ and $(\Xi_\sigma)_{i,\,i} = \sigma$ for $|(\Xi)_{i,\,i}| \leqslant \sigma$ and γ_k is the mean of the elements of $\operatorname{diag}\left(\frac{y}{\left(Hx^{(k)} + b\right)}\right)$. With this approach, the solution of the system (6.29) is obtained by inversion in the Fourier space, using fast transforms.

Algorithm 6.3 details the steps of a projected Newton-like method. A Matlab® version for the restoration of 2D images is downloadable from http://www.dm.unibo.it/~landig/NPTool/NPTool.html.

Algorithm 6.3. Projected-Newton-like method.

Choose the starting point $x^{(0)}$, set the parameters $\rho \in (0, 1)$, $c \in (0, 1/2)$ and $\epsilon > 0$.
FOR $k = 0, 1, 2, \ldots$ DO THE FOLLOWING STEPS:
STEP 1. compute the set $\mathcal{I}_k^+$ and, consequently, the diagonal matrices E_k and F_k;
STEP 2. compute $v^{(k)}$ as solution of the system $U_k v = E_k \nabla\Phi(x^{(k)})$;
STEP 3. compute the search direction $d^{(k)} = E_k v^{(k)} + F_k \nabla\Phi(x^{(k)})$;
STEP 4. set $\alpha_k = 1$; $x^{(k)}(\alpha_k) = P_\Omega(x^{(k)} - \alpha_k d^{(k)})$;
STEP 5. backtracking loop:

IF $\Phi(x^{(k)}) - \Phi(x^{(k)}(\alpha_k)) \geqslant c\left(\alpha_k \sum_{i \notin \mathcal{I}_k^+} \frac{\partial\Phi(x^{(k)})}{\partial x_i} d_i^{(k)} + \sum_{i \in \mathcal{I}_k^+} \frac{\partial\Phi(x^{(k)})}{\partial x_i}(x_i^{(k)} - x_i^{(k)}(\alpha_k))\right)$

go to STEP 6;

ELSE

set $\alpha_k = \rho\alpha_k$; $x^{(k)}(\alpha_k) = P_\Omega(x^{(k)} - \alpha_k d^{(k)})$ and go to STEP 5;

END

STEP 6. set $x^{(k+1)} = x^{(k)}(\alpha_k)$;
STEP 7. if $x^{(k+1)} = x^{(k)}$, then stop, declaring $x^{(k+1)}$ is a stationary point;

In [8], Bardsley and Vogel propose a variant of iteration (6.24), where D_k^{-1} is diagonal with respect to the set of the current active constraints, i.e. $\mathcal{A}_k = \{i\colon x_i = 0\}$; indeed, D_k is defined by

$$(D_k)_{i,\,j} = \begin{cases} \delta_{i,\,j} & \text{if either } i \in \mathcal{A}_k \text{ or } j \in \mathcal{A}_k \\ \dfrac{\partial^2\Phi(x^{(k)})}{\partial x_i \partial x_j} & \text{otherwise;} \end{cases}$$

the vector $d^{(k)}$ is obtained by applying CG method (with a suitable pre-conditioner) to the system $D_k d = E_k \nabla\Phi(x^{(k)})$ and the condition (6.22) is replaced by a less stringent criterion which consists in setting $\alpha_k := \rho^{m_k}$, with $\rho \in (0, 1)$ and m_k the first non-negative integer m such that

$$\Phi(x^{(k)}) > \Phi(x^{(k)}(\rho^m)).$$

The method looks like the algorithm of Moré and Toraldo [94] for bound-constrained quadratic minimization, since each step can be viewed as the minimization of a quadratic approximation of Φ at $x^{(k)}$ restricted to the set $\{x \in \mathbf{R}^N: x_i = 0, i \in \mathcal{A}_k\}$.

6.4 FB methods for non-smooth problems of image reconstruction

In this section, we consider the general problem (6.1) with the additional assumptions that $\Phi(x)$ is convex and the set X^* of the solutions is not empty; the methods discussed in this section apply to the **non-smooth problems** described at the beginning of section 6.1. We can add that, for $b > 0$, $\Phi(x) = f_0(x; y) = \mathrm{KL}(y; Hx + b)$ may have a Lipschitz-continuous gradient on its effective domain and a frequent example of non-smooth regularization $f_1(x)$ is TV regularization. The methods described in this section can be applied also to the smooth problems treated in the previous section, by setting $\Phi(x) = f_0(x; y) + \beta f_1(x)$, where $f_1(x)$ is differentiable and $\Psi(x) = i_\Omega(x)$.

As already mentioned, in the context of signal and image analysis, standard FB methods are widely used to address the problem (6.1). Convergence of the sequence $\{x^{(k)}\}$ to a solution is obtained under the assumption that the gradient of Φ is Lipschitz-continuous with parameter M_Φ, $\alpha_k \in [\epsilon, \frac{2}{M_\Phi} - \epsilon]$ and $\lambda_k \in [\epsilon, 1]$, for a given $\epsilon \in \left(0, \min(1, \frac{1}{M_\Phi})\right)$ [45, 48]. For large M_Φ, we have small step-lengths and slow convergence. In [47], the variant (6.12) with a variable metric is introduced; assuming similar bounds on α_k and λ_k and choosing the s.p.d. matrices D_k in $\mathcal{D}_L$, $L \geqslant 1$ so that $\{D_k\}$ asymptotically approaches a constant matrix (see condition (6.16)), the same convergence result of $\{x^{(k)}\}$ to a solution is obtained. Both versions of the FB scheme strictly depend on the parameter M_Φ, which is essential in establishing the bounds on the step-lengths in a practical implementation. In order to improve the slow convergence of the method and reduce the dependence on M_Φ, some variants of equation (6.12) have been recently proposed.

Without claiming to be exhaustive, in the following we describe the main approaches in the context of Poisson data, with special attention to those related to available software.

6.4.1 Specialized versions of FB method: EM–TV and SPIRAL

Starting from the features of the EM–RL method, in [35] the authors propose an iterative method, named EM–TV, to address the TV regularization of Poisson data, using a model which, in our notation, can be formulated as the following constrained minimization problem

$$\min_{x \geqslant 0} f_0(x; y) + \beta \sum_{i=1}^{N} \|\nabla_i x\|, \tag{6.32}$$

where $\nabla_i x$, $i = 1, \ldots, N$ is the discrete gradient at the ith pixel which can be viewed as the product of the matrix ∇_i and the variable x.

The basic iteration of EM–TV can be interpreted as a special version of the variable metric FB algorithm (6.12), where the scaling matrix D_k is defined as $\mathrm{diag}\left(\frac{H^T 1}{x^{(k)}}\right)$ and $\alpha_k = 1$ for all $k \geqslant 0$:

$$g^{(k)} = \mathrm{diag}\left(\frac{x^{(k)}}{H^T 1}\right) H^T \frac{y}{Hx^{(k)} + b} \qquad \text{EM step} \tag{6.33}$$

$$x^{(k+1)} = \arg\min_{x \in \mathbf{R}^N} \beta \sum_{i=1}^{N} \|\nabla_i x\| + \frac{1}{2}\|x - g^{(k)}\|_{D_k}^2 \quad \text{TV step.} \tag{6.34}$$

Here the quotient of vectors is intended component-wise as in the previous chapters. According to equations (6.33)–(6.34), an approximate solution is obtained alternating EM reconstruction steps and weighted TV denoising steps.

The authors derive the method from the optimality conditions for problem (6.32). An optimal solution-Lagrange multiplier pair (x, λ) of problem (6.32) satisfies the following conditions:

$$0 = H^T 1 - H^T \frac{y}{Hx + b} + \beta p - \lambda \tag{6.35}$$

$$0 = \lambda_i x_i \quad i = 1, \ldots, N, \tag{6.36}$$

with $p \in \partial f_1(x)$. By multiplying equation (6.35) by $\frac{x}{H^T 1}$ and using equation (6.36), we can eliminate the Lagrange multiplier λ and obtain a semi-implicit iterative scheme:

$$x^{(k+1)} = \underbrace{\mathrm{diag}\left(\frac{x^{(k)}}{H^T 1}\right) H^T \frac{y}{Hx^{(k)} + b}}_{g^{(k)}} - \beta\, \mathrm{diag}\left(\frac{x^{(k)}}{H^T 1}\right) p^{(k+1)},$$

with $p^{(k+1)} \in \partial f_1(x^{(k+1)})$; we can immediately verify that $x^{(k+1)}$ is the solution of the weighted TV denoising problem (6.34) and that $x^{(k+1)} = \mathrm{prox}_{\beta f_1, D_k}(g^{(k)})$.

Under standard assumptions on the matrix H (see equation (3.31)), given a strictly positive starting vector $x^{(0)} > 0$, the iteration scheme preserves positivity ([109], lemma 4.11). The convergence of a damped EM–TV method is proved in [109], under suitable assumptions on the damping parameter, hardly controllable in the numerical implementation. Furthermore, we observe that the weighted TV step has to be computed by an iterative method (as for example [11, 38]), so that any iteration of EM–TV becomes expensive. Applications in optical nanoscopy are described in [35], while applications to PET are discussed in [109, 110].

Another special version of FB schemes for the problem (6.32) is proposed in [77]. In this case the metric of the backward step is fixed and the authors attempt to improve the performance of the method by introducing a suitable updating rule for the step-length and an acceptance criterion assuring convergence. The method is named sparse Poisson intensity reconstruction algorithm (SPIRAL) and it is a tailored version for Poisson data of the SPARSA code approach [118].

The basic iteration of SPIRAL is given by a forward step for $\Phi(x) = f_0(x; y)$, followed by a backward step for $\Psi(x) = \beta f_1(x) + i_C(x)$, with $C = \{x \geqslant 0\}$:

$$g^{(k)} = x^{(k)} - \alpha_k \nabla\Phi(x^{(k)}) \tag{6.37}$$

$$x^{(k+1)} = \mathrm{prox}_{\alpha_k\Psi}(g^{(k)}) = \arg\min_{x\in C} \beta f_1(x) + \frac{1}{2\alpha_k}\|x - g^{(k)}\|^2. \tag{6.38}$$

The step-length α_k is updated via a sequence of two steps:

- the initial value of α_k is chosen by a modified Barzilai–Borwein rule [9] and it is intended to capture the curvature of Φ along the most recent direction at the current iterate $x^{(k)}$:

  $$\alpha_k = \min\left(\alpha_{\max}, \max\left(\alpha_{\min}, \frac{\|x^{(k)} - x^{(k-1)}\|^2}{(x^{(k)} - x^{(k-1)})^T \nabla^2\Phi(x^{(k)})(x^{(k)} - x^{(k-1)})}\right)\right),$$

  where $0 < \alpha_{\min} \leqslant \alpha_{\max}$. In particular, for $\Phi(x) = f_0(x; y)$, from the equation (4.12) of the Hessian, we have

  $$\alpha_k = \min\left(\alpha_{\max}, \max\left(\alpha_{\min}, \frac{\|x^{(k)} - x^{(k-1)}\|^2}{\left\|\mathrm{diag}\left(\frac{\sqrt{y}}{Hx^{(k)} + b}\right)H(x^{(k)} - x^{(k-1)})\right\|^2}\right)\right) \tag{6.39}$$

  where again square root and quotient of vectors are intended component-wise; consequently, the evaluation of α_k is not expensive, since $Hx^{(k)}$ and $Hx^{(k-1)}$ are already available from the gradient computation;
- the objective function $F = \Phi + \Psi$ has to satisfy the following acceptance criterion at the new iterate:

  $$F(x^{(k+1)}) \leqslant \max_{i = \max(k+1-M,0),\ldots,k} \left(F(x^{(i)}) - \frac{c}{2\alpha_k}\|x^{(k+1)} - x^{(k)}\|^2\right) \tag{6.40}$$

  where $c \in (0, 1)$ is a constant, usually chosen to be close to 0, and M is a positive integer; if α_k does not satisfy the criterion, it is repeatedly decreased by a factor $\rho \in (0, 1)$ until the solution $x^{(k+1)}$ satisfies condition (6.40).

As already mentioned, with respect to the general form of the classic FB method, SPIRAL method tries to obtain a satisfactory efficiency by a suitable choice of the parameter α_k, adjusting its value in such a way that the condition (6.40), that ensures a sufficient non-monotone decrease of the objective function, is satisfied. The details of SPIRAL are described in Algorithm 6.4. A Matlab® version of this method is downloadable from http://drz.ac/code/spiraltap/.

The main computational burden of an iteration of this algorithm is:

- the evaluations of $\nabla^2\Phi(x^{(k)})$ and $\nabla\Phi(x^{(k)})$ (for $\Phi(x) = f_0(x; y)$ they require only matrix–vector products involving H and H^T);
- the solution of the inner sub-problem (step 3), that must be repeated if the condition (6.40) is not immediately satisfied (step 4).

Convergence results for the method are obtained under the assumptions that the gradient of $\Phi(x) = f_0(x; y)$ is Lipschitz-continuous (this means $b > 0$), the regularization term is a convex function, the whole objective function is coercive and the constraint region is the non-negative orthant [77]. Although the convergence of $\{x^{(k)}\}$ is guaranteed for any initial point $x^{(0)} > 0$, the choice of the starting iterate is an important practical aspect. A recommended choice for an effective initialization is an appropriately scaled vector $H^T y$, or, in emission tomography, a filtered back-projection estimate.

We remark that the convergence results for the scheme (6.37)–(6.38) require that the inner sub-problem involved in the backward step is exactly solved. Nevertheless, also in view of the imposed constraints, this arises only for special penalty terms. Consequently, the complexity of the algorithm 6.4 is strongly dependent on the computation of the proximal operator in equation (6.38), since it can be repeatedly required at any iteration.

Algorithm 6.4. Sparse Poisson intensity reconstruction algorithm (SPIRAL).

In [77] computational techniques to obtain this evaluation are described for some penalty terms. For example, when $f_1(x)$ is the discrete version of TV, an

---

Choose the starting point $x^{(0)}$, set the parameters $\rho \in (0, 1)$, $c \in (0, 1)$, fix a positive integer M and $\alpha_{\max} \geqslant \alpha_{\min} > 0$.

FOR $k = 0, 1, 2, \ldots$ DO THE FOLLOWING STEPS:

STEP 1. compute $\alpha_k = \min\left(\alpha_{\max}, \max\left(\alpha_{\min}, \dfrac{\| x^{(k)} - x^{(k-1)} \|^2}{\| \operatorname{diag}\left(\frac{\sqrt{y}}{Hx^{(k)} + b}\right) H(x^{(k)} - x^{(k-1)}) \|^2}\right)\right)$;

STEP 2. compute $g^{(k)} = x^{(k)} - \alpha_k \nabla\Phi(x^{(k)})$;

STEP 3. compute $x^{(k+1)} = \arg\min_{x\in C} \left(\beta f_1(x) + \frac{1}{2\alpha_k}\|x - g^{(k)}\|^2\right)$;

STEP 4. backtracking loop:

IF $F(x^{(k+1)}) \leqslant \max_{i=\max(k+1-M,\, 0), \ldots, k}\left(F(x^{(i)}) - \frac{c}{2\alpha_k}\|x^{(k+1)} - x^{(k)}\|^2\right)$

go to STEP 5;

ELSE

set $\alpha_k = \alpha_k \rho$;

compute $g^{(k)} = x^{(k)} - \alpha_k \nabla\Phi(x^{(k)})$;

compute $x^{(k+1)} = \arg\min_{x\in C} \left(\beta f_1(x) + \frac{1}{2\alpha_k}\|x - g^{(k)}\|^2\right)$;

set $\lambda_k = \rho\lambda_k$ and go to STEP 4;

END

STEP 5. if a stopping criterion is satisfied then stop, declaring $x^{(k+1)}$ is a stationary point;

---

approximate solution of the inner sub-problem (6.38) can be obtained by an iterative solver (for example [11]), since it becomes a problem of non-negative denoising of $g^{(k)}$ with TV regularization.

6.4.2 Variable metric FB methods with inexact line-search algorithms

In [24], Bonettini *et al* propose a general framework for a variable metric FB scheme (6.12) equipped with a set of tools which assure both the theoretical convergence and the numerical effectiveness of the method. In particular, convergence results are obtained by adopting the generalized line-search strategy proposed in [116]; this rule, combined with the properties of the descent direction, enables one to determine the step-length λ_k; consequently, the assumption on the Lipschitz-continuity of the gradient of Φ is not required and the selection of the other parameters of the algorithm, α_k and D_k, is more flexible. Indeed, it can be based only on considerations intended to improve the practical performance of the method, as, for example, those described for smooth optimization in section 6.3. Furthermore, in the backward step the method allows an inexact solution of the inner sub-problem, preserving the theoretical convergence. Implementable conditions for the computation of the approximate proximal point are provided (see also [20, 117]).

In order to explain these techniques, we recall that, for a generic convex function F and a point $x \in \mathrm{dom}\, F$, the directional derivative of F at x with respect to a direction d, i.e. $F'(x; d) = \lim_{\lambda \downarrow 0} \frac{F(x+\lambda d) - F(x)}{\lambda}$, is well-defined ([107]; see definition A.7 and proposition A.7 in appendix A); furthermore, we can write

$$F'(x; d) \leqslant F(x + d) - F(x).$$

As a consequence, a point x^* is *stationary* for the function F if $x^* \in \mathrm{dom}\, F$ and $F'(x^*; d) \geqslant 0$ for any feasible direction d. When $F'(x; d) < 0$ at a point $x \in \mathrm{dom}\, F$, the vector d is a *descent direction* of F at x.

In order to have a tool for detecting a descent direction for the problem (6.1), the following functions are introduced.

Definition 6.4. *Given* $\alpha \in [\alpha_{\min}, \alpha_{\max}]$, *with* $\alpha_{\min} > 0$, *and a s.p.d. matrix* $D \in \mathcal{D}_L$, $L \geqslant 1$, *the function* $h_{\alpha, D}$: $\mathbf{R}^N \times \mathbf{R}^N \to \bar{\mathbf{R}}$ *is defined by*

$$h_{\alpha, D}(z, x) = \nabla\Phi(x)^T(z - x) + \frac{1}{2\alpha}\|z - x\|_D^2 + \Psi(z) - \Psi(x); \tag{6.41}$$

moreover, the function $\bar{h}_{\alpha, D, \gamma}$: $\mathbf{R}^N \times \mathbf{R}^N \to \bar{\mathbf{R}}$ *with* $\gamma \in [0, 1]$ *is defined by*

$$\bar{h}_{\alpha, D, \gamma}(z, x) = \nabla\Phi(x)^T(z - x) + \frac{\gamma}{2\alpha}\|z - x\|_D^2 + \Psi(z) - \Psi(x). \tag{6.42}$$

For all $x, z \in \mathbf{R}^N$, we have

$$\bar{h}_{\alpha,\,D,\,\gamma}(z,\,x) \leqslant h_{\alpha,\,D}(z,\,x), \tag{6.43}$$

and the equality holds when $\gamma = 1$.

For the assumption on Ψ and the strong convexity of the quadratic term $\frac{1}{\alpha}\|\cdot - x\|_D^2$, both functions $\bar{h}_{\alpha,\,D,\,\gamma}(z,\,x)$ and $h_{\alpha,\,D}(z,\,x)$ are proper, l.s.c. and convex with respect to z, for any $x \in \mathrm{dom}\,\Psi$; furthermore, for a given x, $h_{\alpha,\,D}(z,\,x)$ is a strongly convex function with modulus $\frac{1}{2\alpha_{\max}L}$ with respect to z, i.e. we have for all z_1, z_2:

$$h_{\alpha,\,D}(z_1,\,x) \geqslant h_{\alpha,\,D}(z_2,\,x) + q^T(z_1 - z_2) + \frac{1}{2\alpha_{\max}L}\|z_1 - z_2\|^2, \tag{6.44}$$

where $q \in \partial h_{\alpha,\,D}(z_2,\,x)$. Consequently, for a given $x \in \mathrm{dom}\,\Psi$, $h_{\alpha,\,D}(z,\,x)$ has a unique minimum point, denoted in the following by $p(x;\,h_{\alpha,\,D})$. We observe that $p(x;\,h_{\alpha,\,D})$ is the proximity operator of $\alpha\Psi$ computed at the result of a forward step on Φ:

$$p(x;\,h_{\alpha,\,D}) = \mathrm{prox}_{\alpha\Psi,\,D}(x - \alpha D^{-1}\nabla\Phi(x)). \tag{6.45}$$

The functions $h_{\alpha,\,D}$ and $\bar{h}_{\alpha,\,D}$ have nice features since their behavior is strictly related to the directional derivative of the objective function. These relationships are summarized in the following proposition, proved in a more general framework in [24, 116]. Since for $\gamma = 1$, $h_{\alpha,\,D}(z,\,x) = \bar{h}_{\alpha,\,D,\,\gamma}(z,\,x)$, we only consider $\bar{h}_{\alpha,\,D,\,\gamma}$.

Proposition 6.2. *Let* $\alpha \in [\alpha_{\min},\,\alpha_{\max}]$, *with* $\alpha_{\min} > 0$, $D \in \mathcal{D}_L$, $L \geqslant 1$ *a s.p.d. matrix,* $\gamma \in [0,\,1]$ *and* $\bar{h}_{\alpha,\,D,\,\gamma}$ *the function defined in equation* (6.42). *For a given* $x \in \mathrm{dom}\,\Psi$, *the following statements hold*:

(a) $\bar{h}'_{\alpha,\,D,\,\gamma}(x,\,x;\,d) = F'(x;\,d)$, *for any* $d \in \mathbf{R}^N$;

(b) $\bar{h}_{\alpha,\,D,\,\gamma}(x,\,x) = 0$;

(c) *if* $z \in \mathbf{R}^N$ *and* $\bar{h}_{\alpha,\,D,\,\gamma}(z,\,x) < 0$, *then* $F(x;\,z - x) < 0$;

(d) $\bar{h}_{\alpha,\,D,\,\gamma}(p(x;\,h_{\alpha,\,D}),\,x) \leqslant 0$ *and the equality holds if and only if* $p(x;\,h_{\alpha,\,D}) = x$;

(e) $F'(x,\,p(x;\,h_{\alpha,\,D}) - x) \leqslant 0$ *and the equality holds if and only if* $\bar{h}_{\alpha,\,D,\,\gamma}(p(x;\,h_{\alpha,\,D}),\,x) = 0$.

As a consequence of this proposition, it follows that x^* is a stationary point for the problem (6.1) if and only if it is a fixed point for the operator $p(\cdot;h_{\alpha,\,D})$, i.e. x^* is a root of the equation $\bar{h}_{\alpha,\,D,\,\gamma}(p(\cdot;h_{\alpha,\,D}),\,\cdot) = 0$. It follows that, given a point $x \in \mathrm{dom}\,\Psi$, the vector $d = p(x;\,h_{\alpha,\,D}) - x$ is a descent direction for F at x. Indeed, when $\bar{h}_{\alpha,\,D,\,\gamma}(p(x;\,h_{\alpha,\,D}),\,x) < 0$, then $F'(x;\,d) < 0$, unless $p(x;\,h_{\alpha,\,D}) = x$; in this case x is a solution of the minimization problem.

The second crucial point for a descent method is to ensure a sufficient decrease of the objective function. In order to deal with non-smooth convex objective functions, in [102, 116] a generalization of the monotone Armijo line-search procedure is proposed.

Suppose that a descent direction $d^{(k)} = \bar{w}^{(k)} - x^{(k)}$ is computed for the objective function at the iterate $x^{(k)}$ of an iterative method, with $x^{(k)}, \bar{w}^{(k)} \in \text{dom}\,\Psi$. Given $c, \rho \in (0, 1)$ and $\gamma \in [0, 1]$, by a backtracking strategy one can compute a steplength $\lambda_k := \rho^m$, where m is the first non-negative integer such that:

$$F(x^{(k)} + \lambda_k d^{(k)}) \leqslant F(x^{(k)}) + c\lambda_k \bar{h}_{\alpha_k, D_k, \gamma}(\bar{w}^{(k)}, x^{(k)}). \tag{6.46}$$

This backtracking strategy is well-defined, i.e. it terminates in a finite number of steps, when the two sequences $\{x^{(k)}\}$ and $\{\bar{w}^{(k)}\}$ of points of dom Ψ are such that

$$\bar{h}_{\alpha_k, D_k, \gamma}(\bar{w}^{(k)}, x^{(k)}) < 0 \quad \forall k \geqslant 0. \tag{6.47}$$

Furthermore, when the new iterate $x^{(k+1)}$ is defined so that $F(x^{(k+1)}) \leqslant F(x^{(k)} + \lambda_k d^{(k)})$ and the two sequences $\{x^{(k)}\}$ and $\{\bar{w}^{(k)}\}$ are bounded, the sequence $\{\bar{h}_{\alpha_k, D_k, \gamma}(\bar{w}^{(k)}, x^{(k)})\}$ is bounded and, under the assumption that F is bounded from below, we can prove that $\bar{h}_{\alpha_k, D_k, \gamma}(\bar{w}^{(k)}, x^{(k)}) \to 0$ as $k \to \infty$ [24, 116].

We remark that the modified line-search procedure (6.46) does not need the exact proximal point $w^{(k)} = p(x^{(k)}; h_{\alpha_k, D_k})$: the unique condition required for the descent direction is inequality (6.47), i.e. the strict negativity of $\bar{h}_{\alpha_k, D_k, \gamma}(\bar{w}^{(k)}, x^{(k)})$. This remark allows one to devise a class of variable metric FB methods equipped with an *inexact* computation of the proximal vector $w^{(k)} = p(x^{(k)}; h_{\alpha_k, D_k})$. Nevertheless, to ensure the stationarity of the limit points of $\{x^{(k)}\}$, the following additional condition has to be guaranteed:

$$\lim_{k \in K, k \to \infty} h_{\alpha_k, D_k}(\bar{w}^{(k)}, x^{(k)}) - h_{\alpha_k, D_k}(w^{(k)}, x^{(k)}) = 0, \tag{6.48}$$

where K is a suitable subset of indices. In practice, $\bar{w}^{(k)}$ has to provide better and better approximations of $w^{(k)}$ as $k \to \infty$ ([24], theorem 4.1).

The assumption (6.48) is crucial. If, for example, $N = 1$, $\Phi(x) = \frac{x^2}{2}$, $\Psi(x) = 0$, by setting $c = \rho = \frac{1}{2}$ and $x^{(0)} = 2$, the sequence $x^{(k+1)} = x^{(k)} + \lambda_k(\bar{w}^{(k)} - x^{(k)})$ with $\lambda_k = 1$, $\alpha_k = 1$, $D_k = 1$ and $\bar{w}^{(k)} = x^{(k)} - \frac{1}{2}^{k+1}$ satisfies $h_{\alpha_k, D_k}(\bar{w}^{(k)}, x^{(k)}) = \frac{1}{2}^{k+1}\left(\frac{1}{2}^{k+2} - x^{(k)}\right) < 0$ for all $k \geqslant 0$, but $x^{(k)} = 1 + \frac{1}{2}^{k} \to 1$ as $k \to \infty$, while the unique stationary point is 0. Indeed the condition (6.48) is not satisfied.

In algorithm 6.5, the main steps of an inexact variable metric FB method are detailed. The scheme is named VMILA in [24].

Algorithm 6.5. Inexact variable metric FB algorithm (VMILA).

Choose the starting point $x^{(0)} \in \text{dom}\, \Psi$, set the parameters $c, \rho \in (0, 1)$, $0 < \alpha_{\min} < \alpha_{\max}$, $L \geqslant 1$ and $\gamma \in [0, 1]$.

FOR $k = 0, 1, 2, \ldots$ DO THE FOLLOWING STEPS:

STEP 1. choose the parameter $\alpha_k \in [\alpha_{\min}, \alpha_{\max}]$ and the scaling matrix $D_k \in \mathcal{D}_L$;

STEP 2. select $\bar{w}^{(k)}$ as an approximation of $p(x^{(k)}; h_{\alpha_k, D_k})$ such that $\bar{h}_{\alpha_k, D_k, \gamma}(\bar{w}^{(k)}, x^{(k)}) \leqslant 0$;

if $\bar{h}_{\alpha_k, D_k, \gamma}(\bar{w}^{(k)}, x^{(k)}) = 0$ (equivalently $\bar{w}^{(k)} = x^{(k)}$), then stop, declaring $x^{(k)}$ is a stationary point;

STEP 3. set $d^{(k)} = \bar{w}^{(k)} - x^{(k)}$;

STEP 4. set $\lambda_k = 1$ and $\Delta_k = \bar{h}_{\alpha_k, D_k, \gamma}(\bar{w}^{(k)}, x^{(k)})$;

STEP 5. backtracking loop:

IF $F(x^{(k)} + \lambda_k d^{(k)}) \leqslant F(x^{(k)}) + c\lambda_k \Delta_k$

go to STEP 6;

ELSE

set $\lambda_k = \rho \lambda_k$ and go to STEP 5;

END

STEP 6. set $x^{(k+1)} = x^{(k)} + \lambda_k d^{(k)}$;

In order to complete the description of the approach, we must describe the techniques which can be used for performing step 2, so that condition (6.48) is satisfied. Indeed, the condition (6.48) cannot be checked directly. Two different ways are available for obtaining suitable approximations of the sequence $\{w^{(k)}\}$.

- ϵ_k**-approximation**. When $w^{(k)}$ is the exact minimum point of $h_{\alpha_k, D_k}(w, x^{(k)})$, i.e. $w^{(k)} = \text{prox}_{\alpha_k \Psi, D_k}(x^{(k)} - \alpha_k D_k^{-1} \nabla \Phi(x^{(k)}))$, in view of equation (6.5), the following inclusion follows:

$$\frac{1}{\alpha_k} D_k(g^{(k)} - w^{(k)}) \in \partial \Psi(w^{(k)}) \tag{6.49}$$

where $g^{(k)} = x^{(k)} - \alpha_k D_k^{-1} \nabla \Phi(x^{(k)})$. Following [108, 117], we can relax the condition (6.49) by exploiting the features of the ϵ-subdifferential notion (see definition A.16 and examples in appendix A). In particular, we can require that $\bar{w}^{(k)}$ satisfies the following inclusion

$$\frac{1}{\alpha_k} D_k(g^{(k)} - \bar{w}^{(k)}) \in \partial_{\epsilon_k} \Psi(\bar{w}^{(k)}), \tag{6.50}$$

where ϵ_k is a suitable non-negative value. As a consequence of inclusion (6.50), we have that

$$0 \in \frac{1}{\alpha_k} D_k(\bar{w}^{(k)} - g^{(k)}) + \partial_{\epsilon_k} \Psi(\bar{w}^{(k)}) \subseteq \partial_{\epsilon_k} h_{\alpha_k, D_k}(\bar{w}^{(k)}, x^{(k)}). \tag{6.51}$$

By definition of ϵ-subdifferential, this means

$$h_{\alpha_k, D_k}(w, x^{(k)}) \geqslant h_{\alpha_k, D_k}(\bar{w}^{(k)}, x^{(k)}) - \epsilon_k \quad \forall w \in \mathbf{R}^N.$$

In particular, for $w = w^{(k)}$, which is the exact minimum point of $h_{\alpha_k, D_k}(w, x^{(k)})$, we have

$$h_{\alpha_k, D_k}(\bar{w}^{(k)}, x^{(k)}) - h_{\alpha_k, D_k}(w^{(k)}, x^{(k)}) \leqslant \epsilon_k. \tag{6.52}$$

Now, using equation (6.44), due to the strong convexity of h_{α_k, D_k}, we have

$$\frac{1}{2\alpha_{\max}L}\|w^{(k)} - \bar{w}^{(k)}\|^2 \leqslant h_{\alpha_k, D_k}(\bar{w}^{(k)}, x^{(k)}) - h_{\alpha_k, D_k}(w^{(k)}, x^{(k)}) \leqslant \epsilon_k.$$

As a consequence, by choosing $\{\epsilon_k\}$ such that $\lim_{k\to 0} \epsilon_k = 0$, one assures that condition (6.48) is satisfied.

- **η-approximation**. Following [20], we introduce the set

$$\Pi_\eta(x, h_{\alpha, D}) = \{\bar{w} \in \mathrm{dom}\, \Psi: h_{\alpha, D}(\bar{w}, x) \leqslant \eta h_{\alpha, D}(p(x; h_{\alpha, D}), x)\}, \tag{6.53}$$

where $\eta \in (0, 1]$. If we select $\bar{w}^{(k)} \in \Pi_\eta(x^{(k)}, h_{\alpha_k, D_k})$, then

$$h_{\alpha_k, D_k}(\bar{w}^{(k)}, x^{(k)}) \leqslant \eta h_{\alpha, D}(w^{(k)}, x^{(k)}) \leqslant 0,$$

and the equality $h_{\alpha_k, D_k}(\bar{w}^{(k)}, x^{(k)}) = 0$ holds if and only if $h_{\alpha_k, D_k}(w^{(k)}, x^{(k)}) = 0$, i.e. $w^{(k)} = x^{(k)} = \bar{w}^{(k)}$ and $x^{(k)}$ is a stationary point for the original problem. Then, when an element of $\Pi_\eta(x^{(k)}, h_{\alpha_k, D_k})$ is selected as approximation of $w^{(k)}$, one can prove that any limit point of $\{x^{(k)}\}$ is a stationary point of the problem (6.1) ([24], theorem 4.2). Furthermore, we observe that, when $\bar{w}^{(k)} \in \Pi_\eta(x^{(k)}, h_{\alpha_k, D_k})$, we can write

$$h_{\alpha_k, D_k}(\bar{w}^{(k)}, x^{(k)}) - h_{\alpha_k, D_k}(w^{(k)}, x^{(k)}) \leqslant (\eta - 1)h_{\alpha_k, D_k}(w^{(k)}, x^{(k)}); \tag{6.54}$$

by setting $\epsilon_k = (\eta - 1)h_{\alpha_k, D_k}(w^{(k)}, x^{(k)})$, the inequality (6.54) is equivalent to stating that $0 \in \partial_{\epsilon_k} h_{\alpha_k, D_k}(\bar{w}^{(k)}, x^{(k)})$, with $\epsilon_k = (\eta - 1)h_{\alpha_k, D_k}(w^{(k)}, x^{(k)})$ (see equation (6.52)). Hence an η-approximation $\bar{w}^{(k)}$ of $w^{(k)}$ is also an ϵ_k-approximation.

In summary, for a general variable metric FB method, as algorithm 6.5, when the *inexact* sequence $\{\bar{w}^{(k)}\}$ is computed in such a way that condition (6.48) holds true and $\lim_{k\to\infty} \epsilon_k = 0$ or $\bar{w}^{(k)} \in \Pi_\eta(x, h_{\alpha, D})$, with $\eta \in (0, 1]$, a limit point of $\{x^{(k)}\}$ is a stationary point for the problem (6.1).

We remark that, until now, the assumption of convexity of the function Φ is not used. However, in order to obtain the convergence of the whole sequence $\{x^{(k)}\}$ to a solution of the problem (6.1), the following additional assumptions are required:

- Φ is convex and the solution set X^* of equation (6.1) is not empty;
- $\{D_k\}$ is a sequence of s.p.d. scaling matrices $\{D_k\}$, $D_k \in \mathcal{D}_L$, $L \geqslant 1$, such that the following condition holds

$$D_{k+1} \preccurlyeq (1 + \xi_k)D_k, \qquad \xi_k \geqslant 0, \quad \sum_{k=0}^{\infty} \xi_k < \infty;$$

- the sequence $\{\bar{w}^{(k)}\}$ is chosen in such a way that condition (6.50) holds and one of the following conditions is satisfied:

(a1) the sequence $\{\epsilon_k\}$ is summable;
(a2) $\epsilon_k \leqslant -\tau \bar{h}_{\alpha_k, D_k, \gamma}(\bar{w}^{(k)}, x^{(k)})$, with $\tau > 0$.

This convergence result is obtained in [24], theorems 4.3 and 4.4. The assumptions (a1)–(a2) indicate that the tolerance sequence $\{\epsilon_k\}$ for the approximations of the inner sub-problems can be fixed as an *a priori* summable sequence (case (a1)) or in an adaptive way (case (a2)). Furthermore, when $\nabla\Phi$ is Lipschitz-continuous and condition (a2) holds, it is possible to prove that the rate of convergence of the values of the objective function is asymptotically $\mathcal{O}\left(\frac{1}{k}\right)$ ([24], theorem 4.5).

We observe that algorithm 6.5 is a generalization of SGP, with the additional feature of an inexact proximal step. Consequently, for the choice of the step-length α_k, we can use the same techniques described for SGP, aimed at improving the practical convergence speed of the method without the introduction of significant computational costs. For this reason, diagonal scaling matrices able to capture problem features are to be preferred. As already mentioned, in the context of Poisson data the scaling selection rules suggested by [86] and discussed in section 6.3.1 enable to obtain very efficient numerical results.

6.4.2.1 Techniques for computing an inexact proximal point

The procedure for computing an inexact solution of the inner sub-problem at step 2 of the algorithm 6.5 is plain, but it requires some technicalities. The approximation $\bar{w}^{(k)}$ of the exact solution $w^{(k)}$ can be obtained by an iterative solver equipped with a suitable stopping criterion, ensuring one of the two conditions (a1) or (a2) is satisfied. In both cases, in order to compute an approximation of the minimizer of the inner sub-problem

$$\min_{w} h_{\alpha_k, D_k}(w, x^{(k)}), \tag{6.55}$$

it is convenient to consider its dual form.

By rewriting $\Psi(w)$ in terms of its conjugate function, i.e. $\Psi(w) = \sup_{q \in \mathbf{R}^N}(q^T w - \Psi^*(q))$, in the objective function of the primal form (6.55), the primal-dual form of the problem is obtained; then, since the primal form admits a finite solution, the dual form can be obtained by computing the minimum with respect to w and by substituting it in the primal-dual form.

We describe the procedure by means of a meaningful example. Consider $\Psi(w)$ as the sum of the indicator function of the non-negative orthant C and the discrete total variation of a 2D-image with N pixels, i.e. $\Psi(w) = i_C(w) + \beta \sum_{i=1}^{N} \|\nabla_i w\|$, where ∇_i is the $2 \times N$ matrix of the discrete gradient at the ith pixel and β is the regularization parameter. At each kth iteration, the following inner sub-problem has to be solved:

$$\min_{w\in C} h_{\alpha_k, D_k}(w, x^{(k)}) := \frac{1}{2\alpha_k}\|w - g^{(k)}\|^2_{D_k}$$
$$+\beta\sum_{i=1}^{N}\|\nabla_i w\| - \beta\sum_{i=1}^{N}\|\nabla_i x^{(k)}\| - \frac{\alpha_k}{2}\|D_k^{-1}\nabla\Phi(x^{(k)})\|^2_{D_k}, \tag{6.56}$$

with $g^{(k)} = x^{(k)} - \alpha_k D_k^{-1}\nabla\Phi(x^{(k)})$.

Since $\beta\|\nabla_i w\| = \sup_{q_i\in\mathbf{R}^2} q_i^T\nabla_i w - \iota_{B(0,\beta)}(q_i)$, where $B(0, \beta) = \{q_i \in \mathbf{R}^2: \|q_i\| \leqslant \beta\}$, if we denote by ∇ the $2N \times N$ matrix $(\nabla_1^T, \dots, \nabla_N^T)^T$, we can write

$$\beta\sum_{i=1}^{N}\|\nabla_i w\| = \sup_{q\in\mathbf{R}^{2N}} q^T\nabla w - \iota_{\mathcal{B}_{2N}(\beta)}(q) \tag{6.57}$$

with $\mathcal{B}_{2N}(\beta) = \{q \in \mathbf{R}^{2N}: q = \left(q_1^T, \dots, q_N^T\right)^T, q_i \in B(0, \beta) \subseteq \mathbf{R}^2, i = 1, \dots, N\}$. By substituting equation (6.57) in the objective function of problem (6.56), we obtain its primal-dual form:

$$\min_{w\in C}\max_{q\in\mathbf{R}^{2N}} \mathcal{L}_{\alpha_k, D_k}(w, q, x^{(k)}) := \frac{1}{2\alpha_k}\|w - g^{(k)}\|^2_{D_k}$$
$$+q^T\nabla w - \iota_{\mathcal{B}_{2N}(\beta)}(q) - \beta\sum_{i=1}^{N}\|\nabla_i x^{(k)}\| - \frac{\alpha_k}{2}\|D_k^{-1}\nabla\Phi(x^{(k)})\|^2_{D_k}. \tag{6.58}$$

Now, adding and subtracting $(g^{(k)})^T\nabla^T q + \frac{1}{2\alpha_k}\|\alpha_k D_k^{-1}\nabla^T q\|^2_{D_k} + \frac{1}{2\alpha_k}\|g^{(k)}\|^2_{D_k}$, equation (6.58) can be written as

$$\min_{w\in C}\max_{q\in\mathcal{B}_{2N}(\beta)} \frac{1}{2\alpha_k}\|w - (g^{(k)} - \alpha_k D_k^{-1}\nabla^T q)\|^2_{D_k} - \frac{1}{2\alpha_k}\|g^{(k)} - \alpha_k D_k^{-1}\nabla^T q\|^2_{D_k}$$
$$+\frac{1}{2\alpha_k}\|g^{(k)}\|^2_{D_k} - \beta\sum_{i=1}^{N}\|\nabla_i x^{(k)}\| - \frac{\alpha_k}{2}\|D_k^{-1}\nabla\Phi(x^{(k)})\|^2_{D_k}.$$

Since $\mathcal{L}_{\alpha_k, D_k}(w, q, x^{(k)})$ is a proper convex–concave and continuous function on the non-empty, closed and convex set $C \times \mathcal{B}_{2N}(\beta)$ and $\mathcal{B}_{2N}(\beta)$ is bounded, one can exchange the order of the minimum and maximum (see [107], corollary 37.3.2); by substituting in the primal–dual form the optimal solution $w(q) := P_{C, D_k}(g^{(k)} - \alpha_k D_k^{-1}\nabla^T q)$ of the minimum problem, we obtain the dual form

$$\max_{q\in\mathcal{B}_{2N}(\beta)} h^d_{\alpha_k, D_k}(q, x^{(k)})$$
$$:= \frac{1}{2\alpha_k}\|P_{C, D_k}(g^{(k)} - \alpha_k D_k^{-1}\nabla^T q) - (g^{(k)} - \alpha_k D_k^{-1}\nabla^T q)\|^2_{D_k}$$
$$-\frac{1}{2\alpha_k}\|g^{(k)} - \alpha_k D_k^{-1}\nabla^T q\|^2_{D_k} \tag{6.59}$$
$$+\frac{1}{2\alpha_k}\|g^{(k)}\|^2_{D_k} - \beta\sum_{i=1}^{N}\|\nabla_i x^{(k)}\| - \frac{\alpha_k}{2}\|D_k^{-1}\nabla\Phi(x^{(k)})\|^2_{D_k}.$$

The dual function is continuously differentiable with respect to q on $\mathcal{B}_{2N}(\beta)$ and its gradient at q is

$$\nabla h^d_{\alpha_k, D_k}(q, x^{(k)}) = -\nabla P_{C, D_k}(g^{(k)} - \alpha_k D_k^{-1}\nabla^T q). \tag{6.60}$$

The gradient is Lipschitz-continuous with parameter bounded by $\alpha_k\|D_k^{-1}\|\|\nabla\|^2$.

A similar procedure can be applied to any function Ψ which is the composition of a convex function with a linear operator.

Now, we recall that, by definition of the primal–dual and dual functions, the following inequality holds for all w, q:

$$h^d_{\alpha_k, D_k}(q, x^{(k)}) \leqslant \mathcal{L}_{\alpha_k, D_k}(w, q, x^{(k)}) \leqslant h_{\alpha_k, D_k}(w, x^{(k)}), \tag{6.61}$$

where the equality holds for $w = w^{(k)}$ and q equal to the solution $q^{(k)}$ of the dual problem.

By applying an iterative solver to the dual problem (6.59), we can generate a sequence $\{q^{(k,\ell)}\}$ such that $q^{(k,\ell)} \to q^{(k)}$ as $\ell \to \infty$; the corresponding sequence $\{\bar{w}^{(k,\ell)}\}$, where any vector is related to the dual iterate $q^{(k,\ell)}$ by means of the following equality

$$\bar{w}^{(k,\ell)} = \arg\min_w \mathcal{L}_{\alpha_k, D_k}(w, q^{(k,\ell)}, x^{(k)}),$$

is convergent to the solution $w^{(k)}$ of the primal problem. As a consequence, the sequence $\{h^d_{\alpha_k, D_k}(q^{(k,\ell)}, x^{(k)})\}$ converges, as $\ell \to \infty$, to the maximum $h^d_{\alpha_k, D_k}(q^{(k)}, x^{(k)})$ of the dual functions, which is equal to the required minimum $h_{\alpha_k, D_k}(w^{(k)}, x^{(k)})$ of the primal objective function.

Consider $\eta \in (0, 1]$ and the vector $\bar{w}^{(k,\ell)}$ corresponding to the dual iterate $q^{(k,\ell)}$; an η-approximation of $w^{(k)}$ can be found by stopping the dual iterations when

$$h_{\alpha_k, D_k}(\bar{w}^{(k,\ell)}, x^{(k)}) \leqslant \eta h^d_{\alpha_k, D_k}(q^{(k,\ell)}, x^{(k)}). \tag{6.62}$$

Indeed, for any iterate $q^{(k,\ell)}$ of the inner iteration, the following inequality holds:

$$h^d_{\alpha_k, D_k}(q^{(k,\ell)}, x^{(k)}) \leqslant h^d_{\alpha_k, D_k}(q^{(k)}, x^{(k)}) = h_{\alpha_k, D_k}(w^{(k)}, x^{(k)});$$

furthermore, since $h^d_{\alpha_k, D_k}(q^{(k,\ell)}, x^{(k)}) \to h_{\alpha_k, D_k}(w^{(k)}, x^{(k)})$ as $\ell \to \infty$, for all sufficiently large ℓ, we have that $h^d_{\alpha_k, D_k}(q^{(k,\ell)}, x^{(k)}) > h_{\alpha_k, D_k}(w^{(k)}, x^{(k)})/\eta$ and we can write

$$h_{\alpha_k, D_k}(w^{(k)}, x^{(k)}) < \eta h^d_{\alpha_k, D_k}(q^{(k,\ell)}, x^{(k)}) \leqslant \eta h_{\alpha_k, D_k}(w^{(k)}, x^{(k)}).$$

Since also $h_{\alpha_k, D_k}(\bar{w}^{(k,\ell)}, x^{(k)}) \to h_{\alpha_k, D_k}(w^{(k)}, x^{(k)})$ as $\ell \to \infty$, when condition (6.62) is satisfied for $\ell = \bar{\ell}$, we can set $\bar{w}^{(k)} = \bar{w}^{(k,\bar{\ell})}$, obtaining an η-approximation of $\bar{w}^{(k)}$.

In the case of the constrained discrete TV problem, several iterative solvers for the dual formulation (6.59) are proposed in [11].

Algorithm 6.6. Computation of inexact proximal point $\bar{w}^{(k)}$: η-approximation.

Choose an inner solver $\mathcal{A}$, a starting point $q^{(k,\,0)} \in \text{dom}\,\Psi^*$ and $\eta \in (0,\,1]$;
set $g^{(k)} = x^{(k)} - \alpha_k D_k^{-1} \nabla\Phi(x^{(k)})$.
FOR $\ell = 0,\,1,\,2,\ldots$ DO THE FOLLOWING STEPS:
STEP 1. set $\bar{w}^{(k,\,l)} = \arg\min_w \mathcal{L}_{\alpha_k,\,D_k}(w,\,q^{(k,\,\ell)},\,x^{(k)})$;
STEP 2. condition control:
    IF $h_{\alpha_k,\,D_k}(\bar{w}^{(k,\,l)},\,x^{(k)}) \leqslant \eta h^d_{\alpha_k,\,D_k}(q^{(k,\,l)},\,x^{(k)})$ and $\bar{h}_{\alpha_k,\,D_k,\,\gamma}(\bar{w}^{(k,\,l)},\,x^{(k)}) < 0$
        set $\bar{w}^{(k)} = \bar{w}^{(k,\,l)}$ and EXIT.
    ELSE
        compute the next iterate $q^{(k,\,\ell+1)} = \mathcal{A}(q^{(k,\,\ell)})$;
    END
END

Algorithm 6.6 shows the described procedure.

Now, following [117], we describe how we can compute an ϵ_k-approximation of $w^{(k)}$. This technique is based on the definition of primal–dual gap function, given by

$$\mathcal{G}_{\alpha_k,\,D_k}(w,\,q,\,x^{(k)}) = h_{\alpha_k,\,D_k}(w,\,x^{(k)}) - h^d_{\alpha_k,\,D_k}(q,\,x^{(k)}). \tag{6.63}$$

In view of inequality (6.61), this function is non-negative for all w, q and it is equal to zero for $w = w^{(k)}$ and $q = q^{(k)}$. Given $\epsilon_k \geqslant 0$, if there exists a vector q such that

$$\mathcal{G}_{\alpha_k,\,D_k}(\bar{w}^{(k,\,\ell)},\,q,\,x^{(k)}) = h_{\alpha_k,\,D_k}(\bar{w}^{(k,\,\ell)},\,x^{(k)}) - h^d_{\alpha_k,\,D_k}(q,\,x^{(k)}) \leqslant \epsilon_k, \tag{6.64}$$

with $\bar{w}^{(k,\,\ell)} = \arg\min_w \mathcal{L}_{\alpha_k,\,D_k}(w,\,q^{(k,\,\ell)},\,x^{(k)})$, then $\bar{w}^{(k,\,\ell)}$ is an ϵ_k-approximation of $w^{(k)}$, satisfying equations (6.50)–(6.52) [24, 117]. Thus, using an iterative solver for solving the dual inner sub-problem, we can generate a sequence of vectors $\{q^{(k,\,\ell)}\}$, stopping at the iterate $\bar{\ell}$ that satisfies condition (6.64).

In order to have an ϵ_k-approximation of $w^{(k)}$, in algorithm 6.6 the condition control at step 2 can be replaced by:

$$\mathcal{G}_{\alpha_k,\,D_k}(\bar{w}^{(k,\,\ell)},\,q^{(k,\,\ell)},\,x^{(k)}) \leqslant \epsilon_k \text{ and } \bar{h}_{\alpha_k,\,D_k,\,\gamma}(\bar{w}^{(k,\,l)},\,x^{(k)}) < 0.$$

A Matlab® version of the algorithm 6.5 combined with algorithm 6.6 is downloadable from http://www.oasis.unimore.it/site/home/software.html.

6.4.3 An inertial variable metric FB method

As highlighted in [12, 17, 95], the rate of convergence of the values of the objective function for standard and variable metric FB schemes is only linear, i.e. $\mathcal{O}\left(\frac{1}{k}\right)$. Nevertheless, in the previous sections, we observed that, in spite of the theoretical convergence rate, suitable selection of step-length parameter α_k and/or variable metrics, capturing the features of the considered problem, combined with conditions assuring the decrease of the objective function, can allow one to reach very good performance by improving the convergence behavior of the first iterations.

A well-known approach to overcome the theoretical bound on the rate of convergence of the FB scheme consists in adding an extrapolation step to the basic iteration, yielding a multi-step algorithm, also called heavy ball or inertial method [101]. Inertial methods become very popular in the last decade, thanks to the Nesterov work [95]; they are further developed in [11, 12], where the authors propose the following variant, known also as fast iterative shrinkage-thresholding algorithm (FISTA) for the problem (6.1):

$$\bar{x}^{(k)} = x^{(k)} + \gamma_k(x^{(k)} - x^{(k-1)}), \tag{6.65}$$

$$x^{(k+1)} = \text{prox}_{\alpha_k\Psi}(\bar{x}^{(k)} - \alpha_k \nabla\Phi(\bar{x}^{(k)})) \tag{6.66}$$

where γ_k and α_k are non-negative parameters. In addition to the standard assumptions for the problem (6.1), in this case the gradient of Φ has to be a Lipschitz-continuous function.

In [12, 17], the convergence of the method is investigated by showing that, for suitable sequences of parameters $\{\alpha_k\}$ and $\{\gamma_k\}$, the rate of convergence of the values of the objective function $\{F(x^{(k)})\}$ is $\mathcal{O}\left(\frac{1}{k^2}\right)$. Under additional assumptions on $\{\alpha_k\}$ and $\{\gamma_k\}$, in [7] the authors show that the rate of convergence of the accelerated FB methods is $o\left(\frac{1}{k^2}\right)$ and there is no $h > 2$ such that the order is $\mathcal{O}\left(\frac{1}{k^h}\right)$ for every Φ and Ψ. Under the same assumptions, the convergence of the sequence of iterates $\{x^{(k)}\}$ can be obtained [39], while in [108, 111] the authors address the analysis of inertial schemes where the proximal point (6.66) is inexactly computed.

A drawback in the method (6.65)–(6.66) is that it may be unfeasible when dom Φ does not coincide with the whole space $\mathbf{R}^N$, since the extrapolated point $\bar{x}^{(k)}$ does not necessarily belong to dom Φ. In particular, this is the case of $\Phi = f_0(x; y)$, whose effective domain is a proper subset of the whole space $\mathbf{R}^N$. In [26], a variable metric variant of equations (6.65)–(6.66) addresses the problem (6.1) also when dom $\Phi \neq \mathbf{R}^N$, under the assumption dom $\Phi \supseteq \Upsilon \supseteq$ dom Ψ, where Υ is a non-empty, closed, convex set where $\nabla\Phi$ is Lipschitz-continuous:

$$\bar{x}^{(k)} = P_{\Upsilon,\, D_k}(x^{(k)} + \gamma_k(x^{(k)} - x^{(k-1)})), \tag{6.67}$$

$$x^{(k+1)} = \mathrm{prox}_{\alpha_k\Phi,\, D_k}(\bar{x}^{(k)} - \alpha_k D_k^{-1}\nabla\Phi(\bar{x}^{(k)})). \tag{6.68}$$

Indeed, the inertial step is modified by introducing a projection of the extrapolated vector on Υ with respect to the current metric, so that it is assured that $\bar{x}^{(k)}$ belongs to a subset of dom Φ where $\nabla\Phi$ exists and it is a Lipschitz-continuous function. The second step (6.68) is a variable metric FB iteration related to a parameter α_k, that can be adaptively computed via a backtracking procedure. In order to preserve the theoretical rate of convergence of FISTA in a variable metric setting, suitable choices for the sequences $\{\gamma_k\}$ and $\{D_k\}$ have to be performed:

- as for SGP or VMILA, the sequence of scaling matrices $\{D_k\}$ has to be suitably chosen in the compact set $\mathcal{D}_L$ of the s.p.d. matrices of order N, $L \geqslant 1$, and assumption (6.16) has to be satisfied; as already mentioned, this means that the sequence $\{D_k\}$ asymptotically approaches a constant matrix (see remark 6.1 about a practical rule for choosing $\{D_k\}$);
- as pointed out in [17], the extrapolation parameter γ_k must have the form

$$\gamma_k = \frac{t_{k-1} - 1}{t_k}, \quad \gamma_0 = 0, \tag{6.69}$$

where $\{t_k\}$ is a given sequence of parameters, satisfying the condition

$$t_{k-1}^2 + t_k - t_k^2 \geqslant 0, \quad t_k \geqslant 1. \tag{6.70}$$

An example of a sequence $\{t_k\}$ and corresponding $\{\gamma_k\}$ satisfying equations (6.69)–(6.70) is the following one

$$t_k = \begin{cases} 1 & k = -1, 0 \\ \dfrac{k + a}{a} & k \geqslant 1 \end{cases} \qquad \gamma_k = \begin{cases} 0 & k = 0 \\ \dfrac{k - 1}{k + a} & k \geqslant 1, \end{cases}$$

with $a \geqslant 2$. Indeed, for all $k \geqslant 0$ and $a \geqslant 2$, we have

$$\begin{aligned} t_{k-1}^2 + t_k - t_k^2 &= \frac{(k - 1 + a)^2}{a^2} + \frac{(k + a)}{a} - \frac{(k + a)^2}{a^2} \\ &= \frac{(k + a)(a - 2) + 1}{a^2} \geqslant 0. \end{aligned}$$

The choice $a = 2$ is proposed in [12]; it allows one to prove, also in the case of variable metric, that the sequence $\{F(x^{(k)})\}$ converges to a minimum of the problem as $\mathcal{O}\left(\frac{1}{k^2}\right)$. The more general case $a \geqslant 2$ is considered in [7, 39]; in particular, the assumption $a > 2$ is introduced in [39] to prove the convergence of the sequence $\{x^{(k)}\}$ to a minimum point and in [7] to obtain that $\{F(x^{(k)})\}$ is $o\left(\frac{1}{k^2}\right)$ rather than $\mathcal{O}\left(\frac{1}{k^2}\right)$. These results can be extended to a variable metric setting [28].

The scheme of the method, named SFBEM, is detailed in Algorithm 6.7. When $\Upsilon = \mathbf{R}^N$, FISTA is recovered by setting $D_k = I$ for all $k \geqslant 0$.

Algorithm 6.7. Scaled inertial FB method with backtracking (SFBEM).

Choose $\alpha_0 > 0$, $\rho \in (0, 1)$, $x^{(0)} \in \Upsilon$. Set $x^{(-1)} = x^{(0)}$ and define a sequence of non-negative numbers $\{\gamma_k\}$ and a sequence of scaling matrices $\{D_k\}$, with $D_k \in \mathcal{D}_L$, $L \geqslant 1$.

FOR $k = 0, 1, 2, \ldots$ DO THE FOLLOWING STEPS:

STEP 1. extrapolation: $\bar{x}^{(k)} = P_{\Upsilon, D_k}(x^{(k)} + \gamma_k(x^{(k)} - x^{(k-1)}))$;

STEP 2. set $i_k = 0$, $\alpha_k = \alpha_{k-1}$;

STEP 3. compute

$x_+^{(k)} = \arg\min_{x \in \mathbf{R}^N} \Psi(x) + \frac{1}{2\alpha_k}\|x - (\bar{x}^{(k)} - \alpha_k \nabla\Phi(\bar{x}^{(k)}))\|_{D_k}^2$

STEP 4. backtracking loop:

IF $\Phi(x_+^{(k)}) \leqslant \Phi(\bar{x}^{(k)}) + \nabla\Phi(\bar{x}^{(k)})^T(x_+^{(k)} - \bar{x}^{(k)}) + \frac{1}{2\alpha_k}\|x_+^{(k)} - \bar{x}^{(k)}\|_{D_k}^2$;

go to STEP 5;

ELSE

set $i_k = i_k + 1$; $\alpha_k = \rho^{i_k}\alpha_{k-1}$;

compute

$x_+^{(k)} = \arg\min_{x \in \mathbf{R}^N} \Psi(x) + \frac{1}{2\alpha_k}\|x - (\bar{x}^{(k)} - \alpha_k \nabla\Phi(\bar{x}^{(k)}))\|_{D_k}^2$

and go to STEP 4;

END

STEP 5. set $x^{(k+1)} = x_+^{(k)}$;

STEP 6. if a stopping criterion is satisfied then stop, declaring $x^{(k+1)}$ is a stationary point;

The main computational burden of an iteration of SFBEM is the evaluation of $\nabla\Phi(x^{(k)})$ (for $\Phi(x) = f_0(x; y)$ only matrix–vector products involving H and H^T are required) and the solution of the inner sub-problem (step 3) that has to be repeated if the condition at step 4 is not immediately satisfied. As in SPIRAL, the convergence results are obtained by assuming that the solutions of the inner sub-problem involved in the backward step are exactly computed. Techniques similar to the ones described for the inexact computation of the proximal step of VMILA can be introduced also in SFBEM [28, 111, 117]; nevertheless, numerical experience shows that the extrapolated method is more sensitive to computational errors than the standard FB method and, consequently, improvements in efficiency might be lost whenever the computational errors do not decay sufficiently fast.

As far as the choice of the metric is concerned, its selection usually may improve the practical convergence speed of the method without introducing significant computational costs. For this reason diagonal scaling matrices have to be preferred. Moreover, as discussed in the previous sections, a good choice of the scaling matrix is mainly based on the idea of capturing some problem features.

As already mentioned for SGP and VMILA, the adoption of a convenient variable metric is closely connected to the objective function to be minimized and possible constraints on the solution. Therefore, since the choice of the metric is strongly problem-dependent, it should be evaluated case-by-case. In the case of Poisson data inversion, we refer to the discussion already made for the SGP method (section 6.3.1). For HS regularization of a 2D image corrupted by Poisson noise, a Matlab® version of SFBEM is downloadable from http://www.oasis.unimore.it/site/home/software.html. For this application, $\Phi(x) = f_0(x; y) + \beta f_1(x)$ and $\Psi(x) = \iota_C(x)$.

6.5 The alternating direction method of multipliers (ADMM)

A class of popular methods which can be referred to the proximal splitting algorithms is the class of the augmented Lagrangian techniques and, in particular, the alternating direction method of multipliers (ADMM). ADMM is developed in the seventies [70, 71]; its convergence is proved in the nineties [61] and its convergence rate is the topic of many recent papers. Indeed, ADMM is a state-of-the-art solver for a wide variety of optimization problems in machine learning, signal processing, and many other areas, as discussed in the comprehensive review [32]. ADMM is a version of the augmented Lagrangian methods [14, 80, 105] especially tailored to fit an objective function with a separable structure. The method has been deeply investigated, showing that it is equivalent to other schemes well-known in literature, such as split Bregman method in the context of ℓ_1 regularized problems [73, 112] or Douglas–Rachford splitting method applied to the dual formulation of the original problem in the framework of splitting methods for maximal monotone operators [45, 61, 64, 112]. In this section we summarize the ideas underlying the method and its main features. Then, we show how the method is used in the context of inverse imaging with Poisson data.

6.5.1 The basic method

The method applies to the solution of general separable problems of the form:

$$\begin{aligned} &\min_{x\in\mathbf{R}^n, z\in\mathbf{R}^m} \quad F_1(x) + F_2(z), \\ &\text{subject to } \mathrm{A}x + Bz = c, \end{aligned} \tag{6.71}$$

where $F_1: \mathbf{R}^n \to \bar{\mathbf{R}}$, $F_2: \mathbf{R}^m \to \bar{\mathbf{R}}$ are proper, convex, l.s.c. functions, $A \in \mathbf{R}^{t\times n}$, $B \in \mathbf{R}^{t\times m}$ and $c \in \mathbf{R}^t$. We assume that the problem is feasible and admits a finite solution.

Following the basic idea of the method of multipliers, we can consider the equivalent problem

$$\min_{x\in\mathbf{R}^n, z\in\mathbf{R}^m} \mathcal{F}_{p,\gamma}(x, z) := F_1(x) + F_2(z) + \frac{\gamma}{2}\|Ax + Bz - c\|^2, \quad \text{subject to Ax} + Bz = c, \tag{6.72}$$

where γ is a positive penalty parameter and $\|\cdot\|$ is the standard ℓ_2 norm. The Lagrangian function of the problem (6.72) is given by

$$\mathcal{L}_\gamma(x, z, p) = F_1(x) + F_2(z) + p^T(Ax + Bz - c) + \frac{\gamma}{2}\|Ax + Bz - c\|^2, \tag{6.73}$$

and it is named augmented Lagrangian of the problem (6.71). We observe that $\mathcal{L}_0(x, z, p)$ is the Lagrangian of the original problem (6.71) and the two Lagrangian functions differ for a quadratic term that is equal to zero for any admissible vector (x, z). The associated dual function of $\mathcal{F}_{p,\gamma}(x, z)$ is

$$\mathcal{F}_{d,\gamma}(p) = \inf_{x,z} \mathcal{L}_\gamma(x, z, p). \tag{6.74}$$

The benefit of including the penalization term is that $\mathcal{F}_{d,\gamma}(p)$ can be shown to be differentiable under rather mild conditions on the original problem. Indeed, when the uniqueness of the solution of the problem $\inf_{x,z}\mathcal{L}_\gamma(x, z, p)$ for fixed p holds (for example $\mathcal{L}_\gamma(x, z, p)$ is strictly convex with respect to (x, z) for fixed p), $\mathcal{F}_{d,\gamma}(p)$ is well-defined and is differentiable:

$$\begin{aligned}\nabla\mathcal{F}_{d,\gamma}(p) &= \underbrace{\nabla_{(x,z)}\mathcal{L}_\gamma(x(p), z(p), p)}_{=0}\nabla_p(x(p), z(p)) + \nabla_p\mathcal{L}_\gamma(x(p), z(p), p) \\ &= Ax(p) + Bz(p) - c.\end{aligned}$$

In [80, 105] the Augmented Lagrangian is introduced in the method of multipliers, defined by the following iteration

$$\begin{aligned}(x^{(k+1)}, z^{(k+1)}) &= \arg\min_{x,z} \mathcal{L}_\gamma(x, z, p^{(k)}), \\ p^{(k+1)} &= p^{(k)} + \gamma(Ax^{(k+1)} + Bz^{(k+1)} - c),\end{aligned} \tag{6.75}$$

which can be interpreted as a gradient ascent method, with constant step-length γ, applied to the dual problem $\max_p\mathcal{F}_{d,\gamma}(p)$. Convergence and numerical properties of the method are deeply investigated (see the monograph [14]). ADMM is introduced to exploit the separable structure of the objective function and preserve the convergence features of the multiplier methods. It has the following scheme:

$$x^{(k+1)} = \arg\min_x \mathcal{L}_\gamma(x, z^{(k)}, p^{(k)}), \tag{6.76}$$

$$z^{(k+1)} = \arg\min_z \mathcal{L}_\gamma(x^{(k+1)}, z, p^{(k)}), \tag{6.77}$$

$$p^{(k+1)} = p^{(k)} + \gamma(Ax^{(k+1)} + Bz^{(k+1)} - c). \tag{6.78}$$

The algorithm consists of a x-minimization step (6.76), a z-minimization step (6.77) and a dual variable update (6.78). As in the method of multipliers, the dual variable update uses the parameter γ as a step-length.

ADMM can be viewed as a version of the method of multipliers where a single Gauss–Seidel step over x and z is used instead of the usual joint minimization, exploiting the separability of the objective function.

The so-called *scaled form* is obtained when an auxiliary scaled dual vector $q = \frac{1}{\gamma}p$ is introduced and, consequently, the linear and the quadratic term in equations (6.77)–(6.76) can be assembled:

$$\begin{aligned} x^{(k+1)} &= \arg\min_x \left(F_1(x) + \frac{\gamma}{2}\|Ax + Bz^{(k)} - c + q^{(k)}\|^2 \right), \\ z^{(k+1)} &= \arg\min_z \left(F_2(z) + \frac{\gamma}{2}\|Ax^{(k+1)} + Bz - c + q^{(k)}\|^2 \right), \\ q^{(k+1)} &= q^{(k)} + Ax^{(k+1)} + Bz^{(k+1)} - c. \end{aligned} \tag{6.79}$$

Many convergence results for ADMM can be found in the literature; following [32], we report the convergence properties of the method under the following assumptions:

- F_1 and F_2 are proper, convex and l.s.c. functions such that the sub-problems arising in the x-update and z-update are solvable;
- the un-augmented Lagrangian $\mathcal{L}_0$ has a saddle point, i.e. there exists (x^*, z^*, p^*), not necessarily unique, for which the following inequalities hold for all (x, z, p)

$$\mathcal{L}_0(x^*, z^*, p) \leqslant \mathcal{L}_0(x^*, z^*, p^*) \leqslant \mathcal{L}_0(x, z, p^*);$$

  it follows that $\mathcal{L}_0(x^*, z^*, p^*)$ is finite for any saddle point and that (x^*, z^*) is a solution of the primal problem (6.71) ($Ax^* + Bz^* = c$ and $F_1(x^*) < \infty$, $F_2(z^*) < \infty$); consequently, p^* is the solution of the dual one.

Under these assumptions, we obtain the following results:

- residual convergence: $r^{(k)} = Ax^{(k)} + Bz^{(k)} - c \to 0$ as $k \to \infty$;
- convergence of the values of the objective function: $F_k = F_1(x^{(k)}) + F_2(z^{(k)}) \to \bar{F}$ as $k \to \infty$, where $\bar{F}$ is the optimal value of equation (6.71);
- dual variable convergence: $p^{(k)} \to p^*$ as $k \to \infty$.

Furthermore, under suitable assumptions assuring that the sub-problems (6.76)–(6.77) of ADMM admits a unique solution, we can state also the convergence of the sequences $\{x^{(k)}\}$ and $\{z^{(k)}\}$ to a minimizer of (6.71). This result holds true, for instance, when the matrices A and B have full column-rank or when F_1 is strictly convex and B has full column-rank and so on (see for instance [18]).

In order to devise stopping criteria for ADMM, we write the optimality conditions of the problem (6.71) and of the sub-problems (6.77)–(6.76). In particular, we have that (x^*, z^*, p^*) is a solution of equation (6.71) when the following conditions are satisfied:

$$Ax^*+Bz^*-c = 0, \text{ primal feasibility}$$

$$\begin{cases} 0 \in \partial F_1(x^*) + A^T p^*, \\ 0 \in \partial F_2(z^*) + B^T p^*, \end{cases} \text{dual feasibility.}$$

Furthermore, since $z^{(k+1)}$ minimizes $\mathcal{L}_\gamma(x^{(k+1)}, z, p^{(k)})$, in view of equation (6.78), we have:

$$0 \in \partial F_2(z^{(k+1)}) + B^T p^{(k)} + \gamma B^T(Ax^{(k+1)} + Bz^{(k+1)} - c) = \partial F_2(z^{(k+1)}) + B^T p^{(k+1)}; \tag{6.80}$$

on the other hand, since $x^{(k+1)}$ minimizes $\mathcal{L}_\gamma(x, z^{(k)}, p^{(k)})$, we have:

$$\begin{aligned} 0 &\in \partial F_1(x^{(k+1)}) + A^T p^{(k)} + \gamma A^T(Ax^{(k+1)} + Bz^{(k)} - c) \\ &= \partial F_1(x^{(k+1)}) + A^T p^{(k+1)} - \gamma A^T B(z^{(k+1)} - z^{(k)}). \end{aligned} \tag{6.81}$$

From the comparison between the dual feasibility condition at the solution and equations (6.80)–(6.81), we have that the vector $s^{(k+1)} := \gamma A^T B(z^{(k+1)} - z^{(k)})$ can be viewed as a residual for the dual feasibility condition as well as $r^{(k+1)}$ is a residual for the primal feasibility condition at $(k + 1)$-iteration. This suggests that a reasonable stopping criterion for ADMM is that the primal and dual residuals must be small [32]:

$$\begin{aligned} &\|r^{(k)}\| \leqslant \epsilon_{\text{pri}}, \quad \|s^{(k)}\| \leqslant \epsilon_{\text{dual}} \\ &\epsilon_{\text{pri}} = \epsilon_{\text{abs}}\sqrt{t} + \epsilon_{\text{rel}}\max(\|Az^{(k)}\|, \|Bx^{(k)}\|, \|c\|), \\ &\epsilon_{\text{dual}} = \epsilon_{\text{abs}}\sqrt{n} + \epsilon_{\text{rel}}\|A^T p^{(k)}\|, \end{aligned}$$

where $\epsilon_{\text{abs}} > 0$ is an absolute tolerance and $\epsilon_{\text{rel}} > 0$ is a relative tolerance.

From the practical point of view, ADMM exhibits a very slow convergence to a highly accurate solution; however, ADMM can reach a modest accuracy, sufficient for many applications, within a reasonable number of iterations. Nevertheless, we point out that the choice of the parameter γ has a high impact on the performance of the method. A frequently used generalization consists in using different penalty parameters γ_k at each iteration. The goal is to improve the convergence in practice, and making the effectiveness of the method less dependent on the initial choice of γ. A residual balancing scheme, proposed in [79], is based on the following updating rule:

$$\gamma_{k+1} = \begin{cases} (1 + \tau_k)\gamma_k & \text{if } \|r^{(k)}\| > \nu\|s^{(k)}\| \\ \dfrac{\gamma_k}{1 + \tau_k} & \text{if } \|s^{(k)}\| > \nu\|r^{(k)}\| \\ \gamma_k & \text{otherwise} \end{cases} \tag{6.82}$$

where $\nu > 1$ and $\{\tau_k\}$ is a sequence of non-negative values. The rule is based on the observation that an increase of γ_k strengthens the penalty term, with the result of smaller primal residuals but larger dual ones; conversely, a decrease of γ_k leads to larger primal and smaller dual residuals. The convergence is preserved if $\sum_{k=0}^{\infty}\tau_k < \infty$ and, consequently, γ_k becomes fixed after a finite number K of iterations (possibly very large); a typical choice for $\{\tau_k\}$ is $\tau_k = 1$, for $k \leqslant K$, $\tau_k = 0$, for $k > K$ and $\nu = 10$ [32]. However, we observe that, since the residuals depend on the (arbitrary) scaling of the problem, the rule (6.82) can produce very different performances in different cases, since the behavior of the method is determined also by the choice of ν and τ_k. Other strategies are recently investigated. We mention those in [119], based on the equivalence of ADMM with the Douglas–Rachford splitting method combined with Barzilai–Borwein techniques.

Another point to be considered is that the performance of ADMM depends also on the z- and x-minimization steps; even if closed formulas exist, they can require expensive solutions of linear systems. Furthermore, even if the convergence of the method still holds if the inner sub-problems are not solved exactly [61], the control of the lack of exactness is not effective, being dependent on the exact minimizers. In [62, 78, 96] some variants of inexact ADMM, based on implementable conditions, are proposed.

6.5.2 ADMM for image deconvolution of Poisson data

In recent papers on image deconvolution of Poisson data, ADMM is applied to the solution of the MAP problem (4.30). Using the notation of chapter 4, the objective function can be written as the sum of three terms: $f_0(x; y) = \mathrm{KL}(y; Hx + b)$, with $H \in \mathbf{R}^{M\times N}$; a regularization term $\beta f_1(x) := \psi(Rx)$, given by the composition of a convex function ψ with a linear operator $R \in \mathbf{R}^{N_1\times N}$; the indicator function $\iota_C(x)$ of the non-negative orthant. Therefore, a MAP estimate is a solution of the problem

$$\min_x \mathrm{KL}(y; Hx + b) + \psi(Rx) + \iota_C(x). \tag{6.83}$$

Following the suggestions in [36, 66, 113], we can reformulate this problem as follows:

$$\min_{z,x} \mathrm{KL}(y; z_1) + \psi(z_2) + \iota_C(z_3),$$
$$\text{subject to} \begin{cases} Hx &= z_1 - b \\ Rx &= z_2 \\ x &= z_3. \end{cases} \tag{6.84}$$

Comparing with problem (6.71) we observe that the formulated problem has just that form if $F_1(x) = 0$, $F_2(z) = \mathrm{KL}(y; z_1) + \psi(z_2) + \iota_C(z_3)$ with $z = (z_1, z_2, z_3)$, and the linear constraint $Ax + Bz = c$ is obtained by setting

$$A = \begin{pmatrix} H \\ R \\ I_N \end{pmatrix}, \quad B = -I_{M+N_1+N}, \quad c = \begin{pmatrix} -b \\ 0 \\ 0 \end{pmatrix}. \tag{6.85}$$

Then, taking into account that F_2 has a separable structure, the basic iteration (6.79) of the scaled ADMM, known in the literature as PIDAL or PIDSplit+ algorithm, assumes the following form:

$$\begin{aligned} x^{(k+1)} = \arg\min_{x \in R^N} & \|Hx + b - z_1^{(k)} + q_1^{(k)}\|^2 + \|Rx - z_2^{(k)} + q_2^{(k)}\|^2 \\ & + \|x - z_3 + q_3^{(k)}\|^2, \end{aligned} \tag{6.86}$$

$$z_1^{(k+1)} = \arg\min_{z_1 \in R^M} \mathrm{KL}(y; z_1) + \frac{\gamma}{2}\|Hx^{(k+1)} + b - z_1 + q_1^{(k)}\|^2, \tag{6.87}$$

$$z_2^{(k+1)} = \arg\min_{z_2 \in R^S} \psi(z_2) + \frac{\gamma}{2}\|Rx^{(k+1)} - z_2 + q_2^{(k)}\|^2, \tag{6.88}$$

$$z_3^{(k+1)} = \arg\min_{z_3 \in R^N} \iota_C(z_3) + \frac{\gamma}{2}\|x^{(k+1)} - z_3 + q_3^{(k)}\|^2, \tag{6.89}$$

$$\begin{aligned} q_1^{(k+1)} &= q_1^{(k)} + Hx^{(k+1)} - z_1^{(k+1)} + b, \\ q_2^{(k+1)} &= q_2^{(k)} + Rx^{(k+1)} - z_2^{(k+1)}, \\ q_3^{(k+1)} &= q_3^{(k)} + x^{(k+1)} - z_3^{(k+1)}. \end{aligned}$$

The matrices A and B have full column-rank, so that the convergence of the primal sequences is assured. Indeed, any sub-problem admits only one solution. The sequence of approximate solutions of the problem (6.83) is $\{z_3^{(k)}\}$, which are in C since the sub-problem (6.89) requires a projection on C:

$$z_3^{(k+1)} = P_C(x^{(k+1)} + q_3^{(k)}).$$

The sub-problems (6.86)–(6.87) can be solved explicitly, the first by solving the following linear system

$$(H^T H + R^T R + I_N)x = H^T(z_1^{(k)} - q_1^{(k)} - b) + R^T(z_2^{(k)} - q_2^{(k)}) + z_3^{(k)} - q_3^{(k)}, \tag{6.90}$$

the second by the closed formula of the proximal operator of $\sigma\mathrm{KL}(y; z_1)$ at u, with $u = t^{(k)} = Hx^{(k+1)} + b + q_1^{(k)}$ and $\sigma = \frac{1}{\gamma}$:

$$\mathrm{prox}_{\sigma\mathrm{KL}(y;z_1)}(u) = \frac{1}{2}\left(u - \sigma + \sqrt{(u - \sigma)^2 + 4\sigma y}\right), \tag{6.91}$$

where the vector operators $\sqrt{\cdot}$ and $(\cdot)^2$ are intended component-wise. The computation complexity for solving system (6.90) (step 1) as well as for the sub-problem (6.88) (step 5) are strictly dependent on the application.

The implementation of the ADMM method is detailed in algorithm 6.8.

Algorithm 6.8. Poisson image deconvolution by augmented Lagrangian method (PIDAL).

Choose $\gamma > 0$; set $q_1^{(0)} = 0$, $q_2^{(0)} = 0$, $q_3^{(0)} = 0$; set initial values for $z_1^{(0)}, z_2^{(0)}, z_3^{(0)}$ with $z_3^{(0)} \in C$.

FOR $k = 0, 1, 2, \ldots$ DO THE FOLLOWING STEPS:

STEP 1. compute

$$x^{(k+1)} = (H^T H + R^T R + I_N)^{-1}(H^T(z_1^{(k)} - q_1^{(k)} - b) + R^T(z_2^{(k)} - q_2^{(k)}) + z_3^{(k)} - q_3^{(k)});$$

STEP 2. compute $t^{(k)} = Hx^{(k+1)} + b + q_1^{(k)}$;

STEP 3. compute $z_1^{(k+1)} = \frac{1}{2\gamma}\left(\gamma t^{(k)} - 1 + \sqrt{(\gamma t^{(k)} - 1)^2 + 4\gamma y}\right)$;

STEP 4. compute $v^{(k)} = Rx^{(k+1)} + q_2^{(k)}$;

STEP 5. compute $z_2^{(k+1)} = \mathrm{prox}_{\frac{1}{\gamma}\psi}(v^{(k)})$;

STEP 6. compute $w^{(k)} = x^{(k+1)} + q_3^{(k)}$;

STEP 7. compute $z_3^{(k+1)} = P_C(w^{(k)})$;

STEP 8. set $q_1^{(k+1)} = t^{(k)} - z_1^{(k+1)}$;

STEP 9. set $q_2^{(k+1)} = v^{(k)} - z_2^{(k+1)}$;

STEP 10. set $q_3^{(k+1)} = w^{(k)} - z_3^{(k+1)}$;

STEP 11. if a stopping criterion is satisfied then stop, declaring $z_3^{(k+1)}$ is a stationary point;

For TV regularization of Poisson data [113], we have $\psi(Rx) = \beta\sum_{i=1}^{N}\|\nabla_i x\|$ and the matrix $R = \nabla = (\nabla_1^T, \ldots, \nabla_N^T)^T$ represents the discrete gradient operator in all pixels of the unknown image, so that $R^T R$ can be diagonalized by the discrete cosine transform. If H models a periodic convolution, the solution of the system (6.90) can be obtained by fast transform algorithms with $\mathcal{O}(N \ln N)$ operations. Furthermore, the sub-problem (6.88) is separable and can be solved by a closed formula. Indeed, for each i-pixel, $i = 1, \ldots, N$, given $v^{(k)} = Rx^{(k+1)} + q_2^{(k)}$, we can compute the proximal operator required at step 5 by the following shrinkage operator:

$$(z_2^{(k+1)})_i = \mathrm{shrink}\left(\frac{\beta}{\gamma}, v_i^{(k)}\right) = \frac{v_i^{(k)}}{\|v_i^{(k)}\|}\max\left(\|v_i^{(k)}\| - \frac{\beta}{\gamma}, 0\right). \tag{6.92}$$

As concerns the selection of the parameter γ, a typical value is $\gamma = \frac{\beta}{50}$. Nevertheless, we stress that the performance of the method is crucially dependent on the choice of this parameter.

Similar applications are described in [36, 66], with a penalty term of the form $\beta\|Rx\|_1$, where now the operator R provides the representation of x in some suitable wavelet basis (or other multi-scale system), i.e. the regularization is provided by the ℓ_1 norm of the representation coefficients of x.

Remark 6.3. *ADMM method can be used also for HS regularization* (4.35) *by setting* $\psi(Rx) = \beta\sum_{i=1}^{N}\left(\left\|\begin{pmatrix}\nabla_i x\\ \delta\end{pmatrix}\right\| - \delta\right)$ *and* $R = \begin{pmatrix}\nabla\\ 0\end{pmatrix}$. *In such a case, the right-hand side of the equation* $Rx = z_2$ *in (6.84) is replaced by* $z_2 + (0^T, \delta 1_N^T)^T$.

6.6 Primal–dual methods

A family of methods for solving a convex minimization problem with a non-smooth term composed with a linear operator is based on its primal–dual formulation. In particular, we focus on the following problem, thoroughly investigated is the framework of inverse problems:

$$\min_{x\in\mathbf{R}^N} \mathcal{F}_p(x) \equiv F_1(x) + F_2(Ax) \tag{6.93}$$

where F_1: $\mathbf{R}^N \to \bar{\mathbf{R}}$, F_2: $\mathbf{R}^M \to \bar{\mathbf{R}}$ are convex, proper, l.s.c. functions and A is a $M \times N$ matrix.

From definition 6.2 of conjugate function, by introducing the conjugate functions of $F_2(Ax)$ and $F_1(x)$, it is easy to obtain the primal–dual formulation of the problem (6.93):

$$\begin{aligned}&\min_{x\in\mathbf{R}^N} F_1(x) + \max_{p\in\mathbf{R}^M} p^T Ax - F_2^*(p)\\ &\quad= \min_{x\in\mathbf{R}^N}\max_{p\in\mathbf{R}^M} \mathcal{L}(x, p) \equiv F_1(x) - F_2^*(p) + p^T Ax,\end{aligned} \tag{6.94}$$

and the corresponding dual formulation

$$\max_{p\in\mathbf{R}^M} \mathcal{F}_d(p) \equiv -F_2^*(p) - F_1^*(-A^T p). \tag{6.95}$$

We assume that there exists a pair (x^*, p^*) of optimal solutions of the primal and dual problems; this is also a saddle point for the function $\mathcal{L}(x, p)$ so that the primal optimal and the dual optimal coincide and the following conditions hold (for details see Fenchel's duality theorem A.25):

$$\begin{aligned}&-A^T p^* \in \partial F_1(x^*),\\ &Ax^* \in \partial F_2^*(p^*).\end{aligned}$$

Therefore, for all $(x, p) \in \mathbf{R}^N \times \mathbf{R}^M$, we have

$$\begin{aligned}F_1(x^*) - F_2^*(p) + p^T Ax^* \leqslant F_1(x^*) - F_2^*(p^*) + (p^*)^T Ax^* \leqslant F_1(x) - F_2^*(p^*)\\ + (p^*)^T Ax,\end{aligned} \tag{6.96}$$

and $\mathcal{F}_p(x^*) = \mathcal{L}(x^*, p^*) = \mathcal{F}_d(p^*)$.

A well-known method for solving equation (6.93) is the Arrow–Hurwicz method [3], first used for image processing in [121] and named primal–dual hybrid gradient

(PDHG) algorithm. The well-known Chambolle–Pock (CP) algorithm can be considered as an improved version of this scheme, obtained by introducing an extrapolation step on the primal (or the dual) variable [41]. The PDHG method is studied also in a more general framework: connections to other known algorithms are established [64] and several variants have been developed. In particular, we mention the schemes where the step-length parameters are *a priori* selected sequences [30, 121] or a variable metric is introduced [22].

6.6.1 Chambolle–Pock method

The basic iteration of a class of methods introduced in [41] for solving problem (6.93) is given by the following three steps:

$$p^{(k+1)} = \text{prox}_{\tau F_2^*}(p^{(k)} + \tau A\bar{x}^{(k)}) \tag{6.97}$$

$$x^{(k+1)} = \text{prox}_{\sigma F_1}(x^{(k)} - \sigma A^T p^{(k+1)}) \tag{6.98}$$

$$\bar{x}^{(k+1)} = x^{(k+1)} + \theta(x^{(k+1)} - x^{(k)}) \tag{6.99}$$

where $\bar{x}^{(0)} = x^{(0)}$ and $p^{(0)}$ are starting vectors, τ, σ are positive parameters and $\theta \in [0, 1]$. The choice $\theta = 0$ yields the classic Arrow–Hurwicz algorithm, while $\theta = 1$ gives the CP algorithm. Each iteration basically consists in alternating a proximal gradient ascent in the dual variable and a proximal gradient descent in the primal variable. Additionally, the algorithm performs an over-relaxation step in the primal variable. Note that, by exchanging the updates for $p^{(k+1)}$ and $x^{(k+1)}$, the extrapolation step can be performed also on the p variable. A fundamental assumption for the convergence and the performance of the method is that the proximity operators of the functions F_1 and F_2^* can be obtained by closed-form solutions.

Under the assumptions that F_1 and F_2 are convex, proper, l.s.c. functions such that there exists a saddle point of problem (6.94) and $0 < \sigma\tau\|A\|^2 < 1$, the sequences $\{x^{(k)}\}$ and $\{p^{(k)}\}$ are convergent to x^* and p^*, respectively, as $k \to \infty$, and (x^*, p^*) is a saddle point of problem (6.94) ([41], theorem 1). Furthermore, when F_1 and F_2^* are strongly convex functions, for a suitable choice of the parameters τ and σ, linear convergence of the sequence $\{(x^{(k)}, p^{(k)})\}$ can be obtained. In addition, when F_1 (or F_2^*) is a strongly convex function with parameter M_{F_1} (see definition A.5), an accelerated version of the CP algorithm can be devised. In this scheme, starting from $\tau_0\sigma_0\|A\|^2 < 1$, the relaxation parameter θ_k in equation (6.99) is obtained by an updating rule and the primal and dual step-lengths are dynamically computed at any kth iteration, $k = 0, 1, \ldots$, as follows:

$$\theta_k = \frac{1}{\sqrt{1 + 2M_{F_1}\sigma_k}}, \quad \sigma_{k+1} = \theta_k\sigma_k, \quad \tau_{k+1} = \theta_k\tau_k.$$

A crucial question for the performance of the CP method (6.97)–(6.99) is the estimate of $\|A\|$; when this estimate is not available or is very large, preconditioning techniques can be applied (see for example [40]), by replacing the parameter σ and τ with two s.p.d. matrices Σ and T that satisfy the condition $\|T^{1/2}A\Sigma^{1/2}\|^2 < 1$. In [40], the authors suggest to select Σ and T as diagonal matrices with diagonal entries defined by

$$
\begin{aligned}
(\Sigma)_{j,j} &= c_\Sigma \frac{1}{\sum_{i=1}^{M} |(A)_{(i,j)}|^\zeta}, \; j = 1, \dots, N \\
(T)_{i,i} &= c_T \frac{1}{\sum_{j=1}^{N} |(A)_{(i,j)}|^{2-\zeta}}, \; i = 1, \dots, M,
\end{aligned}
\tag{6.100}
$$

with $\zeta \in [0, 2]$ and c_Σ, $c_T \in (0, 1)$. Typical choices are $\zeta = 1$ and $c_\Sigma = c_T = 0.99$.

To avoid the choice of the two parameters σ and τ, which can strongly affect the effectiveness of CP iteration, the preconditioned form of CP method is used also when $\|A\|$ is known.

6.6.1.1 CP method for image reconstruction with Poisson data

For the image reconstruction problem with Poisson data (6.83), the application of the CP algorithm is not trivial. We consider two different instances of the problem: the denoising case, where $H = I_N$, and the deblurring case.

For the denoising problem, we set $F_1(x) = \mathrm{KL}(y; x + b) + \iota_C(x)$ and $F_2(Ax) = \psi(Rx)$, with $A = R$ (here $M = N_1$). Starting from a suitable vector $x^{(0)} = x^{(-1)} \in C$ (for example $P_C(y)$) and $p^{(0)} = 0$, the iteration of the CP method can be written as

$$
p^{(k+1)} = \arg\min_{p \in \mathbf{R}^{N_1}} \psi^*(p) + \frac{1}{2\tau} \|p - (p^{(k)} + \tau R(2x^{(k)} - x^{(k-1)}))\|^2 \tag{6.101}
$$

$$
x^{(k+1)} = \arg\min_{x \in C} \mathrm{KL}(y; x + b) + \frac{1}{2\sigma} \|x - (x^{(k)} - \sigma R^T p^{(k+1)})\|^2 . \tag{6.102}
$$

where σ and τ have to be chosen in such a way that $\sigma\tau\|R\|^2 < 1$ and the extrapolation step on x is included in equation (6.101). The solution of the sub-problem (6.102) is easily obtained by using the closed formula (6.91), with $u = x^{(k)} - \sigma R^T p^{(k+1)}$ while the computation of $p^{(k+1)}$ in equation (6.101) depends on the regularization functional. For example, for TV regularization of a 2D-image, we have $\psi(Rx) = \beta \sum_{i=1}^{N} \|\nabla_i x\|$ with $R = \nabla = (\nabla_1^T, \dots, \nabla_N^T)^T$. The conjugate of ψ is

$$
\psi^*(p) = \max_{p \in \mathbf{R}^{2N}} (\nabla x)^T p - \iota_{\mathcal{B}_{2N}(\beta)}(p), \tag{6.103}
$$

where $\mathcal{B}_{2N}(\beta) = \{p \in \mathbf{R}^{2N} : p = \left(p_1^T, \dots, p_N^T\right)^T, p_i \in B(0, \beta) \subseteq \mathbf{R}^2, i = 1, \dots, N\}$. Consequently, by setting $v^{(k)} = p^{(k)} + \tau\nabla(2x^{(k)} - x^{(k-1)})$, the ith two-dimensional block of $p^{(k+1)}$ is given by the formula:

$$
(p^{(k+1)})_i = \frac{\beta (v^{(k)})_i}{\max\{\beta, \|(v^{(k)})_i\|\}} \; i = 1, \dots, N. \tag{6.104}
$$

The vector $p^{(k+1)}$ of equation (6.101) can also be derived from the Moreau identity (6.10), using $\mathrm{prox}_{\frac{1}{\tau}\psi}\left(\frac{1}{\tau}(p^{(k)} + \tau\nabla(2x^{(k)} - x^{(k-1)}))\right)$ and the shrinkage operator (see equation (6.92)). Furthermore, the parameters of the CP method have to be chosen so that $\sigma\tau\|\nabla\|^2 < 1$.

In the case of image deblurring, different techniques can be used for applying the CP method to the minimization problem (6.83) (see for example [1, 59, 114]). In order to observe the close analogy with ADMM, we apply the CP method as detailed in algorithm 6.9, with the extrapolation step on the dual variable. In the objective function (6.83), we set $F_1(x) = \iota_C(x)$ and $F_2(Ax) = \mathrm{KL}(y; Hx + b) + \psi(Rx)$, with $A = (H^T\ R^T)^T$ $(m = M + N_1)$. The variable p can be partitioned into two compatible sub-vectors $p_1 \in \mathbf{R}^M$ and $p_2 \in \mathbf{R}^{N_1}$. Starting from a suitable $x^{(0)} \in C$ (for example $P_C(y)$) and $p^{(0)} = p^{(-1)} = 0$, the iteration of the CP method is given by

$$\begin{aligned}
x^{(k+1)} &= P_C(x^{(k)} - \sigma(H^T(2p_1^{(k)} - p_1^{(k-1)}) + R^T(2p_2^{(k)} - p_2^{(k-1)}))),\\
p_1^{(k+1)} &= \arg\min_{p_1\in\mathbf{R}^M} \mathrm{KL}^*(y; p_1) + \frac{1}{2\tau}\|p_1 - (p_1^{(k)} + \tau Hx^{(k+1)})\|^2,\\
p_2^{(k+1)} &= \arg\min_{p_2\in\mathbf{R}^S} \psi^*(p_2) + \frac{1}{2\tau}\|p_2 - (p_2^{(k)} + \tau Rx^{(k+1)})\|^2.
\end{aligned}$$

By means of Moreau identity (6.10) and equation (6.91) for the proximity operator of KL, $p_1^{(k+1)}$ can be obtained as follows:

$$\begin{aligned}
u_1^{(k)} &= p_1^{(k)} + \tau Hx^{(k+1)},\\
p_1^{(k+1)} &= u_1^{(k)} - \tau\,\mathrm{prox}_{\frac{1}{\tau}\mathrm{KL}(y;z_1+b)}\left(\frac{1}{\tau}u_1^{(k)}\right)\\
&= u_1^{(k)} + \tau b - \frac{1}{2}\left(u_1^{(k)} + \tau b - 1 + \sqrt{\left(1 - u_1^{(k)} - \tau b\right)^2 + 4\tau y}\right),
\end{aligned} \tag{6.105}$$

where the operations between vectors are intended component-wise. Similarly, $p_2^{(k+1)}$ can be obtained as follows

$$\begin{aligned}
u_2^{(k)} &= p_2^{(k)} + \tau Rx^{(k+1)},\\
p_2^{(k+1)} &= u_2^{(k)} - \tau\,\mathrm{prox}_{\frac{1}{\tau}\psi}\left(\frac{1}{\tau}u_2^{(k)}\right).
\end{aligned} \tag{6.106}$$

We observe that, when $\psi(Rx)$ is the discrete TV regularization of a 2D-image, in view of equation (6.103), any 2D block of $p_2^{(k+1)}$ can be directly obtained as $\left(p_2^{(k+1)}\right)_i = \dfrac{\beta\left(u_2^{(k)}\right)_i}{\max\{\beta, \|\left(u_2^{(k)}\right)_i\|\}}$, $i = 1, \dots, N$.

Simple algebraic computations in equations (6.105) and (6.106) show that, for τ equal to the γ parameter of ADMM, $\frac{1}{\tau}p_1^{(k+1)}$ and $\frac{1}{\tau}p_2^{(k+1)}$ have formally the same expressions of $q_1^{(k+1)}$ and $q_2^{(k+1)}$, computed at steps 8 and 9 of algorithm 6.8. The main difference between the CP iteration and the implementation of ADMM is in the step required for obtaining $x^{(k+1)}$. Indeed, in the CP iteration the solution of the linear system at step 1 of algorithm 6.8 is replaced by a forward step with step-length σ, based on an approximation of the subgradient of $F_2(Ax)$ at the current iterate and on a projection on C that preserves the feasibility of x. Thus, CP method is very inexpensive since, at each iteration, it only requires the computation of matrix–vector products and of $\text{prox}_{\frac{1}{\tau}\psi}(u_2^{(k)})$. Nevertheless, as in the case of ADMM, the performance of the method depends on a suitable choice of the two parameters σ and τ.

Algorithm 6.9. Poisson image deconvolution by CP method.

Choose σ and τ such that $\sigma\tau\|H^T H + R^T R\| < 1$; $p_1^{(0)} = p_1^{(-1)} = 0$ and $p_2^{(0)} = p_2^{(-1)} = 0$; set $x^{(0)} \in C$.

FOR $k = 0, 1, 2, \ldots$ DO THE FOLLOWING STEPS:

STEP 1. compute $x^{(k+1)} = P_C(x^{(k)} - \sigma(H^T(2p_1^{(k)} - p_1^{(k-1)}) + R^T(2p_2^{(k)} - p_2^{(k-1)})))$;

STEP 2. compute $u_1^{(k)} = p_1^{(k)} + \tau H x^{(k+1)}$;

STEP 3. compute $p_1^{(k+1)} = u_1^{(k)} + \tau b - \frac{1}{2}\left(u_1^{(k)} + \tau b - 1 + \sqrt{\left(1 - u_1^{(k)} - \tau b\right)^2 + 4\tau y}\right)$;

STEP 4. compute $u_2^{(k)} = p_2^{(k)} + \tau R x^{(k+1)}$;

STEP 5. compute $p_2^{(k+1)} = u_2^{(k)} - \tau\,\text{prox}_{\frac{1}{\tau}\psi}(\frac{1}{\tau}u_2^{(k)})$;

STEP 6. if a stopping criterion is satisfied then stop, declaring $x^{(k+1)}$ is a stationary point;

6.6.2 PDHG methods, ϵ-subgradient iteration and variable metric

The primal–dual hybrid gradient (PDHG) method with *a priori* selected primal and dual step-length parameters $\{\tau_k\}$ and $\{\sigma_k\}$, is thoroughly investigated in [65, 121]. The basic iteration of PDHG for the solution of problem (6.93) is given by:

$$p^{(k+1)} = \text{prox}_{\tau F_2^*}(p^{(k)} + \tau_k A x^{(k)}), \tag{6.107}$$

$$x^{(k+1)} = \text{prox}_{\sigma F_1}(x^{(k)} - \sigma_k A^T p^{(k+1)}). \tag{6.108}$$

For a suitable choice of the step-length sequences, Zhu and Chan show with an extensive experimentation that, in the case of TV regularization of data affected by Gaussian noise, the method can achieve in practice a very fast convergence [121]. Convergence results of this iterative scheme can be obtained by interpreting the primal step (6.108) as an ϵ-subgradient method:

$$x^{(k+1)} = x^{(k)} - \sigma_k u^{(k)}, \tag{6.109}$$

where $u^{(k)} \in \partial_{\epsilon_k}(F_1 + F_2 \circ A)(x^{(k)})$, $\epsilon_k \geqslant 0$. In the following we show that equation (6.108) is a special case of iteration (6.109), so that the convergence of the scheme can be obtained thanks to the conditions for the convergence of the ϵ-subgradient method, detailed for a general framework in proposition A.19.

Indeed, in view of proposition A.16, from the dual step (6.107), since $x^{(k)} \in \text{dom}\,(F_2 \circ A)$ and $p^{(k+1)} \in \text{dom}\, F_2^*$, it follows that $A^T p^{(k+1)} \in A^T \partial_{\epsilon_2^k} F_2(Ax^{(k)})$ where

$$\epsilon_2^k = F_2(Ax^{(k)}) - (p^{(k+1)})^T Ax^{(k)} + F_2^*(p^{(k+1)}). \tag{6.110}$$

Furthermore, from the last equality and equation (6.107), by setting $\text{diam}(\text{dom}\, F_2^*) = \sup_d \{d = \|p_1 - p_2\|, p_1, p_2 \in \text{dom}\, F_2^*\}$, we have that

$$\begin{aligned}
\epsilon_2^k &\leqslant F_2(Ax^{(k)}) - (p^{(k+1)})^T Ax^{(k)} + F_2^*(p^{(k+1)}) + \frac{1}{2\tau_k}\|p^{(k+1)} - p^{(k)}\|^2 \\
&= \min_p \left(F_2(Ax^{(k)}) - p^T Ax^{(k)} + F_2^*(p) + \frac{1}{2\tau_k}\|p - p^{(k)}\|^2 \right) \\
&\leqslant F_2(Ax^{(k)}) + \min_p (F_2^*(p) - p^T Ax^{(k)}) + \frac{1}{2\tau_k}\text{diam}(\text{dom}\, F_2^*)^2 \\
&= \frac{1}{2\tau_k}\text{diam}(\text{dom}\, F_2^*)^2;
\end{aligned} \tag{6.111}$$

under the assumption $\text{diam}(\text{dom}\, F_2^*) < \infty$ and $\lim_{k\to\infty}\tau_k = \infty$, it is immediate to show that $\lim_{k\to\infty} \epsilon_2^k = 0$. On the other hand, using the inclusion (6.5) due to the primal step (6.108), we have that

$$g_{F_1}^{(k)} = \frac{x^{(k)} - x^{(k+1)}}{\sigma_k} - A^T p^{(k+1)} \in \partial F_1(x^{(k+1)}), \tag{6.112}$$

or

$$x^{(k+1)} = x^{(k)} - \sigma_k (g_{F_1}^{(k)} + A^T p^{(k+1)}). \tag{6.113}$$

From the definition of subgradient, we can write:

$$\begin{aligned}
F_1(x) &\geqslant F_1(x^{(k+1)}) + \left(g_{F_1}^{(k)}\right)^T (x - x^{(k+1)}) \\
&= F_1(x^{(k)}) + \left(g_{F_1}^{(k)}\right)^T (x - x^{(k)}) - \underbrace{(F_1(x^{(k)}) - F_1(x^{(k+1)}) - (g^{(k)})^T (x^{(k)} - x^{(k+1)}))}_{\epsilon_1^k \geqslant 0};
\end{aligned}$$

then we conclude that $g_{F_1}^{(k)} \in \partial_{\epsilon_1^k}(F_1)(x^{(k)})$ with

$$\epsilon_1^k = F_1(x^{(k)}) - F_1(x^{(k+1)}) - \left(g_{F_1}^{(k)}\right)^T (x^{(k)} - x^{(k+1)}) \geqslant 0. \tag{6.114}$$

Consequently, we obtain that the primal step (6.113) (or equation (6.108)) coincides with the ϵ-subgradient iteration (6.109), where $u^{(k)} = g_{F_1}^{(k)} + A^T p^{(k+1)}$; from equations (6.110) and (6.114), $u^{(k)} \in \partial_{\epsilon_1^k}(F_1)(x^{(k)}) + A^T \partial_{\epsilon_2^k} F_2(Ax^{(k)}) \subset \partial_{\epsilon_k}(F_1 + F_2 \circ A)$, with $\epsilon_k = \epsilon_1^k + \epsilon_2^k$ (see proposition A.14). As a consequence of this interpretation of the PDHG–primal implicit iteration as ϵ-subgradient method, in view of proposition A.19, suitable choices of a divergent sequence $\{\tau_k\}$ and a diminishing, divergent series, square summable sequence $\{\sigma_k\}$ enable one to obtain the following convergence result.

Proposition 6.3 [30, *theorem 3.*] *Let the set* X^* *of the solutions of problem* (6.93) *be bounded. Assume* $F_1(x)$ *locally Lipschitz-continuous and* $\mathrm{diam}(\mathrm{dom}\, F_2^*) < \infty$. *Assume* $\{x^{(k)}\}$ *be the sequence generated by iteration* (6.107)–(6.108), *where the following conditions on the primal and dual step-lengths hold*:

$\lim_{k\to\infty} \tau_k = \infty$,
$\sum_{k=0}^{\infty} \sigma_k = \infty$,
$\sum_{k=0}^{\infty} \sigma_k^2 < \infty$,
$\sum_{k=0}^{\infty} \frac{\sigma_k}{\tau_k} < \infty$.

Then, setting $\bar{F}$ *as the minimum of* $F_1(x) + F_2(Ax)$, *we have*:

- *the sequence* $\{x^{(k)}\}$ *is bounded and, consequently, the sequence* $\{g_{F_1}^{(k)}\}$ *is bounded*;
- $\lim_{k\to\infty} \epsilon_k = 0$;
- $\sum_{k=0}^{\infty} \epsilon_k \sigma_k < \infty$;
- $\lim_{k\to\infty} F_1(x^{(k)}) + F_2(Ax^{(k)}) = \bar{F}$ *and the sequence* $\{x^{(k)}\}$ *converges to a solution of problem* (6.93).

As observed in sections 6.5.2 and 6.6.1, in some applications such as image deblurring from Poisson data, the presence of the term $\mathrm{KL}(y; Hx + b)$ requires a reformulation of the problem which consists in including this term as part of F_2. Indeed, if $\mathrm{KL}(y; Hx + b)$ is included in F_1, the proximal operator of this function cannot be exactly computed.

In order to avoid this reformulation, in [22, 30] the authors propose a version of the PDHG method especially tailored for the case of problem (6.93) with $F_1(x) = f(x) + i_\Omega(x)$; Ω is a non-empty convex set and a closed formula for the proximity operator of $f(x)$ does not exist. In particular, in the primal step (6.108) the proximity operator of F_1 at $x^{(k)} - \sigma_k(g_{F_1}^{(k)} + A^T p^{(k+1)})$ is replaced by the proximity operator of $\iota_\Omega(x)$ computed at a similar vector; in addition, instead of the ϵ-subgradient of F_1, the ϵ-subgradient of $f(x)$ at $x^{(k)}$ is used. In the framework of a variable metric, the basic iteration of this primal-explicit version of scaled PDHG, named SPDHG, is given by:

$$p^{(k+1)} = \mathrm{prox}_{\tau F_2^*}(p^{(k)} + \tau_k A x^{(k)}), \tag{6.115}$$

$$x^{(k+1)} = P_{\Omega,\, D_k}(x^{(k)} - \sigma_k D_k^{-1}(g_f^{(k)} + A^T p^{(k+1)})), \tag{6.116}$$

where is a sequence of s.p.d. matrices, $D_k \in \mathcal{D}_L$, $L \geqslant 1$, and $g_f^{(k)} \in \partial_{\epsilon_1^k} f(x^{(k)})$ with $\lim_{k\to\infty} \epsilon_1^k = 0$. As for the PDHG version (6.107)–(6.108), we have that $A^T p^{(k+1)} \in \partial_{\epsilon_2^k} F_2(x^{(k)})$, with ϵ_2^k given by equation (6.110), so that, under the assumptions diam(dom F_2^*) $< \infty$ and $\lim_{k\to\infty}\tau_k = \infty$, we have $\lim_{k\to\infty} \epsilon_2^k = 0$.

Furthermore, by setting $u^{(k)} = g_f^{(k)} + A^T p^{(k+1)}$, we have $u^{(k)} \in \partial_{\epsilon_1^k}(f)(x^{(k)}) + A^T \partial_{\epsilon_2^k} F_2(Ax^{(k)}) \subset \partial_{\epsilon_k}(f + F_2 \circ A)$, with $\epsilon_k = \epsilon_1^k + \epsilon_2^k$ and $\lim_{k\to\infty} \epsilon_k = 0$. In such a way the primal step (6.116) can be interpreted as the basic iteration of the scaled projected ϵ-subgradient method; it follows that, under assumptions similar to those of the primal-implicit PDHG method, in view of proposition A.19, the convergence of the scaled primal explicit scheme is obtained.

Proposition 6.4 [22, 30]. *Let the set X^* of the solutions of problem* (6.93) *with $F_1(x) = f(x) + \iota_\Omega(x)$ be bounded and assume* diam(dom F_2^*) $< \infty$. *Assume that $\{x^{(k)}\}$ is the sequence generated by iteration* (6.115)–(6.116) *and that there exists a sequence $\{g_f(x^{(k)})\}$, $g_f(x^{(k)}) \in \partial_{\epsilon_1^k} f(x^{(k)})$, such that $\|g_f(x^{(k)})\| \leqslant g_f$, for some $g_f > 0$, with ϵ_1^k converging to zero at least fast as $\frac{1}{\tau_k}$. Let $\{D_k\}$ be a sequence of positive definite matrices, $D_k \in \mathcal{D}_{L_k}$, with $L_k^2 = 1 + \zeta_k$, for some non-negative sequence of parameters $\{\zeta_k\}$. Finally, let the step-length sequences $\{\tau_k\}$, $\{\sigma_k\}$ and $\{\zeta_k\}$ satisfy the following conditions:*

$$\sigma_k = \mathcal{O}\left(\frac{1}{k^t}\right), \quad \tau_k = \mathcal{O}(k^t), \quad \zeta_k = \mathcal{O}\left(\frac{1}{k^q}\right) \quad \frac{1}{2} < t \leqslant 1, \ q > 1. \tag{6.117}$$

Then the sequence $\{x^{(k)}\}$ converges to a solution of problem (6.93) *and $\lim_{k\to\infty} f(x^{(k)}) + F_2(x^{(k)}) = \bar{F}$, where $\bar{F}$ is the minimum of $\mathcal{F}_p(x)$.*

A crucial point for the effectiveness of the standard and variable metric PDHG methods is the selection of the sequences $\{\tau_k\}$ and $\{\sigma_k\}$. By borrowing ideas from [33, 72], in [22] the authors introduce a *level algorithm* allowing adaptive computation of a dynamic step-length σ_k in the primal step (6.116) of the primal explicit version of SPDHG iteration. Under assumptions on $\{\tau_k\}$ and $\{D_k\}$ similar to those of proposition 6.4, it is proved that lim $\inf_{k\geqslant 0}(f(x^{(k)}) + F_2(x^{(k)})) = \bar{F}$.

Remark 6.4. *The primal explicit version of the SPDHG method can be considered also as a special case of a generalized scaled FB method, where the forward step consists of the computation of an ϵ-subgradient of $f(x) + F_2(x)$ at the current iterate and the backward or proximal step is a projection on Ω with respect to a variable D_k-norm. Features and convergence properties of FB subgradient methods are discussed in* [13] *while in [22] these schemes are generalized by introducing the use of an ϵ-subgradient in the forward step and a variable metric in the backward or proximal step.*

6.6.2.1 PDHG methods for image reconstruction

Consider the image reconstruction problem with Poisson data (6.83). In order to apply the primal implicit PDHG method (6.107)–(6.108), we can proceed as in the case of CP method. The application of the primal explicit version of the SPDHG method (6.115)–(6.116) to this problem does not require a specific formulation; indeed, we can set $f(x) = \mathrm{KL}(y; Hx + b)$, $\iota_\Omega(x) = \iota_C(x)$ and $F_2(Ax) = \psi(Rx)$, with $A = R$ (here $M = N_1$), so that $g_f^{(k)}$ is the exact gradient of $\mathrm{KL}(y; Hx + b)$ at $x^{(k)}$ ($\epsilon_1^k = 0$ for all $k \geqslant 0$). In algorithm 6.10 the main steps of the method are detailed. A crucial question is the choice of a suitable variable metric. As already discussed for SGP and other scaled FB methods, an efficient choice is inspired by SGM in [86]. The key point of this approach consists in finding a subgradient decomposition of the form

$$u^{(k)} = V(x^{(k)}) - U(x^{(k)}), \tag{6.118}$$

with $V(x^{(k)}) > 0$ and $U(x^{(k)}) \geqslant 0$ for all k and then defining D_k as a diagonal scaling matrix whose diagonal entries are the projection of $(V(x^{(k)}))_i/(x^{(k)})_i$ onto the set $[1/\sqrt{1 + \zeta_k}, \sqrt{1 + \zeta_k}]$. This strategy has the advantage of agreeing with the non-negativity constraints and strongly depends on the form of the subgradient $u^{(k)}$. For a practical implementation of this strategy, we have to find a decomposition of the vector $u^{(k)} = \nabla\mathrm{KL}(y; Hx^{(k)} + b) + A^T p^{(k+1)}$ as the difference of two non-negative terms. As concerns the first term, the gradient of f has the natural decomposition $\nabla f(x) = H^T\mathbf{1} - H^T\frac{y}{Hx^{(k)} + b}$, where the operator / between vectors is intended component-wise; under standard assumptions on H, we have $H^T\mathbf{1} > 0$ and $H^T\frac{y}{Hx^{(k)} + b} \geqslant 0$ for all $x \geqslant 0$.

The decomposition of the vector $A^T p^{(k+1)}$ depends on the choice of the regularization term. In [22], the authors indicate a procedure for obtaining a decomposition of TV regularization of a $\sqrt{N} \times \sqrt{N}$ image, where $\psi(Rx) = \beta\sum_{i=1}^{N}\|\nabla_i x\|$ and $R = \nabla$. In the following we summarize the main steps of the procedure. If the dual variable is partitioned as $p = \left(p_1^T \;\; p_2^T \;\; \cdots \;\; p_N^T\right)^T$, $p_i \in \mathbf{R}^2$, the dual step (6.115) can be written as follows

$$\begin{aligned} \bar{p}^{(k)} &= p^{(k)} + \tau_k \nabla x^{(k)}, \\ p^{(k+1)} &= S_k \bar{p}^{(k)}, \end{aligned}$$

where S_k is a diagonal $2N \times 2N$ matrix with the following diagonal entries

$$(S_k)_{2\ell-1,\, 2\ell-1} = (S_k)_{2\ell,\, 2\ell} = s_\ell^{(k)} = \frac{\beta}{\max\{\beta, \|(\bar{p}^{(k)})_i\|\}}, \quad \ell = 1, \ldots, N. \tag{6.119}$$

If the method is initialized with $p^{(0)} = 0$, the dual variable can be written as

$$\begin{aligned}
p^{(0)} &= 0,\\
p^{(1)} &= \tau_0 S_0 \nabla x^{(0)},\\
p^{(2)} &= S_1(\tau_0 S_0 \nabla x^{(0)} + \tau_1 \nabla x^{(1)}),\\
p^{(3)} &= S_2(\tau_0 S_1 S_0 \nabla x^{(0)} + \tau_1 S_1 \nabla x^{(1)} + \tau_2 \nabla x^{(2)}),\\
&\vdots\\
p^{(k+1)} &= \sum_{j=0}^{k} \tau_j S_j^k \nabla x^{(j)},
\end{aligned}$$

where we set $S_j^k = S_k S_{k-1} \dots S_j = \prod_{i=j}^k S_i$. As a consequence, the ϵ-subgradient of $F_2 \circ \nabla$ used in equation (6.115) can be expressed as

$$\nabla^T p^{(k+1)} = \sum_{j=0}^{k} \tau_j \nabla^T S_j^k \nabla x^{(j)}. \tag{6.120}$$

We observe that, by dropping the indices j and k, each term in the summation at the right-hand side of equation (6.120) has the form $\nabla^T S \nabla$, where S is a diagonal matrix of order $2N$ with positive entries such that $S_{2\ell,\,2\ell} = S_{2\ell-1,2\ell-1} = s_\ell$, $\ell = 1, \cdots, N$. Given $x \geqslant 0$, this term can be decomposed as follows

$$\nabla^T S \nabla x = V_S x - U_S x$$

where

$$\begin{aligned}
(V_S x)_{i,j} &= (2s_{i,j} + s_{i,j-1} + s_{i-1,j}) x_{i,j} \geqslant 0\\
(U_S x)_{i,j} &= s_{i,j}(x_{i+1,j} + x_{i,j+1}) + s_{i,j-1} x_{i,j-1} + s_{i-1,j} x_{i-1,j} \geqslant 0
\end{aligned}$$

with the correspondence $s_\ell \equiv s_{i,j}$, $j = \lfloor(\ell - 1)/\sqrt{N}\rfloor + 1, i = \ell - \lfloor(\ell - 1)/\sqrt{N}\rfloor \cdot \sqrt{N}$. In such a way, the positive part of $\nabla^T p^{(k+1)}$ is $\sum_{j=0}^k \tau_j V_{S_j^k} x^{(j)}$; then $V(x^{(k)})$ of equation (6.118) can be written as follows

$$V(x^{(k)}) = H^T \mathbf{1} + \sum_{j=0}^{k} \tau_j V_{S_j^k} x^{(j)}. \tag{6.121}$$

Even if it looks quite complex, the term $\sum_{j=0}^k \tau_j V_{S_j^k} x^{(j)}$ can be easily computed in a recursive way, by introducing three auxiliary vectors. Setting $w^{(-1)} = \bar{w}^{(-1)} = \tilde{w}^{(-1)} = 0$, for any kth iteration, $k \geqslant 0$, the three vectors have to be updated by the following rule:

$$w_{i,j}^{(k)} = (w_{i,j}^{(k-1)} + \tau_k x_{i,j}^{(k)}) s_{i,j}^{(k)}, \tag{6.122}$$

$$\bar{w}_{i,j}^{(k)} = (\bar{w}_{i,j}^{(k-1)} + \tau_k x_{i,j}^{(k)}) s_{i-1,j}^{(k)}, \tag{6.123}$$

$$\tilde{w}_{i,j}^{(k)} = (\tilde{w}_{i,j}^{(k-1)} + \tau_k x_{i,j}^{(k)}) s_{i,j-1}^{(k)}, \tag{6.124}$$

where $s_{i,j}^{(k)}$ is given in equation (6.119) and $i, j = 1, \dots, \sqrt{N}$. Finally we have

$$V(x^{(k)}) = H^T\mathbf{1} + w^{(k)} + 2w^{(k)} + \bar{w}^{(k)} + \tilde{w}^{(k)}$$

and, consequently,

$$(D_k)_{i,i} = \max\left\{\frac{1}{L_k}, \min\left\{L_k, \frac{(V(x^{(k)}))_i}{x_i^{(k)}}\right\}\right\}. \tag{6.125}$$

Algorithm 6.10. Poisson image deconvolution by SPDHG method—primal explicit version.

Choose positive sequences $\{\sigma_k\}$ and $\{\tau_k\}$, $\sigma_k = \mathcal{O}\left(\frac{1}{k^t}\right)$, $\tau_k = \mathcal{O}(k^t)$, $\frac{1}{2} < t \leqslant 1$; define a sequence of (diagonal) scaling matrices $\{D_k\}$, with $D_k \in \mathcal{D}_{L_k}$, $L_k^2 = 1 + \zeta_k$, $\zeta_k \geqslant 0$, $\zeta_k = \mathcal{O}\left(\frac{1}{k^q}\right)$, $q > 1$; set $p(0) = 0$ and $x^{(0)} \in C$.

FOR $k = 0, 1, 2, \dots$ DO THE FOLLOWING STEPS:

STEP 1. compute $p^{(k+1)} = \text{prox}_{\tau\psi^*}(p^{(k)} + \tau_k R x^{(k)})$;

STEP 2. compute $u^{(k)} = \nabla \text{KL}(y; Hx^{(k)} + b) + R p^{(k+1)}$;

STEP 3. compute $x^{(k+1)} = P_{C,D_k}(x^{(k)} - \sigma_k D_k^{-1} u^{(k)})$;

STEP 4. if a stopping criterion is satisfied then stop, declaring $x^{(k+1)}$ is a stationary point;

6.6.3 An alternating extragradient method (AEM) for TV image reconstruction

The computation of a saddle point of $\mathcal{L}(x, p)$ in problem (6.94) can be viewed also as a special case of a monotone variational inequality problem, where we have to find $(x^*, p^*) \in \text{dom } F_1 \times \text{dom } F_2^*$ such that

$$\begin{pmatrix} \partial_x \mathcal{L}(x^*, p^*) \\ -\partial_p \mathcal{L}(x^*, p^*) \end{pmatrix}^T \begin{pmatrix} x - x^* \\ p - p^* \end{pmatrix} \geqslant 0 \quad \forall (x, p) \in \text{dom } F_1 \times \text{dom } F_2^* \subset \mathbf{R}^N \times \mathbf{R}^M. \tag{6.126}$$

When $F_1(x) = f(x) + \iota_\Omega(x)$, with $f(x)$ continuously differentiable on its domain and $\Omega \subset \text{dom} f$, $F_2^*(p) = \iota_{\bar{\Omega}}(p)$, with Ω and $\bar{\Omega}$ non-empty, closed and convex subsets, the saddle point problem can be formulated as follows

$$\min_{x \in \Omega} \max_{p \in \bar{\Omega}} f(x) + p^T A x \tag{6.127}$$

and the corresponding variational inequality assumes this simple form:

$$\begin{pmatrix} \nabla f(x^*) + A^T p^* \\ -Ax^* \end{pmatrix}^T \begin{pmatrix} x - x^* \\ p - p^* \end{pmatrix} \geq 0 \quad \forall (x, p) \in C \times \bar{\Omega}. \tag{6.128}$$

Problems with these features arise from TV and HS regularization, where $f(x)$ is a data-fidelity term, for example $f_0(x; y) = \mathrm{KL}(y; Hx + b)$. Indeed, the primal–dual formulation of problem (6.83), with a TV or HS regularization, can be written as

$$\min_{x \in C} \max_{p \in \bar{C}} \mathrm{KL}(y; Hx + b) + p^T z(x), \tag{6.129}$$

where C is the non-negative orthant and the subset $\bar{C}$ and the term $z(x)$ depend on the regularization function. For the TV regularization $\beta \sum_{i=1}^N \|\nabla_i x\|$ of a 2D-image, we have that $\bar{C}$ is $\mathcal{B}_{2N}(\beta) = \bigcup_{i=1}^{N} (B(0, \beta))_i \subset \mathbf{R}^{2N}$, i.e. the union of the set of N balls centered at the origin of $\mathbf{R}^2$ with radius β, and $z(x) \equiv \nabla x \in \mathbf{R}^{2N}$.

When the regularization term is the HS regularization on a 2D-image, i.e.

$$\beta \sum_{i=1}^{N} \left(\left\| \begin{matrix} \nabla_i x \\ \delta \end{matrix} \right\| - \delta \right),$$

with $\delta > 0$, we have that $\bar{C}$ is $\mathcal{B}_{3N}(\beta) = \bigcup_{i=1}^{N} (B(0, \beta))_i \subset \mathbf{R}^{3N}$, i.e. the union of the set of N balls centered at the origin of $\mathbf{R}^3$ with radius β and $z(x) \equiv ((\nabla x)^T, \delta 1^T)^T \in \mathbf{R}^{3N}$. For both regularization terms, the saddle point problem (6.129) is related to a smooth convex–concave objective function.

Thanks to the equivalence between the problems (6.127) and (6.128), in [29] the authors propose an alternating extragradient method (AEM) especially tailored for general smooth saddle point problems as

$$\min_{x \in \Omega} \max_{p \in \bar{\Omega}} F(x, p), \tag{6.130}$$

where $F(x, p)$: $\mathbf{R}^n \times \mathbf{R}^m \to \bar{\mathbf{R}}$ is a continuously differentiable convex–concave function; for example, $F(x, p) = f(x) + p^T Ax$ in equation (6.127).

The method is a generalization of [2] and it allows for an adaptive choice of the step-length which does not make explicit use of the Lipschitz constant of the gradient, but it is based on its local approximation, as in the Khobotov's method [82, 91]. The basic iteration is given by the following formulas

$$\bar{p}^{(k)} = P_{\bar{\Omega}}(p^{(k)} + \alpha_k \nabla_p F(x^{(k)}, p^{(k)})) \tag{6.131}$$

$$x^{(k+1)} = P_{\Omega}(x^{(k)} - \alpha_k \nabla_x F(x^{(k)}, \bar{p}^{(k)})) \tag{6.132}$$

$$p^{(k+1)} = P_{\bar{\Omega}}(p^{(k)} + \alpha_k \nabla_p F(x^{(k+1)}, p^{(k)})), \tag{6.133}$$

where the step-length α_k is chosen in a bounded interval $[\alpha_{\min}, \alpha_{\max}]$, with $0 < \alpha_{\min} < \alpha_{\max}$, so that the following conditions are satisfied

$$\begin{cases} 1 - 2\alpha_k a_k - 2\alpha_k^2 b_k^2 & \geqslant \epsilon \\ 1 - 2\alpha_k c_k & \geqslant \epsilon \end{cases} \tag{6.134}$$

where $0 < \epsilon < 1$ is constant and

$$\begin{aligned} a_k &= \frac{\|\nabla_x F(x^{(k+1)}, \bar{p}^{(k)}) - \nabla_x F(x^{(k)}, \bar{p}^{(k)})\|}{\|x^{(k+1)} - x^{(k)}\|}, \\ b_k &= \frac{\|\nabla_p F(x^{(k+1)}, p^{(k)}) - \nabla_p F(x^{(k)}, p^{(k)})\|}{\|x^{(k+1)} - x^{(k)}\|}, \\ c_k &= \frac{\|\nabla_p F(x^{(k+1)}, p^{(k)}) - \nabla_p F(x^{(k+1)}, \bar{p}^{(k)})\|}{\|p^{(k)} - \bar{p}^{(k)}\|}. \end{aligned} \tag{6.135}$$

If there exists a saddle point for the convex–concave smooth function F on the closed, convex set $\Omega \times \bar{\Omega}$, the sequence $\{(x^{(k)}, p^{(k)})\}$ generated by the algorithm (6.131)–(6.133), with $\alpha_k \in [\alpha_{\min}, \alpha_{\max}]$ such that equation (6.134) holds, converges to a saddle point of F in $\Omega \times \bar{\Omega}$ ([29], theorem 1).

Therefore, in order to have the well-definiteness and the convergence of the sequence generated by equation (6.131)–(6.133) to a saddle point of F, for all $k \geqslant 0$ a value α_k, bounded away from zero and satisfying (6.134), must exist. A sufficient condition assuring that, for any fixed $\epsilon \in (0, 1)$, there exists a positive number $\bar{\alpha}_\epsilon > 0$ such that for all $\alpha \in [0, \bar{\alpha}_\epsilon]$ the inequalities (6.134) hold for all $k \geqslant 0$, is the local Lipschitz-continuity of the gradients of F. More precisely, it is required that for every compact subset $\Upsilon \subset \Omega \times \bar{\Omega}$ there exists a positive constant M_Υ such that

$$\begin{aligned} \|\nabla_x F(x, p) - \nabla_x F(\bar{x}, p)\| &\leqslant M_\Upsilon \|x - \bar{x}\| \\ \|\nabla_p F(x, p) - \nabla_p F(\bar{x}, p)\| &\leqslant M_\Upsilon \|x - \bar{x}\| \\ \|\nabla_p F(x, p) - \nabla_p F(x, \bar{p})\| &\leqslant M_\Upsilon \|p - \bar{p}\| \end{aligned} \tag{6.136}$$

for all $(x, p), (\bar{x}, p), (x, \bar{p}) \in \Upsilon$.

Algorithm 6.11. Alternating extragradient method (AEM).

Choose the starting point $(x^{(0)}, p^{(0)}) \in \Omega \times \bar{\Omega}$, set the parameters ρ, $\epsilon \in (0, 1)$, $\alpha_{\max} > 0$.

FOR $k = 0, 1, 2, \ldots$ DO THE FOLLOWING STEPS:

STEP 1. choose $\alpha \leqslant \alpha_{\max}$;

STEP 2. compute tentative points

$$\bar{p}^{+} = P_{\bar{\Omega}}(p^{(k)} + \alpha \nabla_p F(x^{(k)}, p^{(k)})),$$
$$x^{+} = P_{\Omega}(x^{(k)} - \alpha \nabla_x F(x^{(k)}, \bar{p}^{+}))$$

STEP 3. compute

$$a = \frac{\| \nabla_x F(x^{+}, \bar{p}^{+}) - \nabla_x F(x^{(k)}, \bar{p}^{+}) \|}{\| x^{+} - x^{(k)} \|},$$
$$b = \frac{\| \nabla_p F(x^{+}, p^{(k)}) - \nabla_p F(x^{(k)}, p^{(k)}) \|}{\| x^{+} - x^{(k)} \|},$$
$$c = \frac{\| \nabla_p F(x^{+}, p^{(k)}) - \nabla_p F(x^{+}, \bar{p}^{(k)}) \|}{\| p^{(k)} - \bar{p}^{+} \|}$$

STEP 4. compute

$$\bar{\alpha} = \begin{cases} \min\left\{ \frac{\sqrt{a^2 + 2b^2(1-\epsilon)} - a}{2b^2}, \frac{1-\epsilon}{2c} \right\} & \text{if } b > 0,\ c > 0 \\ \min\left\{ \frac{1-\epsilon}{2a}, \frac{1-\epsilon}{2c} \right\} & \text{if } a > 0,\ c > 0,\ b = 0 \\ \frac{\sqrt{a^2 + 2b^2(1-\epsilon)} - a}{2c^2} & \text{if } b > 0,\ c = 0 \\ \frac{1-\epsilon}{2c} & \text{if } a = 0,\ c > 0,\ b = 0 \\ \frac{1-\epsilon}{2a} & \text{if } a > 0,\ c = 0,\ b = 0 \\ \alpha & \text{otherwise} \end{cases}$$

STEP 5. Check convergence condition:

IF $\alpha \leqslant \bar{\alpha}$ THEN

$\alpha_k = \alpha$;
$\bar{p}^{(k)} = \bar{p}^{+}$;
$x^{(k+1)} = x^{+}$;

ELSE

$\alpha = \min(\bar{\alpha}, \rho\alpha)$;
go to STEP 2;

ENDIF

STEP 6. compute $p^{(k+1)} = P_{\bar{\Omega}}(p^{(k)} + \alpha_k \nabla_p F(x^{(k+1)}, p^{(k)}))$;

STEP 7. if a stopping criterion is satisfied then stop, declaring $x^{(k+1)}$ is a stationary point;

A practical implementation of the method (6.131)–(6.133) is given in algorithm 6.11. The method can also be written by exchanging the role of the primal and dual variables x, p. We remark that AEM consists of successive gradient ascent (6.131) and descent (6.132) steps followed by an extragradient step (6.133).

Under suitable assumptions on the gradient of F, the convergence condition $\alpha \leqslant \bar{\alpha}$ at step 5 is algorithmically equivalent to conditions (6.134), since the loop between step 2 and step 5 terminates in a finite number of trials. Indeed, for a given

$\epsilon > 0$, since α is reduced each time at least by a fixed amount ρ, after a finite number of steps, its value will be smaller than $\bar{\alpha}_\epsilon$ so that the convergence condition (6.134) is fulfilled. This also implies that the sequence $\{\alpha_k\}$, generated by algorithm 6.11, is bounded away from zero with $\alpha_{\min} \geqslant \rho\bar{\alpha}_\epsilon$.

Numerical experience shows that the sequence $\{\alpha_k\}$ generated by algorithm 6.11 tends to approach a fixed value. For this reason, at step 1, it is convenient to adopt a quite conservative choice for the tentative value of α at the iterate $k \geqslant 1$, as the smallest value between $\alpha_{\max}$ and the mean value value of the last M values of α

$$\alpha \leftarrow \min\left\{\frac{1}{\min(k, M)} \sum_{j=1}^{\min(k,M)} \alpha_{k-j}, \quad \alpha_{\max}\right\} \tag{6.137}$$

where M is a fixed integer. With this choice, a reduction of the step-length for fulfilling the convergence condition is not frequently required.

Now we consider the discrete primal–dual version (6.129) of the objective function (6.83), using respectively HS and TV regularization. The function $F(x, p) = \mathrm{KL}(y; Hx + b) + p^T z(x)$ is convex with respect to x and concave with respect to p. The set Ω is the non-negative orthant C and the domain $\bar{\Omega} = \bar{C} = \mathcal{B}_{3N}(\beta)$ or $\bar{\Omega} = \bar{C} = \mathcal{B}_{2N}(\beta)$ is a compact set. Furthermore, under the standard assumption (3.31) on H, $\mathbf{1}$ does not belong to the null space of H and, consequently, the objective function of equation (6.83) is coercive and there exists a minimum value. Then, the strong duality theorem A.24 (see appendix A) ensures the existence of a solution of problem (6.129).

We observe that for $F(x, p) = \mathrm{KL}(y; Hx + b) + p^T z(x)$ we have

$$\begin{aligned} \nabla_x F(x, p) &= H^T 1 - H^T \frac{y}{Hx + b} + \nabla^T \tilde{p} \\ \nabla_p F(x, p) &= z(x) \end{aligned}$$

where again quotient of vectors is intended component-wise, $\tilde{p} \equiv p$ in the TV model while $\tilde{p}$ represents the first $2N$ entries of the dual variable p in the HS model. Since $\nabla_p F(x, p) - \nabla_p F(x, \bar{p}) = 0$ for all $x \in C$, $p, \bar{p} \in \bar{C}$, we have $c_k = 0$ for all k. The Lipschitz constant of $\nabla_p F(\cdot, p)$ is equal to the norm of the matrix ∇, while the Lipschitz-continuity of $\nabla_x F(\cdot, p)$ holds in every compact subset of the domain of the KL divergence. Then, AEM is well-defined and generates a sequence of vectors convergent to a saddle point of the function in (6.129).

As concerns the computational complexity of AEM, we observe that any iteration requires matrix–vector products, one projection on Ω and two projections on $\bar{\Omega}$. When a backtracking procedure is started, each step requires additional matrix–vector products and two further projections on Ω and $\bar{\Omega}$, respectively. A Matlab® version of the algorithm 6.11 for smoothed and unsmoothed total variation functional (6.83) is downloadable from http://www.unife.it/prin/software.

6.7 Majorization–minimization approach

The majorization–minimization (MM) approach is a methodology frequently used in inverse problems applications for the design of suitable iterative methods. The literature is very wide. We only provide a few references concerning medical imaging, both in the case of Poisson data [55, 85] and in the case of least-squares approximation [56]. A very successful application is that proposed by Daubechies, Defrise and De Mol [53] for Tikhonov regularization with a sparsity constraint. For a tutorial on MM methods see [81].

The MM approach is also a very general strategy in numerical optimization, based on the *optimization transfer principle*. MM techniques represent a class of iterative methods, whose basic idea is to substitute at each iteration $k \geqslant 0$ the general objective function $F(x)$ to be minimized by a *surrogate function* $\phi^{(k)}(x)$, usually dependent on the current iterate $x^{(k)}$ (for example $\phi^{(k)}(x) = \phi(x, x^{(k)})$). Figure 6.4 shows an example of $F(x)$ and a surrogate function at $x^{(k)}$ and $x^{(k+1)}$. This function must be such that its minimization is easy; moreover, it must satisfy the *majorization or surrogate conditions*:

$$\begin{aligned} \phi^{(k)}(x^{(k)}) &= F(x^{(k)}), \\ \phi^{(k)}(x) &\geqslant F(x) \quad \text{for } x \in \operatorname{dom} F. \end{aligned} \tag{6.138}$$

In the presence of a constrained minimization problem, the conditions (6.138) have to be satisfied for any x in the feasible region Ω. When F and $\phi^{(k)}$ are differentiable, it follows that this additional condition holds:

$$\nabla F(x^{(k)}) = \nabla \phi^{(k)}(x^{(k)}). \tag{6.139}$$

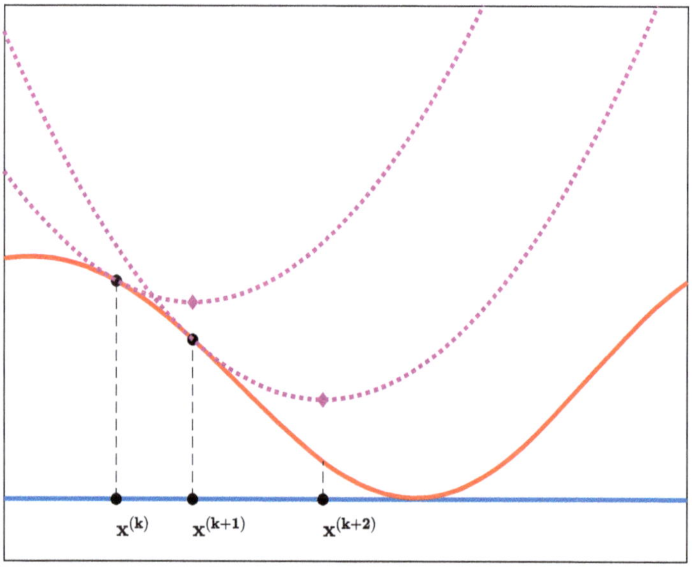

Figure 6.4. An example of surrogate function for $F(x)$ (red line) at $x^{(k)}$ and $x^{(k+1)}$. We observe that $x^{(k+1)} = \arg\min \phi^{(k)}(x)$ and $x^{(k+2)} = \arg\min \phi^{(k+1)}(x)$.

Following the principle described in [100] in the context of the line-search techniques, a MM scheme consists in choosing a suitable $x^{(0)}$ and alternating at each kth iteration the following two steps:

- choose a surrogate function $\phi^{(k)}(x)$;
- set $x^{(k+1)} \in \arg\min \phi^{(k)}(x)$.

As a consequence of equation (6.138), it is immediately found that the update rule for $x^{(k+1)}$ will monotonically decrease F, also when $x^{(k+1)}$ is not the exact minimizer of $\phi^{(k)}(x)$, but satisfies the condition $\phi^{(k)}(x^{(k+1)}) \leqslant \phi^{(k)}(x^{(k)})$.

Remark 6.5. *The FB methods* (6.12) *in section* 6.2 *(without relaxation) could be considered as special instances of MM methods. Indeed, in view of the descent lemma A.3, when* $F(x) = \Phi(x) + \Psi(x)$ *is a convex function and* $\Phi(x)$ *is differentiable with gradient* M_Φ*-Lipschitz-continuous, the function* $\Phi(x^{(k)}) + \nabla\Phi(x^{(k)})^T(x - x^{(k)}) + \frac{1}{2\alpha_k}(x - x^{(k)})^T D_k(x - x^{(k)}) + \Psi(x)$ *can be viewed as a surrogate function* $\phi^{(k)}(x)$ *for* $F(x)$ *with* $\alpha_k < \frac{1}{LM_\Phi}$ *and* $D_k \in \mathcal{D}_L$, $L \geqslant 1$; *then, the proximity forward step is precisely the minimization of this function. The introduction of a greater flexibility in FB methods can lead to differentiate these schemes from the MM approach.*

For the data-fidelity function $f_0(x; y) = \mathrm{KL}(y; Hx)$ (i.e. $b = 0$), which is a sum of terms depending on the scalar product of a row of the matrix H with x, a surrogate function can be introduced by considering each term separately. In such a way the EM–RL method can be considered as a MM method [54].

Indeed, let us consider the part of $f_0(x; y)$ which depends on the variable x, neglecting the other terms, irrelevant for the minimization; we can write:

$$
\begin{aligned}
F(x) &= \sum_{i=1}^{N}\left(y_i\left(-\ln\left(\sum_{j=1}^{N} H_{i,j}x_j\right)\right) + (Hx)_i\right) \\
&= \sum_{i=1}^{N}\left(y_i\left(-\ln\left(\sum_{j=1}^{N} H_{i,j}x_j^{(k)}\frac{x_j}{x_j^{(k)}}\frac{(Hx^{(k)})_i}{(Hx^{(k)})_i}\right)\right) + (Hx)_i\right).
\end{aligned}
\tag{6.140}
$$

Since $H_{i,j} \geqslant 0$, $\frac{\sum_{j=1}^{N} H_{i,j}x_j^{(k)}}{\left(Hx^{(k)}\right)_i} = 1$ and $-\ln(x)$ is a convex function, by applying Jensen's inequality, we obtain

$$
F(x) \leqslant \sum_{i=1}^{N}\left(y_i\sum_{j=1}^{N}\frac{H_{i,j}x_j^{(k)}}{(Hx^{(k)})_i}\left(-\ln\left(\frac{x_j}{x_j^{(k)}}(Hx^{(k)})_i\right)\right) + (Hx)_i\right). \tag{6.141}
$$

The right-hand side can be viewed as a separable convex surrogate function for $F(x)$

$$\phi(x, x^{(k)}) = \sum_{i=1}^{N}\sum_{j=1}^{N}\phi_{i,j}\left(\frac{x_j}{x_j^{(k)}}\right), \tag{6.142}$$

with $\phi_{i,j}(z_j) = y_i\frac{H_{i,j}x_j^{(k)}}{(Hx^{(k)})_i}\left(-\ln(z_j(Hx^{(k)})_i)\right) + H_{i,j}x_j^{(k)}z_j$, $i, j = 1, \dots, N$. The separability of $\phi(x, x^{(k)})$ leads to the multiplicative update:

$$x_j^{(k+1)} = x_j^{(k)}\arg\min_{z_j}\sum_{i=1}^{N}\phi_{i,j}(z_j); \tag{6.143}$$

by computing the minimum one re-obtains the EM–RL method, possibly a very natural way for deriving this algorithm.

The same technique can be used for any convex problem with non-negativity constraints, leading to an iterative scheme with a multiplicative update for x.

Remark 6.6. *From the explicit expression of jth entry of the gradient of* $\phi(x, x^{(k)})$

$$\frac{\partial\phi(x, x^{(k)})}{\partial x_j} = \sum_{i=1}^{N}\left(H_{i,j} - y_i\frac{H_{i,j}x_j^{(k)}}{(Hx^{(k)})_i}\frac{1}{x_j}\right), \tag{6.144}$$

it follows that the Hessian matrix of the surrogate function $\phi(x, x^{(k)})$ *is a diagonal matrix, whose diagonal entries are*

$$\frac{\partial^2\phi(x, x^{(k)})}{\partial x_j^2} = \sum_{i=1}^{N}y_i\frac{H_{i,j}x_j^{(k)}}{(Hx^{(k)})_i}\frac{1}{x_j^2}. \tag{6.145}$$

In view of equation (6.144), *from the first order optimality conditions at* $x^{(k+1)}$ *for the surrogate function* ($\nabla\phi(x^{(k+1)}, x^{(k)}) = 0$), *we have*

$$\sum_{i=1}^{N}H_{i,j} = \sum_{i=1}^{n}y_i\frac{H_{i,j}x_j^{(k)}}{(Hx^{(k)})_i}\frac{1}{x_j^{(k+1)}}, \quad j = 1, \dots, N.$$

Then, by substituting this equality into $\frac{\partial^2\phi(x^{(k+1)}, x^{(k)})}{\partial x_j^2}$ *we obtain*

$$\left(\nabla^2\phi\left(x^{(k+1)}, x^{(k)}\right)\right)_{j,j} = \left(\frac{(H^T\mathbf{1})_j}{x_j^{(k+1)}}\right), \quad j = 1, \dots, N. \tag{6.146}$$

This equality can provide a motivation for the selection of the scaling matrix D_{k+1} *in SGP method (see equation* (6.19)) *when we have to minimize the data-fidelity function* $f_0(x; y)$; *indeed, by neglecting the bounds imposed by the assumptions required for the* *convergence of the method, D_{k+1} can be interpreted as the Hessian of a surrogate function for the KL divergence.*

The MM method can be very useful also for the computation of MAP estimate by solving problem (4.29), because its application is very simple in the presence of a separable regularization term (as T0 or the ℓ_1 norm of x). Indeed, the surrogate function can be obtained by combining the surrogate function of the data-fidelity function $f_0(x; y) = \mathrm{KL}(y; Hx)$ with the penalty term. The computation of the minimum point of $\phi(x, x^{(k)})$ is obtained by a closed formula.

For non-separable regularization terms, there exist different strategies. In the following we report the techniques proposed in [54] and [57].

First, we consider a penalty term of the form

$$\sum_{i=1}^{t} \psi_i((Rx)_i) \tag{6.147}$$

where R is a $t \times N$ matrix and ψ_i is a strictly convex function, twice continuously differentiable and bounded from below; an example can be $\psi_i(z) = \frac{z^2}{2}$ for any $i = 1, \dots, t$ and R a discrete differential operator. In [54] a separable surrogate function for the function (6.147) is devised using a trick similar to the one used to obtain equation (6.142). Consider $\alpha_{i,j}$, $j = 1, \dots, N$, $i = 1, \dots, t$, non-negative real numbers such that $\sum_{j=1}^{N} \alpha_{i,j} = 1$, with $\alpha_{i,j} = 0$ when $R_{i,j} = 0$. We can write

$$\psi_i((Rx)_i) = \psi_i\left(\sum_{j=1;\alpha_{i,j}\neq 0}^{N} \alpha_{i,j}\left(\frac{R_{i,j}x_j}{\alpha_{i,j}} + (Rx^{(k)})_i - \frac{R_{i,j}x_j^{(k)}}{\alpha_{i,j}}\right)\right). \tag{6.148}$$

As before, in view of the convexity of ψ_i, using Jensen's inequality, we obtain

$$\sum_{i=1}^{t} \psi_i((Rx)_i) \leqslant \sum_{i=1}^{t} \sum_{j=1;\alpha_{i,j}\neq 0}^{N} \alpha_{i,j}\psi_i\left(\frac{R_{i,j}x_j}{\alpha_{i,j}} + (Rx^{(k)})_i - \frac{R_{i,j}x_j^{(k)}}{\alpha_{i,j}}\right). \tag{6.149}$$

The right-hand side of the last inequality can be viewed as a surrogate function

$$\phi(x, x^{(k)}) = \sum_{j=1}^{N}\left(\sum_{i=1;\alpha_{i,j}\neq 0}^{t} \alpha_{i,j}\psi_i\left(\frac{R_{i,j}x_j}{\alpha_{i,j}} + (Rx^{(k)})_i - \frac{R_{i,j}x_j^{(k)}}{\alpha_{i,j}}\right)\right)$$

for equation (6.147) and any term of the external sum depends only on x_j, so that

$$\frac{\partial\phi(x, x^{(k)})}{\partial x_j} = \sum_{i=1;\alpha_{i,j}\neq 0}^{t} R_{i,j}\frac{\partial\psi_i\left(\frac{R_{i,j}x_j}{\alpha_{i,j}} + (Rx^{(k)})_i - \frac{R_{i,j}x_j^{(k)}}{\alpha_{i,j}}\right)}{\partial x_j}. \tag{6.150}$$

By combining this surrogate function with a separable surrogate function for the data-fidelity term, it is possible to find a simple updating rule for $x^{(k+1)}$.

A further technique is proposed in [57] to devise a surrogate cost function for the differentiable discrete approximation of the TV penalty term, whose general expression can be written as

$$f_1(x) = \sum_{i=1}^{N} \sqrt{\delta^2 + \sum_{j \in S_i} (x_i - x_j)^2}, \tag{6.151}$$

where $\delta > 0$ is a small constant parameter and $S_i \subseteq \{1, \dots, N\}$ denotes the indices of some neighborhood of pixel/voxel i. For example, in the case of a square 2D image of N pixels, ordered column-wise, the HS regularization function can be interpreted as a special case of equation (6.151) for $S_i = \{i + 1, i + \sqrt{N}\}$.

Following [56], we introduce a surrogate function for $f_1(x)$ at $x^{(k)}$. We set

$$\Delta(x_i) = \sqrt{\delta^2 + \sum_{j \in S_i} (x_i - x_j)^2},$$

and we define the following function

$$\phi(x, x^{(k)}) = \sum_{i=1}^{N} \frac{\delta^2 + \sum_{j \in S_i} \left((x_i - x_j)(x_i^{(k)} - x_j^{(k)}) + \left(x_j^{(k)} - x_j\right)^2 + (x_i^{(k)} - x_i)^2 \right)}{\Delta(x_i^{(k)})}. \tag{6.152}$$

We can immediately observe that $f_1(x^{(k)}) = \phi(x^{(k)}, x^{(k)})$, i.e. the first one of the conditions (6.138) is verified.

As concerns the second one, we observe that

$$\begin{aligned} \phi(x, x^{(k)}) - f_1(x) = \sum_{i=1}^{N} \frac{1}{\Delta(x_i^{(k)})} \Bigg\{ \Bigg(\delta^2 &+ \sum_{j \in S_i} (x_i - x_j)(x_i^{(k)} - x_j^{(k)}) \\ &+ \left(x_j^{(k)} - x_j\right)^2 + (x_i^{(k)} - x_i)^2 \Bigg) \\ &- \sqrt{\delta^2 + \sum_{j \in S_i} (x_i - x_j)^2} \sqrt{\delta^2 + \sum_{j \in S_i} \left(x_i^{(k)} - x_j^{(k)}\right)^2} \Bigg\}; \end{aligned} \tag{6.153}$$

now, using the following inequality in the previous equation

$$
\begin{aligned}
&-\sqrt{\delta^2 + \sum_{j\in S_i}\left(x_i - x_j\right)^2}\sqrt{\delta^2 + \sum_{j\in S_i}\left(x_i^{(k)} - x_j^{(k)}\right)^2} \\
&\qquad \geqslant -\frac{1}{2}\left(\delta^2 + \sum_{j\in S_i}\left(x_i - x_j\right)^2 + \delta^2 + \sum_{j\in S_i}\left(x_i^{(k)} - x_j^{(k)}\right)^2\right),
\end{aligned}
\tag{6.154}
$$

we obtain

$$
\phi(x, x^{(k)}) - f_1(x) \geqslant \sum_{i=1}^{N}\frac{\sum_{j\in S_i}\left(x_i^{(k)} - x_j^{(k)} - x_i - x_j\right)^2}{2\Delta(x_i^{(k)})} \geqslant 0. \tag{6.155}
$$

Therefore, the second condition of a surrogate function holds for $\phi(x, x^{(k)})$ defined in equation (6.152). This function can be written as a sum of terms each depending on a single variable x_i and, as a consequence of this separability property, it can be used in a MM method, in combination with a separable surrogate function for the data-fidelity function. Indeed, the expression (6.152) can be written as the quadratic function

$$
\begin{aligned}
\phi(x, x^{(k)}) &= \sum_{i=1}^{N} w_i(x^{(k)})(x_i - z_i(x^{(k)}))^2 + r_i, \\
w_i(x^{(k)}) &= \frac{1}{\Delta(x_i^{(k)})}\sum_{j\in S_i} 1 + \sum_{j\in \bar{S}_i}\frac{1}{\Delta(x_j^{(k)})}, \\
z_i(x^{(k)}) &= \frac{1}{2w_i(x^{(k)})}\left(\frac{1}{\Delta(x_i^{(k)})}\sum_{j\in S_i}(x_i^{(k)} + x_j^{(k)}) + \sum_{j\in \bar{S}_i}\frac{(x_i^{(k)} + x_j^{(k)})}{\Delta(x_j^{(k)})}\right),
\end{aligned}
\tag{6.156}
$$

where r_i is a term independent of x and $\bar{S}_i$ denotes the adjoint neighborhood of S_i, defined by $\ell \in \bar{S}_i$ if and only if $i \in S_\ell$; for example, for the HS regularization, $\bar{S}_i = \{i - 1, i - \sqrt{N}\}$. The Hessian of $\phi(x, x^{(k)})$ is (except for a constant factor) a diagonal matrix with positive entries equal to a weighted local average of the inverse discrete gradient of $x^{(k)}$ at the pixel i.

Combining the surrogate function (6.142) of the data-fidelity function with that of the penalty term (6.156), we can obtain a surrogate function for applying the MM method to the minimization problem of the function (4.29). Also in this case we can find a closed formula to determine the minimum point of the combined surrogate function at any iteration of the MM scheme.

6.8 Towards non-convex minimization problems

Recent problems in image and signal processing require the minimization of non-convex functions; this problem arises for instance in the case of non-convex regularization terms or in blind deconvolution problems. Another example is provided by a more accurate and nonlinear model of transmission tomography [63, 84]. In this section, we give a few hints about the ideas at the basis of the methods developed to deal with these problems, referring to the scientific literature for further insights. First of all, as already mentioned in section 6.2, we recall that the main convergence results for the gradient projection method do not require the convexity of the objective function; moreover, convergence results for general FB methods have been obtained even for non-convex minimization problems [4, 6], provided that the objective function satisfies the Kurdyka–Lojasiewicz inequality; these results hold true also when a variable metric is used and a closed formula for the proximity operator of Ψ is not available [25, 43].

Definition 6.5. *For any limit point $\bar{x}$ of the sequence $\{x^{(k)}\}$, a function F has the Kurdyka–Lojasiewicz property at $\bar{x}$ if there exists $v \in (0, \infty)$, a neighborhood U of $\bar{x}$ and a continuous concave function ϕ: $[0, v) \to [0, \infty)$ such that*

- $\phi(0) = 0$;
- *ϕ is continuously differentiable on $(0, v)$;*
- *the Kurdyka–Lojasiewicz inequality*

$$\phi'(F(x) - F(\bar{x}))dist(0, \partial F(x)) \geqslant 1$$

holds for all $x \in U \cap \{x: F(\bar{x}) < F(x) < F(\bar{x}) + v\}$.

Several significant classes of functions satisfy the Kurdyka–Lojasiewicz property at each point of their domain, as, for example, all p-norms, real polynomials, indicator functions of semi-algebraic sets and all semi-algebraic and real analytic functions. In particular, the data-fidelity function $f_0(x; y) = \mathrm{KL}(y; Hx + b)$ satisfies this property in its domain. This property can be exploited also in the context of recent iteratively re-weighted algorithms for image processing problems with non-convex (and possibly non-smooth) regularization terms; they are based on the solution of a sequence of convex optimization problems, where the non-convex part is at each iteration approximated by means of a majorizing convex surrogate function [99].

In the following, we limit ourselves to considering the block coordinate descent methods.

6.8.1 Block coordinate descent methods

Consider the minimization problem where the objective function is the sum of block separable terms plus a continuously differentiable one where the variables are coupled

$$\min_{x \in \mathbf{R}^n} F(x) \equiv F_0(x_1, \dots, x_t) + \sum_{i=1}^{t} F_i(x_i); \tag{6.157}$$

here F is assumed to be a coercive function, F_0: $\mathbf{R}^n \to \mathbf{R}$, F_i: $\mathbf{R}^{n_i} \to \mathbf{R}$, $i = 1, \dots, t$, with $n = \sum_{i=1}^{t} n_i$ and the unknown vector $x \in \mathbf{R}^n$ can be partitioned into t blocks,

$x = (x_1, \dots, x_t)$, with $t \geqslant 2$; the formulation (6.157) includes constrained minimization of $F_0(x)$ over a Cartesian product of convex sets $\Omega = \Omega_1 \times \cdots \times \Omega_t$, since $F_i(x_i)$ can be the indicator function of the convex set Ω_i, $i = 1, \dots, t$.

A standard approach for solving equation (6.157) is the block coordinate descent (BCD) method, where at any k-iteration the objective function is minimized with respect to the i_k-block coordinates, with $i_k \in \{1, \dots, t\}$, while the others remain fixed. Thus, starting from $x^{(0)} \in \mathbf{R}^n$, the kth iteration of the BCD method has the following steps:

(a) select $i_k \in \{1, \dots, t\}$;

(b) $x_{i_k}^{(k+1)} = \arg\min_{x_{i_k}}(F_0(x_1^{(k)}, \dots, x_{i_{k-1}}^{(k)}, x_{i_k}, x_{i_{k+1}}^{(k)}, \dots, x_t^{(k)}) + F_{i_k}(x_{i_k}))$;

(c) $x^{(k+1)} = (x_1^{(k)}, \dots, x_{i_k-1}^{(k)}, x_{i_k}^{(k+1)}, x_{i_k+1}^{(k)} \dots, x_t^{(k)})$.

When i_k is chosen at step (a) according to the following *cyclic rule*

$$i_k = \left(k - t\left\lfloor\frac{k}{t}\right\rfloor\right) + 1 \quad k \geqslant 0, \tag{6.158}$$

the BCD method is also referred to as the block nonlinear Gauss–Seidel (GS) method or alternating optimization method. When the ith sub-problem at step (b) has a unique solution, for $i = 1, \dots, t$, the limit points of the sequence $\{x^{(k)}\}$ generated by this last method are stationary for equation (6.157) [16, 100]. On the other hand, when the function F is not strictly convex with respect to each block of variables, the method may fail to locate the stationary points; in particular, Powell in [106] produced an example with $t = 3$ where the limit points of the sequence generated by the Gauss–Seidel method are not critical. Convergence results for the block nonlinear GS method have been given under suitable convexity assumptions on the objective function [76, 90, 115]. In particular, in [76] it is shown that, in the case of $t = 2$, also for a non-convex problem, every limit point of the sequence $\{x^{(k)}\}$ is a stationary point of the problem.

Several variants of the basic BCD method have been developed, depending on the assumptions on F_i, $i = 0, 1, \dots, t$, and the difficulties in solving the sub-problems at step (b). In the context of differentiable optimization, the introduction of suitable line-search based schemes at step (b) enables more general convergence results to be obtained without convexity assumptions on the objective function [75, 76] for unconstrained and constrained problems, also when the inner sub-problems are inexactly solved [21, 37]. In some cases, the convergence is obtained under the assumption that the selection of i_k at step (a) follows an essentially cyclic rule, i.e. the blocks are updated in an arbitrary order as far as each of them is updated at least once within a given number T of iterations.

In order to overcome the non-convexity of the objective function with respect to any block of variables, two main variants of the BCD approach are proposed in the literature:

- the *block coordinate proximal gradient* methods where at step (b) the minimization sub-problem is substituted by a proximal operator of $F_0(x_1^{(k)}, \dots, \cdot, \dots, x_t^{(k)}) + F_{i_k}$ at $x_i^{(k)}$ with respect to a suitable metric $\frac{1}{\alpha_k} D_{i_k}$ [6];

- the *block coordinate forward–backward* methods or proximal alternating methods [5, 44], well-known as PALM and BC-VMFB, where the step (b) is given by

$$x_{i_k}^{(k+1)} = \text{prox}_{\alpha_k F_{i_k}, D_{i_k}}(x_{i_k}^{(k)} - \alpha_k D_{i_k}^{-1} \nabla_{i_k}(F_0(x_1^{(k)}, \ldots, x_{i_k}, \ldots, x_t^{(k)}));$$

  in particular, in PALM, $D_{i_k} = I$ while in [44] an inexact version for the computation of the proximal operator is proposed; furthermore, the selection of i_k follows an essentially cyclic rule. The convergence of a unifying scheme of PALM and BC-VMFB is given in [97].

The list and the reported references are far from being exhaustive, as the development of methods for non-convex optimization in signal and image processing are still an active research field.

References

[1] Anthoine S, Aujol J-F, Boursier J and Mélot C 2012 Some proximal methods for CBCT and PET tomography *Inverse Probl. Imaging* **6** 565–98

[2] Antipin A S 2003 Feedback-controlled saddle gradient processes *Autom. Remote Control* **55** 311–20

[3] Arrow K J, Hurwicz L and Uzawa H 1958 Studies in linear and non-linear programming *Stanford Mathematical Studies in the Social Sciences* vol II ed H B Chenery, S M Johnson, S Karlin, T Marschak, R M Solow (Stanford, CA: Stanford University Press)

[4] Attouch H and Bolte J 2009 On the convergence of the proximal algorithm for non-smooth functions involving analytic features *Math. Program.* **116** 5–16

[5] Attouch H, Bolte J, Redond P and Soubeyran A 2010 Proximal alternating minimization and projection methods for nonconvex problems: an approach based on the Kurdyka-Lojasiewicz inequality *Math. Oper. Res.* **35** 438–57

[6] Attouch H, Bolte J and Svaiter B F 2013 Convergence of descent methods for semi-algebraic and tame problems: proximal algorithms, forward-backward splitting, and regularized Gauss-Seidel methods *Math. Program. Ser.* A **137** 91–129

[7] Attouch H and Peypouquet J 2016 The rate of convergence of Nesterov's accelerated forward-backward method is actually faster than $1/k^2$ *SIAM J. Optim.* **26** 1824–34

[8] Bardsley J M and Vogel C R 2003 A non-negatively constrained convex programming method for image reconstruction *SIAM J. Sci. Comput.* **25** 1326–43

[9] Barzilai J and Borwein J M 1988 Two-point step size gradient methods *IMA J. Numer. Anal.* **8** 141–48

[10] Bauschke H H and Combettes P L 2011 *Convex Analysis and Monotone Operator Theory in Hilbert Spaces* (Berlin: Springer) https://link.springer.com/book/10.1007%2F978-3-319-48311-5

[11] Beck A and Teboulle M 2009 Fast gradient-based algorithms for constrained total variation image denoising and deblurring problems *IEEE Trans. Image Process* **18** 2419–34

[12] Beck A and Teboulle M 2009 A fast iterative shrinkage-thresholding algorithm for linear inverse problems *SIAM J. Imaging Sci.* **2** 183–202

[13] Bello Cruz J Y 2017 On proximal sub-gradient splitting method for minimizing the sum of two non-smooth convex functions *Set-Valued Var. Anal.* **25** 245–63

[14] Bertsekas D P 1982 *Constrained Optimization and Lagrange Multiplier Methods* (New York: Academic) https://doi.org/10.1002/net.3230150112

[15] Bertsekas D P 1982 Projected Newton methods for optimization problems with simple constraints *SIAM J. Control Optim.* **20** 221–46

[16] Bertsekas D P 1999 *Nonlinear Programming* 2nd edn (Belmont, MA: Athena Scientific)

[17] Bertsekas D P 2015 *Convex Optimization Algorithms* (Belmont, MA: Athena Scientific) 2015

[18] Bertsekas D P and Tsitsiklis J 1988 *Parallel and Distributed Computation: Numerical Methods* (Englewood Cliffs, NJ: Prentice-Hall) http://hdl.handle.net/1721.1/3719

[19] Birgin E G, Martinez J M and Raydan M 2000 Non-monotone spectral projected gradient methods on convex sets *SIAM J. Optim.* **10** 1196–211

[20] Birgin E G, Martinez J M and Raydan M 2003 Inexact spectral projected gradient methods on convex sets *IMA J. Numer. Anal.* **23** 539–59

[21] Bonettini S 2011 Inexact block coordinate descent methods with application to the nonnegative matrix factorization *IMA J. Numer. Anal.* **31** 1431–52

[22] Bonettini S, Benfenati A and Ruggiero V 2016 Scaling techniques for ϵ-subgradient methods *SIAM J. Optim.* **26** 1741–72

[23] Bonettini S, Landi G, Loli Piccolomini E and Zanni L 2013 Scaling techniques for gradient projection-type methods in astronomical image deblurring *Int. J. Comput. Math.* **90** 9–29

[24] Bonettini S, Loris I, Porta F and Prato M 2016 Variable metric inexact line-search based methods for nonsmooth optimization *SIAM J. Optim.* **26** 891–921

[25] Bonettini S, Loris I, Porta F, Prato M and Rebegoldi S 2017 On the convergence of a linesearch based proximal-gradient method for nonconvex optimization *Inverse Probl.* **55** 055005

[26] Bonettini S, Porta F and Ruggiero V 2016 A variable metric forward-backward method with extrapolation *SIAM J. Sci. Comput.* **38** A2558–84

[27] Bonettini S and Prato M 2015 New convergence results for the scaled gradient projection method *Inverse Probl.* **31** 095008

[28] Bonettini S, Rebegoldi S and Ruggiero V 2018 Inertial variable metric techniques for the inexact forward-backward algorithm *SIAM J. Sci. Comput.* **40** A3180–210

[29] Bonettini S and Ruggiero V 2011 An alternating extragradient method for total variation based image restoration from Poisson data *Inverse Probl.* **27** 095001

[30] Bonettini S and Ruggiero V 2012 On the convergence of primal-dual hybrid gradient algorithms for total variation image restoration *J. Math. Imaging Vis.* **44** 236–53

[31] Bonettini S, Zanella R and Zanni L 2009 A scaled gradient projection method for constrained image deblurring *Inverse Probl.* **25** 015002

[32] Boyd S, Parikh N, Chu E, Peleato B and Eckstein J 2011 Distributed optimization and statistical learning via the alternating direction method of multipliers *Found. Trends Mach. Learn.* **3** 1–122

[33] Brännlund U, Kiwiel K C and Lindberg P O 1995 A descent proximal level bundle method for convex nondifferentiable optimization *Oper. Res. Lett.* **17** 121–6

[34] Bredies K 2009 A forward-backward splitting algorithm for the minimization of non-smooth convex functionals in Banach space *Inverse Probl.* **25** 015005

[35] Brune C, Sawatzky A and Burger M 2009 Bregman-EM-TV Methods with application to optical nanoscopy *Scale Space and Variational Methods in Computer Vision. SSVM 2009* ed X C Tai, K Mørken, M Lysaker and K A Lie (Lecture Notes in Computer Science, vol 5567) (Berlin, Heidelberg: Springer) pp 235–46

[36] Carlavan M and Blanc-Féraud L 2012 Sparse Poisson noisy image deblurring *IEEE Trans. Image Process* **21** 1834–46

[37] Cassioli A, Di Lorenzo D and Sciandrone M 2013 On the convergence of inexact block coordinate descent methods for constrained optimization *Eur. J. Oper. Res.* **231** 274–81

[38] Chambolle A 2005 Total variation minimization and a class of binary MRF models *Energy Minimization Methods in Computer Vision and Pattern Recognition. EMMCVPR 05* (Lecture Notes in Computer Sciences vol 3757) (Berlin: Springer) pp 136–52

[39] Chambolle A and Dossal C 2015 On the convergence of the iterates of the fast iterative shrinkage/thresholding algorithm *J. Optim. Theory Appl.* **166** 968–82

[40] Chambolle A and Pock H 2011 Diagonal preconditioning for first order primal-dual algorithms in convex optimization *Int. Conf. on Computer Vision (Barcelona)* 1762–9

[41] Chambolle A and Pock T 2011 A first-order primal-dual algorithm for convex problems with applications to imaging *J. Math. Imaging Vis.* **40** 120–45

[42] Chen G H G and Rockafellar R T 1997 Convergence rates in forward-backward splitting *SIAM J. Optim.* **7** 421–44

[43] Chouzenoux E, Pesquet J-C and Repetti A 2014 Variable metric forward-backward algorithm for minimizing the sum of a differentiable function and a convex function *J. Optim. Theory Appl.* **162** 107–32

[44] Chouzenoux E, Pesquet J-C and Repetti A 2016 A block coordinate variable metric forward-backward algorithm *J. Glob. Optim.* **66** 457–85

[45] Combettes P L and Pesquet J-C 2011 Proximal splitting methods in signal processing *Fixed-point Algorithms for Inverse Problems in Science and Engineering, Springer Optimization and Its Applications* ed H H Bauschke, R S Burachik, P L Combettes, V Elser, D R Luke and H Wolkowicz (New York, NY: Springer) pp 185–212

[46] Combettes P L and Vũ B C 2013 Variable metric quasi-Féjer monotonicity *Nonlinear Anal. Theory Methods Appl.* **78** 17–31

[47] Combettes P L and Vũ B C 2014 Variable metric forward-backward splitting with applications to monotone inclusions in duality *Optimization* **63** 1289–318

[48] Combettes P L and Wajs V R 2005 Signal recovery by proximal forward-backward splitting *SIAM Multisc. Model. Simul.* **4** 1168–200

[49] Dai Y H and Fletcher R 2005 On the asymptotic behaviour of some new gradient methods *Math. Program.* **103** 541–59

[50] Dai Y H and Fletcher R 2005 Projected Barzilai-Borwein methods for large-scale box-constrained quadratic programming *Numer. Math.* **100** 21–47

[51] Dai Y H and Fletcher R 2006 New algorithms for singly linearly constrained quadratic programming problems subject to lower and upper bounds *Math. Program.* **106** 403–21

[52] Dai Y H, Hager W H, Schittkowski K and Zhang H 2006 The cyclic Barzilai-Borwein method for unconstrained optimization *IMA J. Numer. Anal.* **26** 604–27

[53] Daubechies I, Defrise M and De Mol C 2004 An iterative thresholding algorithm for linear inverse problems with a sparsity constraint *Commun. Pure Appl. Math.* **57** 1413–57

[54] De Pierro A R 1995 A modified expectation maximization algorithm for penalized likelihood estimation in emission tomography *IEEE Trans. Med. Imag.* **14** 132–7

[55] De Pierro A R 1995 On the convergence of an EM-type algorithm for penalized likelihood estimation in emission tomography *IEEE Trans. Med. Imag.* **14** 762–5

[56] Defrise M, Kinahan P E and Michel C J 2005 Image reconstruction algorithms in PET *Positron Emission Tomography* ed D L Bailey, D W Townsend, P E Valk and M N Maisey (Berlin: Springer) pp 63–91

[57] Defrise M, Vanhove C and Liu X 2011 An algorithm for total variation regularization in high-dimensional linear problems *Inverse Probl.* **27** 065002

[58] di Serafino D, Ruggiero V, Toraldo G and Zanni C J 2018 On the steplength selection in gradient methods for unconstrained optimization *Appl. Math. Comput.* **318** 176–95
[59] Dupé F-X, Fadili M J and Starck J-L 2011 Linear inverse problems with various noise models and mixed regularizations *1st Int. Workshop on New Computational Methods for Inverse Problems (May 2011, Cachan, France)*
[60] Eckstein J 1989 Splitting methods for monotone operators with applications to parallel optimization *PhD Thesis* Department of Civil Engineering, Massachusetts Institute of Technology, Cambridge, MA Report LIDS-TH-1877 Laboratory for Information and Decision Sciences, MIT http://hdl.handle.net/1721.1/14356
[61] Eckstein J and Bertsekas D 1992 On the Douglas-Rachford splitting method and the proximal point algorithm for maximal monotone operators *Math. Program.* **55** 293–318
[62] Eckstein J and Yao W 2017 Approximate ADMM algorithms derived from Lagrangian splitting *Comput. Optim. Appl.* **68** 363–405
[63] Elbakri I A and Fessler J A 2002 Statistical image reconstruction for polyenergetic X-ray computed tomography *IEEE Trans. Med. Imaging* **21** 89–99
[64] Esser E 2009 Applications of Lagrangian-based alternating direction methods and connections to split Bregman *Technical Report CAM Report* 09–31 UCLA ftp://ftp.math.ucla.edu/pub/camreport/cam09-31.pdf
[65] Esser E, Zhang X and Chan T F 2010 A general framework for a class of first order primal-dual algorithms for convex optimization in imaging science *SIAM J. Imag. Sci.* **3** 1015–46
[66] Figueiredo M A T and Bioucas-Dias J M 2010 Restoration of Poissonian images using alternating direction optimization *IEEE Trans. Image Process.* **19** 3133–45
[67] Fletcher R 2012 A limited memory steepest descent method *Math. Program. Ser.* A **135** 413–36
[68] Frassoldati G, Zanghirati G and Zanni L 2008 New adaptive stepsize selections in gradient methods *J. Ind. Manag. Optim.* **4** 299–312
[69] Friedlander A, Martínez J M, Molina B and Raydan M 1999 Gradient method with retards and generalizations *SIAM J. Numer. Anal.* **36** 275–89
[70] Gabay D and Mercier B 1976 A dual algorithm for the solution of nonlinear variational problems via finite elements approximations *Comput. Math. Appl.* **2** 17–40
[71] Glowinski R and Marroco A 1975 Sur l'approximation, par éléments finis d'ordre un, et la résolution, par pénalisation-dualité d'une classe de problémes de Dirichlet non linéaires *ESAIM: Model. Math. Anal. Numer.* **9** 41–76
[72] Goffin J L and Kiwiel K C 1999 Convergence of a simple subgradient level method *Math. Program.* **85** 207–11
[73] Goldstein T and Osher S 2009 The split Bregman method for ℓ_1 regularized problems *SIAM J. Imaging Sci.* **2** 323–43
[74] Grippo L, Lampariello F and Lucidi S 1986 A nonmonotone line-search technique for Newton's method *SIAM J. Numer. Anal.* **23** 707–16
[75] Grippo L and Sciandrone M 1999 Globally convergent block-coordinate techniques for unconstrained optimization *Optim. Method Softw.* **10** 587–637
[76] Grippo L and Sciandrone M 2000 On the convergence of the block nonlinear Gauss-Seidel method under convex constraints *Oper. Res. Lett* **26** 127–36
[77] Harmany Z T, Marcia R F and Willett R M 2012 This is SPIRAL-TAP: sparse poisson intensity reconstruction algorithms - theory and practice *IEEE Trans. Image Process.* **21** 1084–96
[78] He B, Liao L-Z, Han D and Yang H 2002 A new inexact alternating directions method for monotone variational inequalities *Math. Program.* **92** 103–18

[79] He B, Yang H and Wang S 2000 Alternating direction method with self-adaptive penalty parameters for monotone variational inequalities *J. Optim. Theory Appl.* **106** 337–56

[80] Hestenes M R 1969 Multiplier and gradient methods *J. Optim. Theory Appl.* **4** 302–20

[81] Hunter D R and Lange K 2004 A tutorial on MM algorithms *Am. Stat.* **58** 30–7

[82] Khobotov E N 1987 Modification of the extragradient method for solving variational inequalities and certain optimization problems *USSR Comput. Math. Phys.* **27** 120–7

[83] Landi G and Loli Piccolomini E 2008 A projected Newton-CG method for nonnegative astronomical image deblurring *Numer. Algebra.* **48** 279–300

[84] Lange K and Carson R 1984 EM reconstruction algorithms for emission and transmission tomography *J. Comput. Assist. Tomogr.* **8** 306–16

[85] Lange K and Fessler J A 1995 Globally convergent algorithms for maximum a posteriori transmission tomography *IEEE Trans. Image Process* **4** 1430–8

[86] Lantéri H, Roche M and Aime C 2002 Penalized maximum likelihood image restoration with positivity constraints: multiplicative algorithms *Inverse Probl.* **18** 1397–419

[87] Levitin E S and Polyak B T 1966 Constrained minimization methods *U.S.S.R. Comput. Math. Math. Phys.* **6** 1–50

[88] Lions P L and Mercier B 1979 Splitting algorithms for the sum of two nonlinear operators *SIAM J. Numer. Anal.* **16** 964–79

[89] Luenberger D G 1989 *Linear and Nonlinear Programming* 2nd edn (Reading, MA: Addison-Wesley)

[90] Luo Z-Q and Tseng P 1992 On the convergence of the coordinate descent method for convex differentiable minimization *J. Optim. Theory Appl.* **72** 7–35

[91] Marcotte P 1991 Application of Khobotov's algorithm to variational inequalities and network equilibrium problems *INFOR* **29** 258–70

[92] Martinet B 1970 Brève communication. Régularisation d'inéquations variationnelles par approximations successives. ESAIM: Mathematical Modelling and Numerical Analysis - Modélisation Mathématique et Analyse Numérique vol 4 no. R3 pp 154–158 (http://www.numdam.org/item/M2AN_1970__4_3_154_0/)

[93] Mercier B 1979 *Lectures on Topics in Finite Element Solution of Elliptic Problems* (Lectures on Mathematics vol 63) (Bombay: Tata Institute of Fundamental Research)

[94] Moré J J and Toraldo G 1991 On the solution of large quadratic programming problems with bound constraints *SIAM J. Optim.* **1** 93–113

[95] Nesterov Y 2004 *Introductory Lectures on Convex Optimization : A Basic Course. Applied Optimization* (Boston, Dordrecht, London: Kluwer Academic) https://link.springer.com/book/10.1007%2F978-1-4419-8853-9

[96] Ng M K, Wang F and Yuan X 2011 Inexact alternating direction methods for image eecovery *SIAM J. Sci. Comput.* **33** 1643–68

[97] Ochs P 2016 Unifying abstract inexact convergence theorems and block coordinate variable metric iPiano (arXiv:1602.07283)

[98] Ochs P, Chen Y, Brox T and Pock T 2014 iPiano: inertial proximal algorithm for non-convex optimization *SIAM J. Imaging Sci.* **7** 1388–419

[99] Ochs P, Dosovitskiy A, Brox T and Pock T 2015 On iteratively reweighted algorithms for nonsmooth nonconvex optimization in computer vision *SIAM J. Imaging Sci.* **8** 331–72

[100] Ortega J M and Rheinboldt W C 1970 *Iterative Solution of Nonlinear Equations in Several Variables* (New York: Academic) https://doi.org/10.1002/zamm.19720520813

[101] Polyak B 1987 *Introduction to Optimization* (NY: Optimization Software - Inc.)

[102] Porta F and Loris I 2015 On some steplength approaches for proximal algorithms *Appl. Math. Comput.* **253** 345–62

[103] Porta F, Prato M and Zanni L 2015 A new steplength selection for scaled gradient methods with application to image deblurring *J. Sci. Comput.* **65** 895–919

[104] Porta F, Zanella R, Zanghirati G and Zanni L 2015 Limited-memory scaled gradient projection methods for real-time image deconvolution in microscopy *Commun. Nonlinear Sci. Numer. Simul.* **21** 112–27

[105] Powell M J D 1969 A method for nonlinear constraints in minimization problems *Optimization* ed R Fletcher (New York: Academic) pp 283–98

[106] Powell M J D 1973 On search directions for minimization algorithms *Math. Program.* **4** 193–201

[107] Rockafellar R T 1970 *Convex Analysis* (Princeton: Princeton University Press)

[108] Salzo S and Villa S 2012 Inexact and accelerated proximal point algorithms *J. Convex Anal.* **19** 1167–92

[109] Sawatzky A, Brune C, Kösters T, Wübbeling F and Burger M 2013 EM-TV methods for inverse problems with Poisson noise *Level Set and PDE Based Reconstruction Methods in Imaging* vol LNM 2090 ed M Burger and S Osher (Cham: Springer) pp 71–142

[110] Sawatzky A, Brune C, Wübbeling F, Kösters T, Schafers K and Burger M 2008 Accurate EM-TV algorithm in PET for low SNR *IEEE Nucl. Sci, Symp. Conf. Rec.* 5133–7

[111] Schmidt M, Le Roux N and Bach F 2011 Convergence rates of inexact proximal-gradient methods for convex optimization *Advances in Neural Information Processing Systems 24* ed J Shawe-Taylor, R S Zemel, P L Bartlett, F Pereira and K Q Weinberger pp 1458–66 (Curran Associates, Inc.)

[112] Setzer S 2011 Operator splittings, Bregman methods and frame shrinkage in image processing *Int. J. Comput. Vis.* **92** 265–80

[113] Setzer S, Steidl G and Teuber T 2010 Deblurring Poissonian images by split Bregman techniques *J. Vis. Commun. Image Represent* **21** 193–9

[114] Teuber T, Steidl G and Chan R H 2013 Minimization and parameter estimation for seminorm regularization models with I-divergence constraints *Inverse Probl.* **20** 035007

[115] Tseng P 2001 Convergence of a block coordinate descent method for nondifferentiable minimization *J. Optim. Theory Appl.* **109** 475–94

[116] Tseng P and Yun S 2009 A coordinate gradient descent method for nonsmooth separable minimization *Math. Program.* **117** 387–423

[117] Villa S, Salzo S, Baldassarre L and Verri A 2013 Accelerated and inexact forward-backward algorithms *SIAM J. Optim.* **23** 1607–33

[118] Wright S J, Nowak R D and Figuereido M A T 2009 Sparse reconstruction by separable approximation *IEEE Trans. Signal Process* **57** 2479–93

[119] Xu Z, Figueiredo M and Goldstein T 2017 Adaptive ADMM with spectral penalty parameter selection *Proceedings of the 20th International Conference on Artificial Intelligence and Statistics, Proc. of Machine Learning Research (Fort Lauderdale, FL, USA, 20–22 Apr)* **vol 54**; Singh A and Zhu J pp 718–27

[120] Zhou B, Gao L and Dai Y H 2006 Gradient methods with adaptive step-sizes *Comput. Optim. Appl.* **35** 69–86

[121] Zhu M and Chan T F 2008 An efficient primal-dual hybrid gradient algorithm for total variation image restoration *Technical Report* UCLA, Center for Applied Math ftp://ftp.math.ucla.edu/pub/camreport/cam08-34.pdf

IOP Publishing

Inverse Imaging with Poisson Data
From cells to galaxies
Mario Bertero, Patrizia Boccacci and Valeria Ruggiero

Chapter 7

Numerics

In this chapter we attempt to give a flavor to the iterative methods discussed in the previous chapters by showing them at work on numerical tests. We start with the methods minimizing the data-fidelity functions by demonstrating examples of their semi-convergence property. Next, we compare methods for the minimization of smooth regularized functions and, separately, those for the minimization of non-smooth regularized functions. In both cases we focus on edge-preserving regularization. Finally, we describe a few applications to real data.

7.1 Semi-convergent methods

As discussed in section 5.1, EM–RL has two main properties: the first is that the sequence defined by the iteration converges to a minimizer of the data-fidelity function; the second is that the iteration has the so-called semi-convergence property. As specified in that section, this property is not theoretically proved, at least to our knowledge, as in the case, for instance, of the iterative Landweber method ([2], section 6.1). Roughly speaking, this property means that the iterates of the EM–RL first approach the 'true' solution and then go away. Therefore, the number of iterations plays a role similar to a regularization parameter and early stopping of iteration can provide a satisfactory and approximate solution of the problem of Poisson data inversion. In the case of the least-squares approach to inverse problems specific stopping rules, based on the Morozov discrepancy principle, have been introduced and their regularization effect has been demonstrated. In the case of iterative methods for Poisson data inversion, similar theoretical results are not available. However, some methods, described in chapters 5 and 6, have the semi-convergence property as EM–RL, at least as demonstrated by numerical experiments.

7.1.1 Image deconvolution

We first consider the problem of image deconvolution which is relevant both in microscopy and astronomy. For simplicity, we consider two synthetic objects which

doi:10.1088/2053-2563/aae109ch7

are frequently used as examples of astronomical images. They have small bright spots and diffuse regions or sharp details. Both objects have size 256 × 256 and are shown in the upper panels of figure 7.1; they are named here object A and object B. From these objects two simulated sets of data are obtained by convolving them with the PSFs shown in the lower panels of the same figure; finally the convolved images are perturbed with different levels of Poisson noise (see remark 3.2).

More precisely, the first set of data, considered in [8], consists of three images obtained by convolving object A with an ideal PSF, corresponding to a telescope with a circular mirror, as given in equation (2.18), with $D = 8.25$ m and $\lambda = 2.2$ μm. The sampled PSF is obtained by considering 256 × 256 pixels with a pixel size of 5×10^{-3} arcsec and it is shown in the bottom-left panel of figure 7.1. As concerns the different noise level of the three images, we recall that the noise level increases when the total number of photons decreases. In order to simulate objects with different brightness with respect to the uniform background emission, we consider the three cases specified in the first three rows of table 7.1. The resulting blurred and noisy images are shown in the first row of figure 7.2, ordered from the lowest noise level (on the left) to the highest noise level (on the right).

The second set of data is considered in [20]. Three test problems are generated by considering the same object observed with different observation times. The object has the same brightness with respect to the background, but the total number of photons increases for increasing observation time. The parameters of the three versions of this object are summarized in the last three rows of table 7.1. The three versions are convolved with a PSF and then perturbed with Poisson noise. The PSF used simulates that taken by a ground-based telescope; it was downloaded from http://www.mathcs.emory.edu/~nagy/RestoreTools/index.html. It is shown in the

Figure 7.1. Original objects (upper row) and central parts of the corresponding PSFs in sqrt-scale (lower row).

Table 7.1. Background levels and maximum values of the six test objects before background addition.

Test object	Background	Maximum counting
Object A1	6.76×10^3	2.87×10^5
Object A2	6.76×10^3	4.55×10^4
Object A3	6.76×10^3	2.87×10^3
Object B1	100	25500
Object B2	10	2550
Object B3	1	255

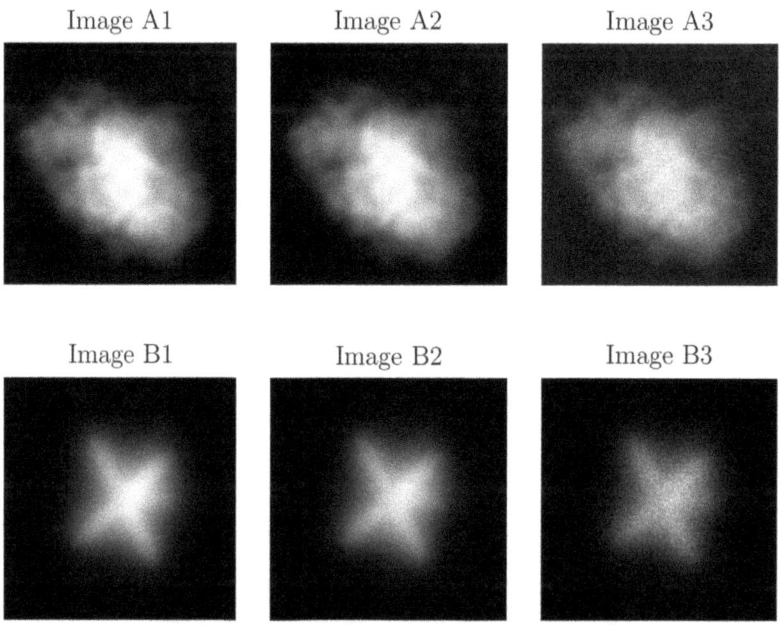

Figure 7.2. Blurred and noisy images. The left, middle and right panels refer to low, medium and high noise levels, respectively.

bottom-right panel of figure 7.1. The resulting images are shown in the lower panels of figure 7.2. For all test problems we assume periodic boundary conditions.

We use the six simulated sets of data to compare the semi-convergence properties of the following methods proposed for the solution of problem (4.13).

- The classic EM–RL method (5.1); in our implementation, the detected data and the background are scaled as follows: $y \to y/\max(y)$, $b \to b/\max(y)$.
- The accelerated EM–RL version proposed in [3]: this exploits a vector extrapolation to determine the point on which the EM–RL iteration is applied; the extrapolation, consisting of a shift along the direction given by the difference between the current iteration and the previous one, introduces a

slight computational cost and, consequently, the cost per iteration of the method is only slightly larger than that of EM–RL; in the following, we denote by EM-MATLAB the implementation of this algorithm available in the *deconvlucy* function of the image processing Matlab® toolbox.

- The SGP method described in algorithm 6.1, equipped with $\mathrm{ABB}_{\min}$ steplength selection detailed in algorithm 6.2, and with diagonal scaling matrices defined in equation (6.19); the variable bounds for their eigenvalues are set as $L_k = \sqrt{1 + \frac{10^{11}}{(k+1)^2}}$ (see equation (6.17)). In algorithm 6.1 a non-monotone line search strategy is used with $M = 10$; the other parameters are set as $\rho = 0.4$, $c = 10^{-4}$; the parameters of algorithm 6.2 are $\alpha_{\min} = 10^{-5}$, $\alpha_{\max} = 10^5$, $\tau_1 = 0.5$. Furthermore, in our implementation, the detected data and the background are scaled as in the case of EM–RL; any iteration of the SGP method requires two matrix–vector products as EM–RL or EM-MATLAB, if the backtracking loop is not required; the projection on the non-negative orthant is inexpensive; as a consequence, the cost per iteration is comparable with that of EM-MATLAB.
- The projected Newton-like method described in algorithm 6.3, equipped with a truncated CG method for an approximate solution of the linear system at step 2 [16]; the CG algorithm is stopped when the inner residual is less than 0.1 or when 30 iterations are performed. In our experiments we use the Matlab® routine NCGP, downloadable from http://www.dm.unibo.it/~landig/NPTool/NPTool.html; the parameters of the algorithm 6.3 are $\epsilon = 10^{-14}$, $\rho = \frac{1}{2}$ and $c = 10^{-4}$ (see equations (6.23) and (6.25)). We observe that each iteration of the CG method requires a matrix–vector product of the Hessian matrix with a vector, with a cost equal to four FFTs, almost equivalent to an iteration of EM–RL or SGP; then, if s is the total number of the performed FFTs, the ratio $R = s/4$ can be a measure of the complexity of NCGP with respect to the number of iterations of EM–RL or SGP.

For EM–RL, SGP and NCGP, the initial guess $x^{(0)}$ is a constant vector with all pixel values equal to $\frac{1}{N}\sum_{i=1}^{N}(y_i - b)$. For EM-MATLAB, the default starting point of *deconvlucy* is given by the detected image; in order to take into account the background and preserve the flux in the recovered image, we set the starting point equal to $y - b$. We observe that, if the linear equality constraint of the flux conservation is explicitly specified in equation (4.13), then the SGP method can manage this additional constraint by computing the projection on the feasible region thanks to a cheap algorithm, as for example the linear-time algorithm proposed in [11]. In the following we denote this version of the SGP method as SGP-EQ. The increase of the cost per iteration is irrelevant: for the size of our test problems, it amounts to a maximum of 5% of the cost of one iteration without this projection. Since the total number of iterations of SGP-EQ is in general lower than that of SGP, then the difference in cost between the two versions is very small and sometimes this is counterbalanced by a slightly better reconstruction. All the numerical simulations are carried out in a Matlab® R2018a (version 9.4.0.813 654) environment on a computer equipped with a processor Intel(R) Core (TM) i7-7560U 2.4 GHz.

In table 7.2 we summarize the numerical results obtained by the considered methods. For each method, the following quantities are reported:

- the minimum relative rms error ρ_K with respect to the original object $\bar{x}$, defined in equation (4.42), the related SSIM index (indicated as SSIM_K in brackets) defined in equation (4.44), and the corresponding iteration number K for which ρ_K is obtained; for NCGP, we report in brackets the ratio $R = s/4$, providing an approximate comparison of the iterations of the method with those of the others;
- μ_K and σ_K, which are the mean and standard deviation of the normalized residuals, defined in equation (4.45), of the recovered image at the iteration K; we recall that, for a statistical goodness of the reconstruction, the value of μ_K should be close to 0 and that of σ_K close to 1;
- the maximum value SSIM_S of the SSIM index, the related relative rms error ρ_S in brackets and the corresponding iteration S for which this value of SSIM is obtained.

In the table an asterisk is used to mark the cases where the minimum of the relative rms error or the maximum of SSIM is not reached within the prefixed maximum number of iterations (5000).

If we look at the values of ρ_K and SSIM_S we conclude that, from this point of view all the methods are essentially equivalent; the main difference concerns the number of iterations. In figure 7.3 we plot the behavior of the relative rms error as a function of the number of iterations for all methods and all test problems.

However, if we look at the values of μ_K, σ_K, we notice that in the case of Image B2 and Image B3 the values of σ_K for the reconstructions provided by EM–RL are much greater than 1. Therefore, for these test problems we also show in figure 7.4 the maps of the normalized residuals, defined in equation (4.45), for EM–RL, EM-MATLAB, SGP and NCGP. Artifacts are clearly visible in the normalized residuals of the reconstructions provided by EM–RL and EM-MATLAB, indicating that the statistical goodness of these reconstructions is lower than that of the reconstructions provided by the other methods. Fainter artifacts are visible also in the other reconstructions because the reconstruction of the aircraft is quite a hard task. However, if we look at details of the reconstructions of image B2 shown in figure 7.5, we discover that, while the reconstructions provided by EM–RL, SGP and SGP-EQ are very similar, in those provided by EM-MATLAB and NGP some distortions of the object are visible. If we come back to table 7.2, we find that these methods (which do not implement the constraint of flux conservation) are those requiring the smallest number of iterations. This result seems to indicate that, in the case of semi-convergence, methods requiring a very small number of iterations could provide unsatisfactory results. A possible interpretation is that their steps in the object space are so long that they cannot adequately approach the correct solution.

Table 7.2. Numerical values of the FOMs characterizing the quality of the reconstructions provided by the methods EM–RL, EM-MATLAB, SGP, SGP-EQ and NCGP for the considered test problems. For the definition of the parameters see the text.

Method	K	ρ_K (SSIM_K)	μ_K	σ_K	S	SSIM_S (ρ_S)
	Image A1					
EM–RL	5000*	0.185 (0.575)	2.74×10^{-6}	0.989	5000*	0.575 (0.185)
EM-MATLAB	355	0.185 (0.575)	-7.76×10^{-6}	0.988	352	0.575 (0.185)
SGP	402	0.185 (0.576)	2.54×10^{-4}	0.988	404	0.576 (0.185)
SGP-EQ	406	0.185 (0.576)	5.85×10^{-5}	0.988	404	0.576 (0.185)
NCGP	43 (1213)	0.187 (0.567)	-2.03×10^{-5}	0.988	24 (614)	0.567 (0.187)
	Image A2					
EM–RL	4047	0.186 (0.572)	1.50×10^{-5}	0.988	3338	0.572 (0.186)
EM-MATLAB	105	0.187 (0.567)	-6.41×10^{-6}	0.986	96	0.567 (0.187)
SGP	135	0.187 (0.571)	2.26×10^{-5}	0.987	126	0.571 (0.187)
SGP-EQ	129	0.187 (0.571)	9.78×10^{-6}	0.987	129	0.571 (0.187)
NCGP	7 (121)	0.188 (0.562)	5.60×10^{-4}	0.987	6 (89)	0.562 (0.188)
	Image A3					
EM–RL	414	0.194 (0.553)	1.41×10^{-12}	0.990	335	0.553 (0.194)
EM-MATLAB	10	0.212 (0.388)	1.91×10^{-3}	0.990	12	0.388 (0.213)
SGP	24	0.194 (0.553)	1.24×10^{-3}	0.991	23	0.553 (0.194)
SGP-EQ	23	0.194 (0.553)	5.66×10^{-5}	0.991	19	0.553 (0.194)
NCGP	2 (8)	0.201 (0.537)	-3.32×10^{-2}	1.01	2 (8)	0.537 (0.201)
	Image B1					
EM–RL	5000*	0.295 (0.860)	1.08×10^{-3}	1.00	5000*	0.860 (0.295)
EM-MATLAB	120	0.332 (0.860)	-2.39×10^{-2}	1.07	215	0.871 (0.347)
SGP	261	0.289 (0.875)	-5.33×10^{-3}	1.00	1030	0.886 (0.310)
SGP-EQ	203	0.289 (0.873)	1.72×10^{-2}	1.00	298	0.887 (0.301)
NCGP	1480 (51378)	0.318 (0.888)	7.74×10^{-3}	0.996	1616 (56 138)	0.888 (0.318)
	Image B2					
EM–RL	2435	0.342 (0.828)	3.42×10^{-14}	18.4	5000*	0.845 (0.359)
EM-MATLAB	85	0.359 (0.840)	-5.01×10^{-3}	1.01	141	0.849 (0.400)
SGP	186	0.340 (0.831)	-6.36×10^{-3}	0.997	999	0.850 (0.405)
SGP-EQ	83	0.335 (0.838)	1.09×10^{-2}	0.997	321	0.853 (0.401)
NCGP	292 (9975)	0.352 (0.855)	8.14×10^{-3}	0.995	347 (11 877)	0.856 (0.356)

	Image B3					
EM–RL	545	0.404 (0.778)	2.84×10^{-13}	5.90	3820	0.804 (0.617)
EM-MATLAB	45	0.416 (0.794)	2.05×10^{-2}	0.991	85	0.814 (0.500)
SGP	51	0.401 (0.788)	6.46×10^{-3}	0.995	121	0.812 (0.536)
SGP-EQ	46	0.403 (0.781)	5.74×10^{-3}	0.995	151	0.850 (0.405)
NCGP	73 (2443)	0.457 (0.811)	6.38×10^{-4}	0.994	34 (1105)	0.813 (0.465)

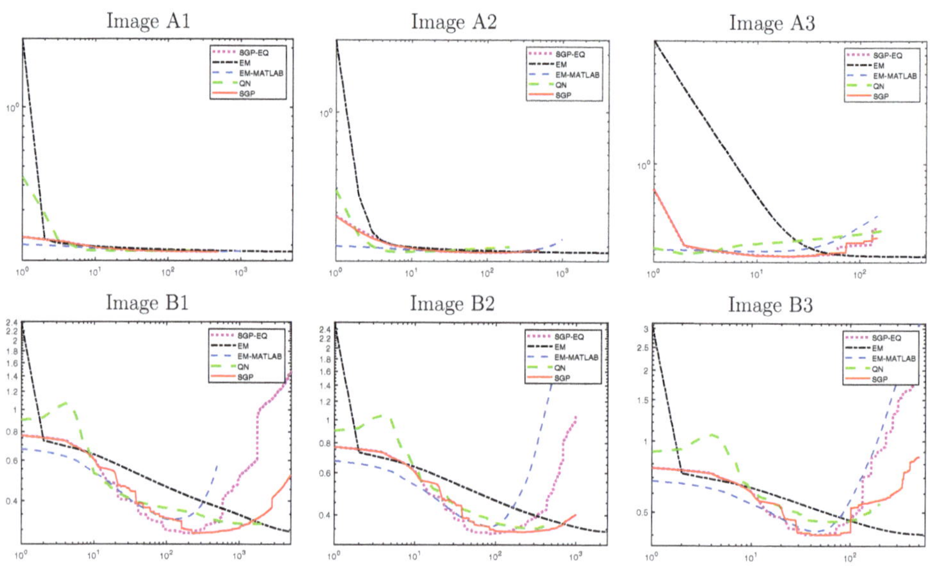

Figure 7.3. Plot of the relative rms error as a function of the number of iterations for EM–RL, EM-MATLAB, SGP, SGP-EQ and NCGP for all the test problems.

7.1.2 Emission tomography

In the case of emission tomography we consider the comparison of two standard methods such as EM–RL and OSEM with one of the methods introduced in chapter 6, namely SGP. In this numerical experiment a version of SGP adapted to object reconstruction from projections is used. The semi-convergence of the first two methods is already well-known from numerical practice while that of SGP has been demonstrated in the case of image deconvolution.

The test problem we consider is the reconstruction of the section of a phantom from its projections, i.e. the data computed for this simulation are simply the projections of the activity of the radio-tracers in each point of the phantom slice. In other terms, we compute the sinogram (see section 3.1.2) without taking into account the many physical effects that influence image formation as, for instance, attenuation or collimator blur.

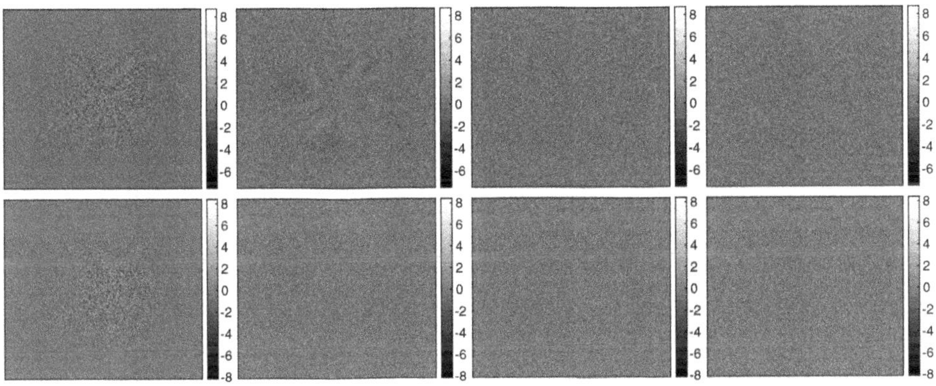

Figure 7.4. Maps of the normalized residuals for the reconstructions corresponding to the minimum relative rms error and obtained by means of EM–RL (left panels), EM-MATLAB (middle-left panels), SGP (middle-right panels) and NCGP (right panels). In the upper row the maps of the normalized residuals of image B2 while in the bottom row those of image B3.

More precisely, we consider a bottom section of the Jaszczak phantom, frequently used to calibrate SPECT scanners. The physical phantom is a Plexiglas cylinder containing rods of different diameters distributed in sectors. It is usually filled with a radioactive solution and placed in the SPECT scanner. In the tomographic reconstruction of the phantom the parts of Plexiglas are seen as 'cold', i.e. black, while the radioactive solution is seen as 'hot', i.e. white.

For our simulation a digital version of Jaszczak phantom (1024×1024 pixels) is used. We compute the sinogram by computing 180 projections uniformly distributed on $[0, \pi)$ which are subsequently corrupted by Poisson noise. The phantom is shown in figure 7.6 (left panel) together with the corresponding sinogram (right panel).

The three iterative methods are applied to the synthetic data; in figure 7.7 we plot, for the three methods, the behavior of the relative rms error ρ_k as a function of the number of iterations. In all cases we find a minimum value of the error and the corresponding reconstruction is taken as the best reconstruction provided by the method. These reconstructions are shown in figure 7.8 where, in the upper-left panel also the reconstruction provided by FBP (section 3.2) is shown. Since this reconstruction contains negative values, which are not replaced by zero, it appears as a lower contrast reconstruction.

For a quantitative comparison of the three iterative methods in table 7.3 we give the number of iterations K corresponding to the minimum relative rms error, the value ρ_k of this error and the values of μ_k and σ_k of the normalized residuals. As a conclusion, the three methods provide similar results while OSEM is the most efficient one.

7.2 Methods for edge-preserving regularization

For the comparison of different optimization methods in the solution of the MAP problem (4.30), we must select some specific regularization function and some specific test objects. We focus on methods for edge-preserving reconstructions

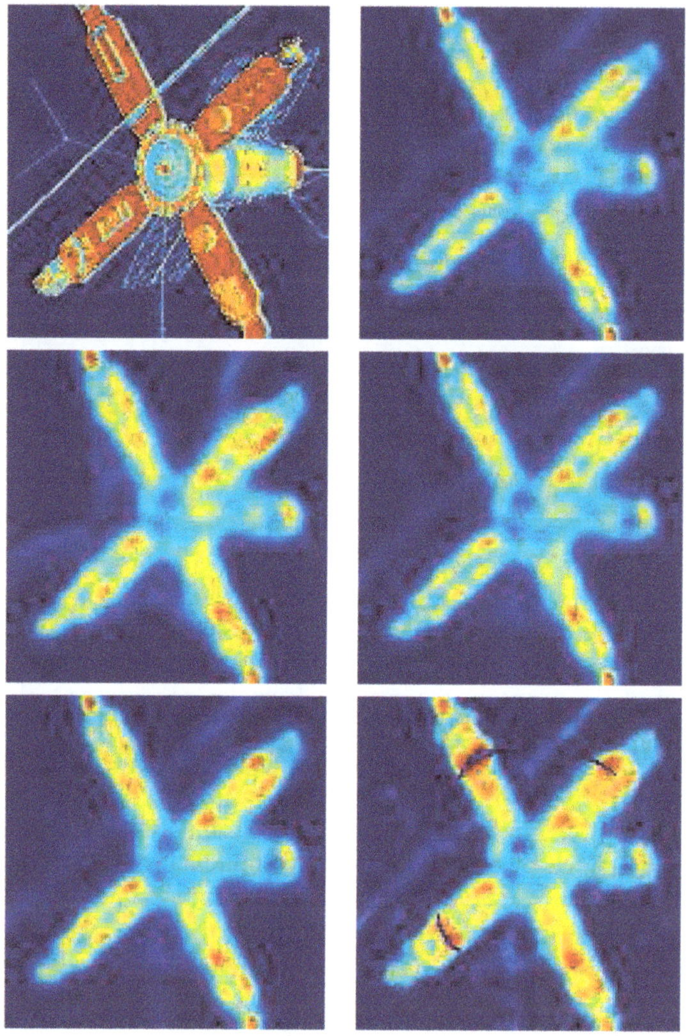

Figure 7.5. Details of the recovered images for the test problem image B2: from top to bottom, in the first row the original object and the result obtained by EM–RL; in the second the result of EM-MATLAB and that of SGP; in the third the result of SGP-EQ and that of NCGP. The same color scale is used in all images to emphasize the differences between them.

because we have at our disposal two regularizers with this property: one is the HS regularizer of equation (4.35), which is differentiable, the other is the TV regularizer of equation (4.36) which is not differentiable. The first contains a threshold parameter δ such that, when it tends to zero, the HS regularizer tends to the TV regularizer. We recall that, for this reason, the HS regularizer is sometimes called the TV-smooth regularizer.

The advantage of this choice is that, in the HS case we can compare methods for the minimization of smooth functions while in the TV case we can compare methods

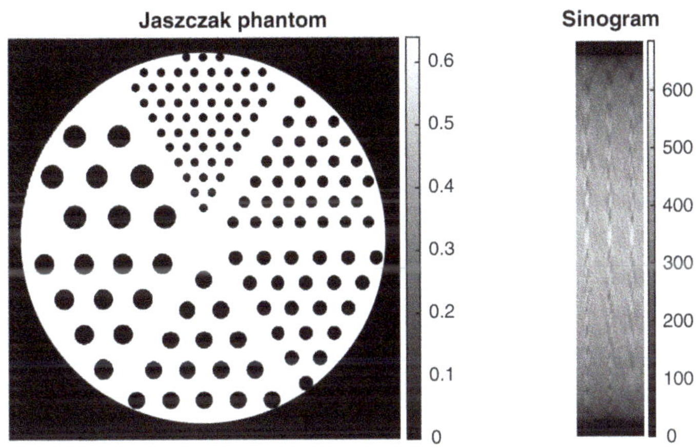

Figure 7.6. Digital version of a section of a Jaszczak phantom (left panel) and the corresponding noisy sinogram, i.e. a simulated set of raw data acquired by a SPECT scanner (right panel).

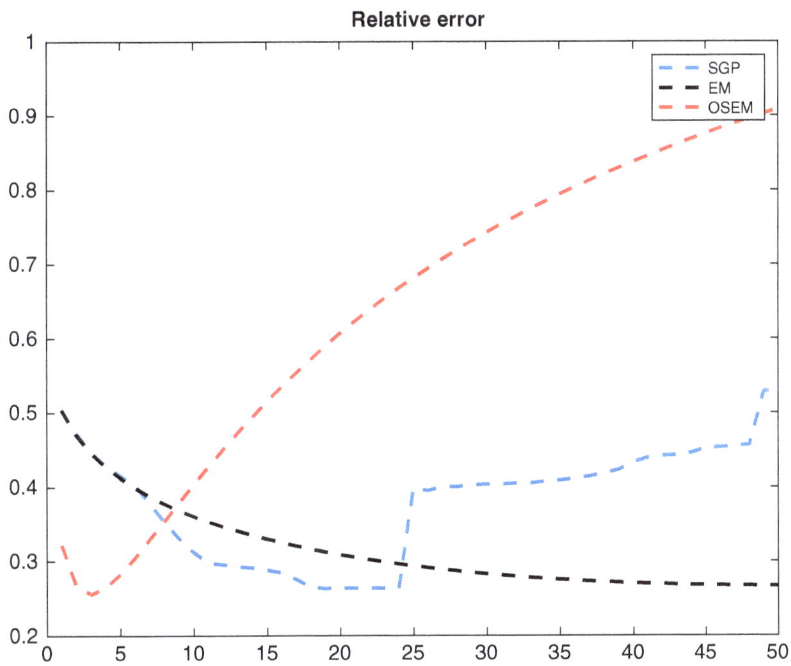

Figure 7.7. Plot of the relative rms error ρ_k as a function of the number of iterations for SGP, EM–RL and OSEM.

for the minimization of non-smooth functions. As a byproduct we also have a comparison between HS and TV.

Our strategy is as follows.

- For generating the test problems, we select a set of test objects and we compute the corresponding images.

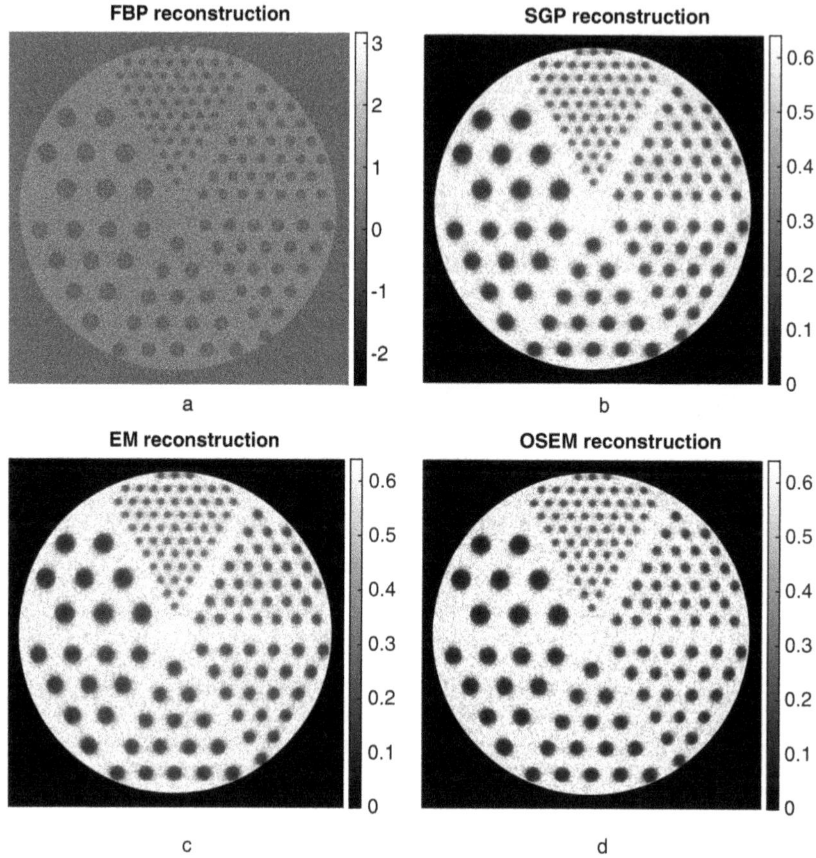

Figure 7.8. Comparison of the reconstructions of the phantom shown in figure 7.6 from its sinogram. (a) Filtered back projection. (b) SGP after 20 iterations. (c) EM–RL after 50 iterations. (d) OSEM (with 18 subsets) after four iterations.

Table 7.3. Quantitative comparison of EM–RL, OSEM (18 subsets), and SGP for the emission tomography simulation. The quantities are defined as in table 7.2.

Method	K	ρ_K	μ_K	σ_K
EM	50	0.27	0.11	1.02
OSEM	4	0.26	−0.04	0.91
SGP	20	0.26	0.04	1.02

- We select a test method not very efficient but stable, applicable to both HS and TV. In the case of HS, we use this method for estimating the best values, according to one particular FOM, of the regularization parameter β and of the thresholding parameter δ for each one of the test objects.

- For the best values of the two parameters we compute by a very large number of iterations of the test method an estimate of the minimum value of the objective function.
- We estimate the efficiency of the different methods by comparing the number of iterations required for reaching the minimum value of the objective function.
- The best value of β estimated in the previous experiments is used for comparing the methods for TV regularization.

All the numerical experiments are carried out in a Matlab® R2018a (version 9.4.0.813 654) environment on a computer equipped with a processor Intel(R) Core (TM) i7-7560U 2.4 GHz.

7.2.1 HS regularization

We consider three test problems, named *Scameraman*, *Sspacecraft* and *micro*. The first two test problems are described in section 4.4. The third test problem *micro* is described in [18] and can be downloaded from https://voices.uchicago.edu/willett/research/software/. The original object of size 128×128 simulates the dendrite spines of a neuron and its values are in the range [0,70]; its total flux is $2.946\,1 \times 10^5$ and the background term b is set to zero. The corresponding simulated image is assumed to be acquired with a confocal microscope. In figure 7.9, the original image and the blurred noisy image are reported; it is evident that the latter is very noisy. As for *Scameraman* and *Sspacecraft*, we assume periodic boundary conditions.

As the test method, we select the AEM method described in section 6.6.3 (algorithm 6.11), since this method is well suited to solve the minimization problem (4.30) with both HS and TV regularization, even if other methods could be used (as for example ADMM). The result of the analysis does not depend on the solver used. The setting of the parameters of algorithm 6.11 is $M = 10$, $\epsilon = 10^{-4}$, $x^{(0)} = \max(0, y)$; the initial guess for the dual variable is set to zero. The method is stopped when the relative ℓ_2 norm between two successive primal–dual iterates is less than a tolerance *tol* or the maximum number of 8000 iterations is exceeded. For *micro*, $\mathrm{tol} = 5 \times 10^{-6}$, while for the other two test problems, $\mathrm{tol} = 10^{-5}$.

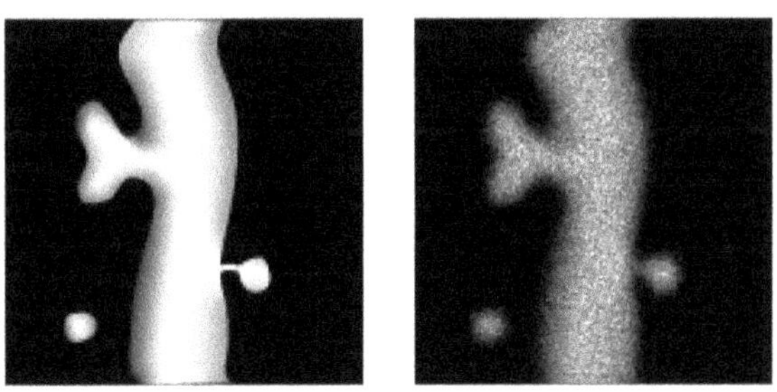

Figure 7.9. Test problem *micro*: original image on the left and blurred noisy image on the right [18].

Next, borrowing the numerical experience from [7], for each one of the deblurring problems selected as a test, we evaluate the following error function on a grid of values in the (β, δ) plain:

$$\rho(\beta, \delta) = \frac{\|x^*_{\beta,\delta} - \bar{x}\|}{\|\bar{x}\|}; \tag{7.1}$$

here $\bar{x}$ is the original object and $x^*_{\beta,\delta}$ is the computed solution of the problem (4.30) with $f_1(x)$ equal to the HS regularizer. When $\delta = 0$, $x^*_{\beta,0}$ is the minimizer of the objective function with TV regularization.

In figure 7.10, the level curves of $\rho(\beta, \delta)$ for the three test problems are shown, whereas in table 7.4 the pairs (β, δ) corresponding to the observed minimum of $\rho(\beta, \delta)$ are given. For a comparison, we report both the minimum relative rms error with HS regularization and the error $\rho(\beta, 0)$ obtained with TV regularization. The latter is obtained by taking in the TV case the value of β of the HS case. In all cases, we remark that when δ is close to 1% of the maximum value of the image, the values of $\rho(\beta, \delta)$ do not significantly change for a fixed β; indeed, for small values of δ, the level curves tend to become almost parallel to the vertical axis and the accuracy of the results depends only on β.

In figure 7.11 we highlight some effects of the different regularizations on the reconstruction of the test object *micro*. The figure shows the reconstruction obtained

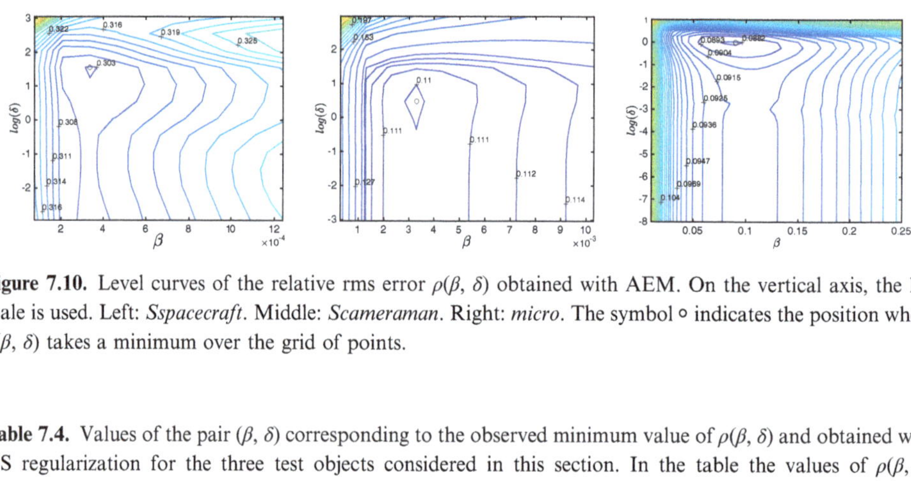

Figure 7.10. Level curves of the relative rms error $\rho(\beta, \delta)$ obtained with AEM. On the vertical axis, the log scale is used. Left: *Sspacecraft*. Middle: *Scameraman*. Right: *micro*. The symbol ○ indicates the position where $\rho(\beta, \delta)$ takes a minimum over the grid of points.

Table 7.4. Values of the pair (β, δ) corresponding to the observed minimum value of $\rho(\beta, \delta)$ and obtained with HS regularization for the three test objects considered in this section. In the table the values of $\rho(\beta, 0)$ corresponding to TV regularization are also given. These results are obtained by means of the AEM method.

Test problem	(β, δ)	$\rho(\beta, \delta)$
Sspacecraft	$(3.35 \times 10^{-4}, 35.89)$	0.30226
	$(3.35 \times 10^{-4}, 0)$	0.30365
Scameraman	$(3.3 \times 10^{-3}, 2.966)$	0.10997
	$(3.3 \times 10^{-3}, 0)$	0.11003
micro	$(9 \times 10^{-2}, 1)$	0.088119
	$(9 \times 10^{-2}, 0)$	0.091192

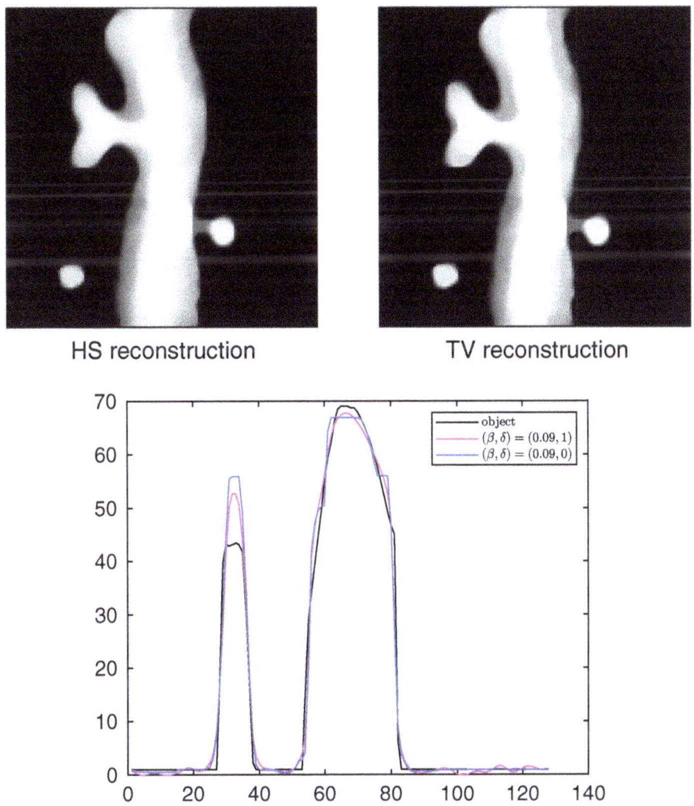

Figure 7.11. Test problem *micro*. Top-left panel: reconstruction obtained with HS regularization and the pair (β, δ) corresponding to the minimum relative rms error. Top-right panel: reconstruction obtained with TV regularization, with the same value of β. Bottom panel: profile of row 64 in the original object (black line), in the HS reconstruction with (β, δ) = (0.09,1) and in the TV reconstruction with $\beta = 0.09$.

by HS regularizer for the observed best pair (β, δ) and that obtained by TV regularization with the same value of β. In addition, the plot of one row of the two reconstructed images is compared with the plot of the corresponding row of the original object. The different behavior of the two regularization terms is evident: the TV regularization emphasizes the edges, but produces a well-known cartoon effect, while the HS regularizer is able to avoid this effect, without excessive deterioration of the edges of the image.

Finally, we use the results obtained with AEM to compare the performance of methods applicable to the minimization of a smooth functional, i.e., the case of HS regularization. We consider the following methods.

- The SGP method described in the algorithms 6.1–6.2, with $\Phi(x) = f_\beta(x; y)$ and $\Psi(x)$ given by the indicator function of the non-negative orthant C. An effective rule for the choice of the sequence of the scaling matrices for this solver is defined by equation (6.20); in particular, for HS regularization the decomposition of the gradient $-\nabla f_1(x) = U_1(x) - V_1(x)$ is derived in section

5.4. In this application we set $L_k = \sqrt{1 + \frac{10^{10}}{(k+1)^2}}$; the other parameters of SGP are $M = 10$, $\rho = 0.4$, $c = 10^{-4}$, $\alpha_{\min} = 10^{-5}$, $\alpha_{\max} = 10^5$, $\tau_1 = 0.5$.

- The SFBEM method, i.e. the scaled inertial forward–backward method with backtracking, described in algorithm 6.7. Also in this case, following [6], we set $\Phi(x) = f_\beta(x; y)$ and $\Psi(x) = \iota_C(x)$, whereas the rule for the scaling matrices is the same as that used for SGP. The initial value for α_0 is set equal to 10^4; the algorithm automatically corrects this parameter at the very first iteration, possibly reducing it with some backtracking steps, but a reduction of α_k never occurs at the successive iterations.
- The algorithm 6.8 implementing the ADMM method. Even if this method is tailored for non-differentiable regularization, it can be also used in the case of a differentiable one, as observed in remark 6.3.

For a comparison of these algorithms, the iterations are stopped when the value of the objective function becomes smaller than that obtained with AEM equipped with the previously described stopping rule. The values of (β, δ) corresponding to the minimum rms error are used.

The execution times of AEM and of the other algorithms are reported in table 7.5. In the case of ADMM we report only the results obtained for the value of the parameter $\gamma \in \{\beta, \frac{\beta}{5}, \frac{\beta}{50}\}$ which provides the best performance. Furthermore, the table reports the execution times of AEM and ADMM for the solution of equation (4.30) with TV regularization. Also in this case, ADMM is stopped when the value of the objective function becomes smaller than that obtained by AEM with the previously described stopping rule. The results of table 7.5 enable us to affirm that, in the case of a smooth edge-preserving regularization, the methods for smooth optimization, as SGP or SFBEM, can be successfully applied, obtaining a good performance in terms of accuracy and computational time. On the other hand, methods tailored for solving non-smooth regularization problems require a greater complexity; consequently, it may be not convenient to use them for solving a smooth minimization problem.

7.2.2 TV regularization

In this section, we consider three test problems of different size and we compare the numerical results obtained by solving, with different methods, the minimization problem in equation (4.30) with TV regularization.

The first test problem is *micro*, the 128×128 object already considered in the previous section; the original object and the blurred noisy image are shown in figure 7.9. The regularization parameter β is set equal to 0.09. The second test problem is *phantom*: the original object 256×256 is generated by the Matlab® function `phantom` and is scaled by a factor 1000; the PSF is a Gaussian function, with standard deviation 3, truncated at the 9×9 central pixels. In this case the values of the original image are in the range [0, 1000] and we introduce a background $b = 10$; the relative distance between the original object and the blurred noisy data in ℓ_2

norm is 0.385 and $y_i \in [1, 852]$. For this test problem, we select $\beta = 0.007$. Figure 7.12 shows the original object and the blurred noisy data. The third test problem, named *tubulins*, is a 512×512 object representing a micro-tubulin network inside a cell [17]. In this case, the values of the original object are in the range [0, 686], whereas those of the blurred and noisy image are in [0,446]; the background is set

Table 7.5. Performance of the methods AEM, ADMM, SGP and SFBEM for the numerical solution of the problem (4.30) with HS regularizer; on the right of the table, the performance of AEM and ADMM is reported also in the case of TV regularizer; for each test problem, *k* denotes the numbers of iterations needed to obtain a value of the objective function smaller than the final value obtained by AEM for the same pair (β, δ); *Time* indicates the execution time in seconds; this value has been obtained as a mean of the execution times corresponding to several runs.

Method	$\rho(\beta, \delta)$	k	*Time*	$\rho(\beta, \delta)$	k	*Time*
	Sspacecraft, HS regularizer with $(\beta, \delta) = (3.35 \times 10^{-4}, 35.89)$			*Sspacecraft*, TV regularizer with $\beta = 3.35 \times 10^{-4}$		
AEM	0.30226	5383	54.3	0.30365	6562	66.6
ADMM	0.30049	1905	29.1	0.30266	1872	29.5
SGP	0.30025	824	8.09			
SFBEM	0.30209	530	7.11			
	Scameraman, HS regularizer with $(\beta, \delta) = (3.3\ 10^{-3}, 2.966)$			*Scameraman*, TV regularizer with $\beta = 3.3 \times 10^{-3}$		
AEM	0.10987	1680	17.2	0.11003	1731	17.6
ADMM	0.10962	242	3.67	0.10990	244	4.15
SGP	0.10920	96	0.909			
SFBEM	0.10924	60	0.917			
	micro, HS regularizer with $(\beta, \delta) = (9 \times 10^{-2}, 1)$			*micro*, TV regularizer with $\beta = 9\ 10^{-2}$		
AEM	0.08812	1786	4.31	0.09120	2414	5.86
ADMM	0.08826	559	1.86	0.09230	675	2.54
SGP	0.08851	82	0.20			
SFBEM	0.08810	98	0.34			

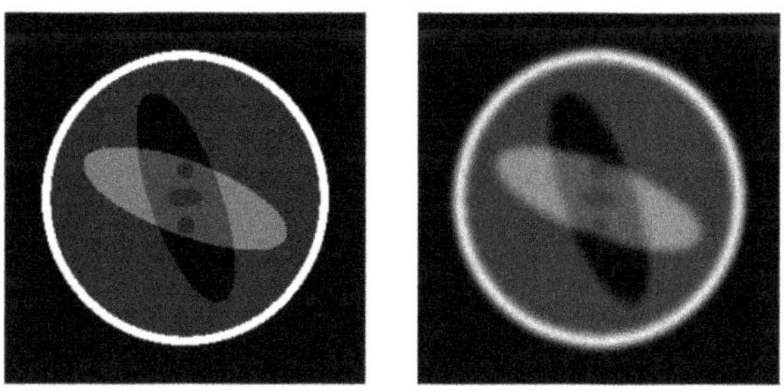

Figure 7.12. Test problem *phantom*: original object on the left and blurred noisy image on the right.

equal to 1 and the relative distance between the original object and the blurred noisy data in ℓ_2 norm is 0.756; β is set equal to 4×10^{-4}. In figure 7.13 the original object and the blurred noisy image are shown. Periodic boundary conditions are assumed in all three cases.

For all test problems, we compute the solution x_β^* of the minimization problem (4.30) by running a huge number of iterations of one of the methods considered in this section; in particular, we use the CP method described in section 6.6.1. Then, for any iterative method, we evaluate not only the relative reconstruction error but also the progress toward the high accuracy estimate of the minimum point x_β^*. If we denote $x^{(k)}$ the result at iteration k of a given minimization problem, in addition to the relative reconstruction error $\rho_k = \frac{\| x^{(k)} - \bar{x} \|}{\| \bar{x} \|}$ of the kth iterate with respect to the original object $\bar{x}$, we compute the relative error with respect to x_β^* and the relative difference between the value $f_\beta(x^{(k)})$ of the objective function and the high accuracy optimal value $f_\beta(x_\beta^*)$:

$$e_k = \frac{\|x^{(k)} - x_\beta^*\|}{\|x_\beta^*\|} \qquad f_k = \frac{f_\beta(x^{(k)}) - f_\beta(x_\beta^*)}{f_\beta(x_\beta^*)}.$$

In our experiments, the TV regularization of the considered test problems is addressed by the following methods.

- The inexact variable metric FB method named VMILA, described in section 6.4.2 and detailed in algorithm 6.5 [5]. The backward step requires the solution of the sub-problem (6.55), which is inexactly solved by applying the classic FISTA method [1] to its dual formulation (see equation (6.60)). The step-length of the inner FISTA is fixed as the reciprocal of an estimate of the Lipschitz constant of the gradient of the dual function; this estimate depends on the outer step-length and on the condition number of D_k. For the details of the inexact computation of the proximal iterate $\tilde{w}^{(k)}$ and of the descent direction we refer to algorithm 6.6, with the inner parameter η equal to 10^{-6}. For the choice of the step-length α_k and of the sequence of scaling matrices $\{D_k\}$, we can use the same techniques described for SGP in section 6.3.1; in particular, D_k is chosen according to the rule of equation (6.19), with

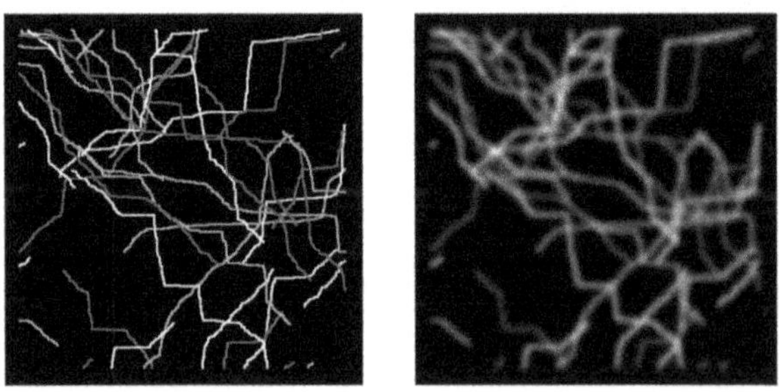

Figure 7.13. Test problem *tubulins*: original object on the left and blurred noisy image on the right [17].

$L_k = \sqrt{1 + \frac{10^{13}}{(k+1)^c}}$. While in SGP a standard choice is $c = 2$, for VMILA it is convenient to set $c > 2$ and in particular in these experiments, $c = 4$ in the case of *micro* and *phantom* and $c = 3$ for *tubulins*. This choice (producing a faster convergence of $\{L_k\}$ to 1 and, consequently, of $\{D_k\}$ to the identity matrix) is strictly related to the computational complexity of the inner solver FISTA; indeed, large values of the condition number of D_k could lead to short inner steps, increasing the inner iteration number, especially when, as the outer iterations proceed, the required inner accuracy increases. Thus, we allow a large condition number for D_k only at the first iterations, when a larger tolerance on the inner sub-problem is required, and we squeeze it very quickly to the identity matrix in the successive iterations, so that the inner iteration number remains bounded. On the other hand, at the first iteration, the choice of large L_k allows more freedom in the selection of D_k, allowing some performing approach such as the split gradient strategy.

- The specialized version of ADMM for Poisson deconvolution, described in section 6.5.2 and detailed in algorithm 6.8 [14, 19]; the value of the parameter γ is set equal to the value in $\{\beta, \frac{\beta}{5}, \frac{\beta}{50}\}$, providing the best performance. We recall that, among the considered methods, this is the only one that requires the solution of a linear system at any iteration; for this application the system can be solved by means of FFT.
- The primal–dual CP method described in section 6.6.1 [10]; in order to avoid the selection of the parameters σ and τ, we use the preconditioned version of the method, where these parameters are replaced by the diagonal positive definite matrices Σ and T, defined in equation (6.100). For TV regularization, the matrix A is given by $(H^T, \nabla^T)^T \in \mathbf{R}^{3N \times N}$, where N is the number of the pixels of the image. In view of the standard assumptions on H (see equations (3.31), (3.32) and (3.33)), since the elements of H are not easily available because only the PSF is given, the parameter ζ in equation (6.100) has to be chosen equal to 1. The method is implemented as detailed in algorithm 6.9 where σ and τ are replaced by the diagonal entries of the matrices Σ and T, i.e. $(\Sigma)_{j,j} = \frac{0.99}{5}$, $j = 1, \ldots, N$ and $(T)_{i,i} = 0.99$, $i = 1, \ldots, N$, $(T)_{i,i} = \frac{0.99}{5}$, $i = N+1, \ldots, 3N$.
- The primal explicit version of the SPDHG method described in section 6.6.2 and detailed in algorithm 6.10 [4]; the sequence of diagonal scaling matrices is defined according to the rule of equation (6.125), with $L_k = \sqrt{1 + \frac{c}{(k+1)^2}}$; moreover, $c = 10^{10}$ for *micro* and $c = 10^{13}$ for the other two test problems, as in VMILA. The sequences $\{\sigma_k\}$ and $\{\tau_k\}$ are chosen as follows

$$\tau_k = c_1 + c_2 k, \qquad \sigma_k = \frac{1}{c_3 + c_4 k}. \tag{7.2}$$

In order to illustrate the effectiveness of the method, the values c_i are manually optimized for each test problem (see table 7.6).

Table 7.6. Parameter setting for SPDHG and SSL methods.

	SPDHG		SSL
	τ_k	σ_k	τ_k
micro	$0.1 + 10^{-3}k$	$(0.5 + 10^{-7}k)^{-1}$	$0.1 + 10^{-3}k$
phantom	$0.01 + 10^{-4}k$	$(0.5 + 10^{-8}k)^{-1}$	$0.9 + 10^{-2}k$
tubulins	$1 + 10^{-3}k$	$(0.5 + 10^{-8}k)^{-1}$	$0.01 + 10^{-2}k$

- A variant of the SPDHG method, based on the scaled ϵ-subgradient level (SSL) proposed in [4]. As remarked in section 6.6.2, in this variant only the sequence $\{\tau_k\}$ is *a priori* selected and, in the primal step of the SPDHG iteration, a level algorithm [9, 15] allows an adaptive computation of the dynamic step-length σ_k. The parameter values of SSL for the three test problems are given in table 7.6. For details of the method, see [4], algorithm 1.
- The AEM method described in section 6.6.3 and detailed in algorithm 6.11; we use the standard setting of the parameters ($M = 10$, $\epsilon = 10^{-4}$).

For all methods, we choose $x^{(0)} = \max(0, y)$; for the primal–dual methods, the initial guess for the dual variable is set to zero. Since the described methods have iterations with a different complexity and different stopping rules, they are compared by considering the numerical results obtained within a prefixed time t. In table 7.7 we report for any test problem the number k of iterations carried out in t seconds, the relative reconstruction error ρ_k, the relative minimization error e_k and the relative error on the value of the objective function f_k; in figure 7.14 we plot the behavior, as a function of time, of the relative error on the function values f_k and of the relative minimization error e_k.

A few comments can be derived from table 7.7 and figure 7.14.

- The relative error on the objective function f_k and the relative minimization error e_k may have a different behavior. Indeed, a faster decrease in the objective function may not correspond to a lower minimization error; this phenomenon could be explained as an effect of the ill-conditioning of the considered problems and of the different paths described by the algorithms toward the solution. However, the values of the relative reconstruction error ρ_k show that, apart from very slow methods, the estimates of the original objects are very similar for all methods, i.e. the effectiveness of the methods in terms of accuracy is comparable. This point can be directly verified, by observing the reconstructed images obtained with different methods in figure 7.15.
- The effectiveness of some methods, such as ADMM, SPDHG, SSL, depends on a suitable setting of relevant parameters. Other methods such as AEM, the

Table 7.7. Numerical results obtained, within a prefixed time t, by several solution methods of the problem of equation (4.30) with TV regularization; k is the number of iterations (for VMILA the total number of inner iterations of the inner solver FISTA is reported in brackets); ρ_k is the relative rms error with respect to the original object; e_k is the relative error with respect to the estimated minimum point x^*_β and f_k is the relative error with respect to the estimated minimum value of the objective function.

Method	k	ρ_k	e_k	f_k
micro	Results after $t = 5$ s			
VMILA	254 (2292)	9.2×10^{-2}	1.8×10^{-3}	2.7×10^{-4}
ADMM	853	9.2×10^{-2}	7.3×10^{-3}	1.5×10^{-5}
CP	1661	9.2×10^{-2}	2.6×10^{-3}	2.3×10^{-5}
SPDHG	1633	9.2×10^{-2}	1.9×10^{-3}	1.3×10^{-3}
SSL	1107	9.1×10^{-2}	1.5×10^{-2}	3.9×10^{-4}
AEM	1389	9.0×10^{-2}	1.9×10^{-2}	9.3×10^{-5}
phantom	Results after $t = 20$ s			
VMILA	623 (3223)	1.4×10^{-1}	2.5×10^{-3}	7.9×10^{-6}
ADMM	1260	1.4×10^{-1}	9.2×10^{-4}	2.0×10^{-4}
CP	2296	1.4×10^{-1}	1.1×10^{-2}	5.2×10^{-5}
SPDHG	1647	1.4×10^{-1}	3.5×10^{-3}	6.9×10^{-3}
SSL	1404	1.4×10^{-1}	4.1×10^{-3}	4.1×10^{-4}
AEM	1854	1.7×10^{-1}	7.9×10^{-2}	4.1×10^{-3}
tubulins	Results after $t = 80$ s			
VMILA	457 (815)	4.4×10^{-1}	6.7×10^{-2}	4.7×10^{-5}
ADMM	703	4.5×10^{-1}	8.1×10^{-2}	2.5×10^{-3}
CP	1160	4.8×10^{-1}	2.0×10^{-1}	5.1×10^{-3}
SPDHG	1180	4.4×10^{-1}	1.1×10^{-1}	1.3×10^{-3}
SSL	820	4.4×10^{-1}	7.3×10^{-2}	4.8×10^{-4}
AEM	938	6.3×10^{-1}	4.9×10^{-1}	2.1×10^{-1}

preconditioned version of CP and, partially, VMILA are more robust with respect to a restricted set of initial parameters. Indeed, AEM and VMILA employ adaptive strategies to self-tune their inner parameters; nevertheless AEM exhibits a slow convergence, while VMILA can achieve a good performance, mainly for large size problems. In this case the practical super-convergence achieved by VMILA can counterbalance the complexity of the inner solver and the matrix–vector products involving H.

- One single iteration of the primal–dual methods, as the preconditioned CP and SPDHG, is very inexpensive; then, in a prefixed time, these methods can

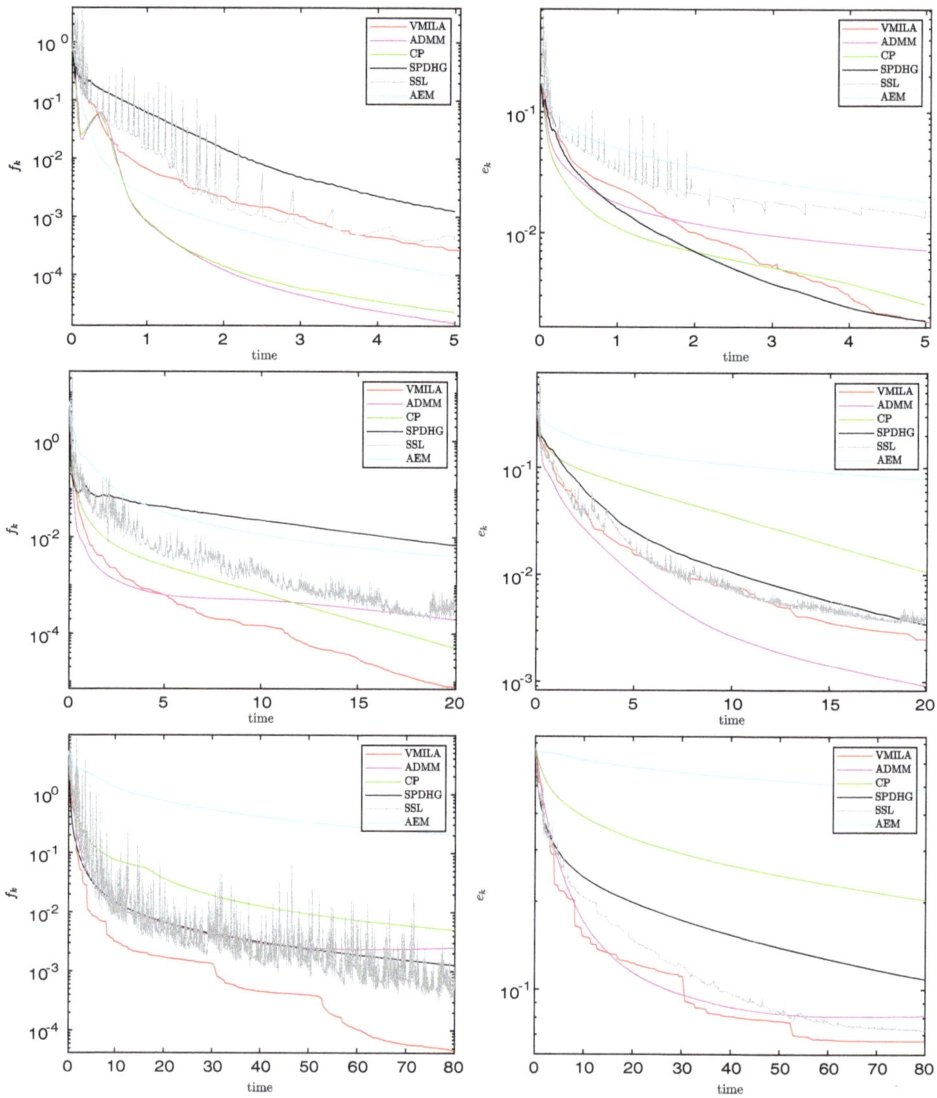

Figure 7.14. Behavior of the relative error f_k on the values of the objective function (left column) and of the relative minimization error e_k (right column) as a function of the time (in seconds) for the three test problems: first row *micro*, second row *phantom*; third row *tubulins*. We observe from table 7.7 that the final values of the objective function are different.

perform many more iterations than other algorithms. Nevertheless, when the matrix–vector products Hx and H^{Tx} are expensive (for example for large N), the time-consuming acceleration strategies used in other methods enable better performance to be obtained. See, for example, the solution of the linear system required at any step of ADMM, or the self-estimate of the step-length in SSL, or the suitable selection of the step-length and the variable metric of VMILA.

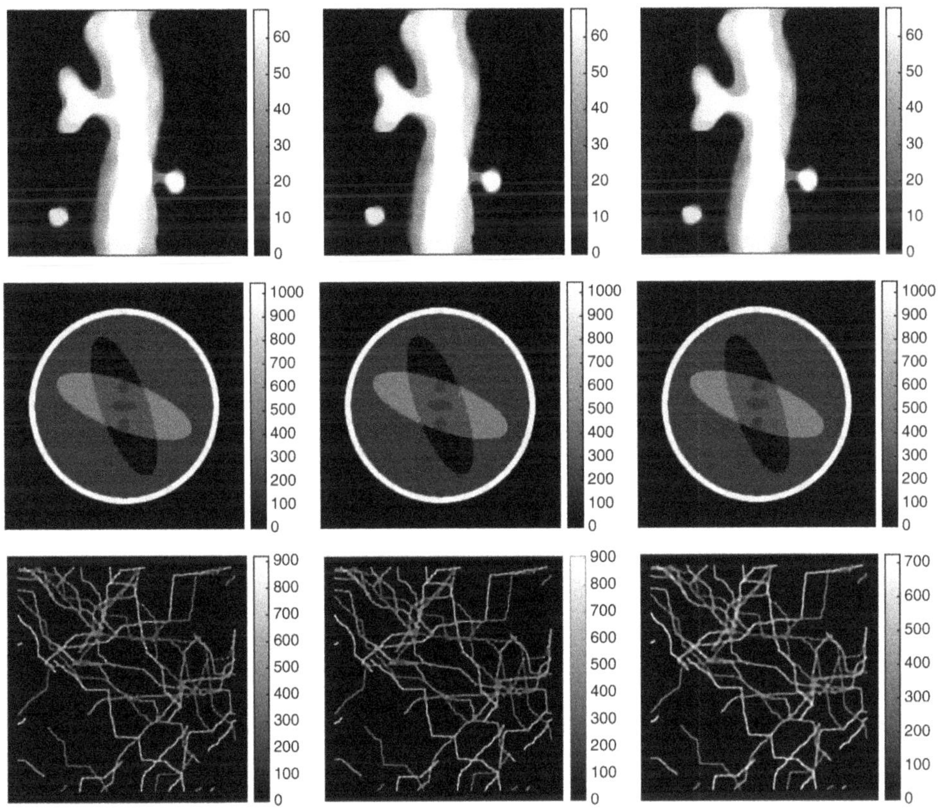

Figure 7.15. Top row: reconstruction of *micro* obtained from left to right by VMILA, ADMM and SPDHG. Middle row: reconstruction of *phantom* obtained from left to right by VMILA, ADMM and SSL. Bottom row: reconstruction of *tubulins* obtained from left to right by VMILA, ADMM and CP.

7.3 Image reconstruction of real data

In this section, examples of deconvolution of real images are shown. The main difficulties consist in the correct modeling of the direct problem, i.e. to determine the PSF (both 2D and 3D), and in the choice of a suitable stopping criterion for the considered algorithm.

7.3.1 3D confocal imaging

The forward problem of image formation in confocal microscopy has been addressed in detail in chapter 2, section 2.1.2. The deconvolution of 3D confocal images can improve the resolution achieved. We use SGP without regularization so that a suitable stopping of the iterations is required.

G Vicidomini (Italian Institute of Technology) kindly provided images of tubulin filaments acquired with a confocal microscope equipped with a lens immersed in oil with a numerical aperture of 1.4. The physical dimensions of the volume are $20 \times 20 \times 6$ μm.

Since the sampling consists of 500 × 500 × 20 pixels, the voxel dimensions are 40 × 40 × 300 nm.

Another parameter of confocal imaging is the 'dwell time' that refers to the time that the laser spot remains in one location within the specimen. The laser beam is continually scanned across and down the sample and the image is constructed from discrete pixels that are compiled as the laser scan proceeds. The pixel intensity results from the time-averaged response of the PMT to the emitted fluorescence as the laser spot moves across this small unit of distance. For the tubulin image, the dwell time is set to 50 μs.

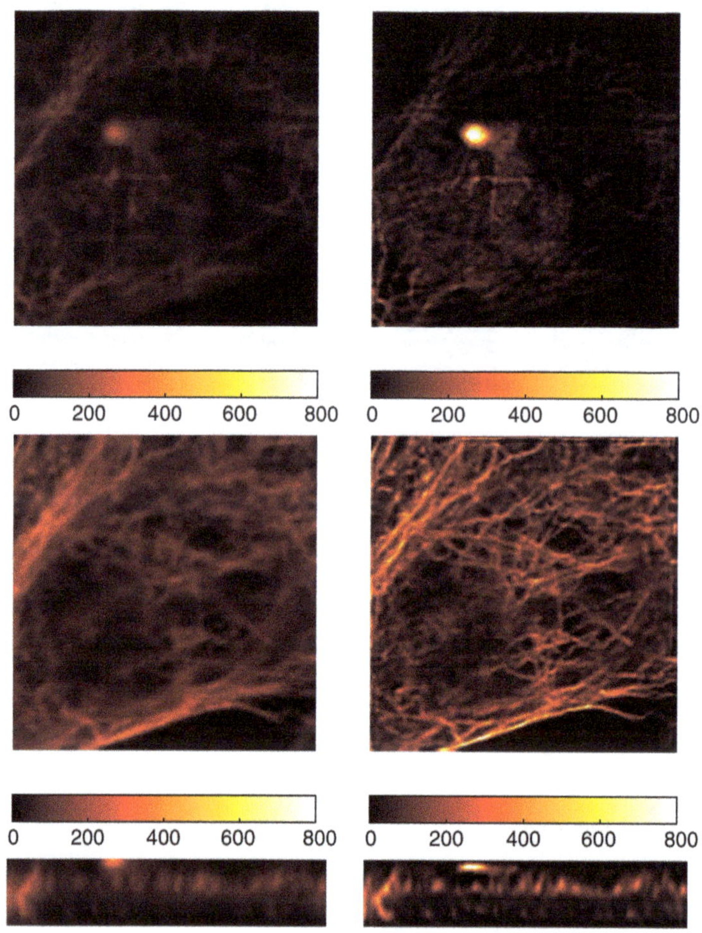

Figure 7.16. 3D deconvolution of a confocal image of tubulin filaments stained with Abberior STAR red and imaged by 60x and NA = 1.4 oil objective lens. Starting from the top: an *xy* section of the original image (left) and its reconstruction (right); another *xy* plane of the original image (left) and its reconstruction (right); an *yz* section of the original section (left) and its reconstruction (right). Courtesy of G Vicidomini—Italian Institute of Technology.

Before applying deconvolution, the PSF must be determined. For simplicity, the PSF was modeled as a 3D Gaussian with FWHM = 300 nm in the x, y plane and FWHM = 1500 nm in the z direction, i.e. the optical axis.

For real data, it is difficult to determine when the iterations of the deconvolution algorithm should be stopped. The discrepancy method is not applicable because the PSF is not exactly known. Few iterations lead to a blurred result, too many iterations produce a result with a few very bright pixels and are therefore unrealistic.

A good deconvolution is a compromise between these two limiting cases and can be evaluated 'on sight'. In figure 7.16 we present the 3D deconvolution of tubulin filaments image after 10 iterations of the SGP algorithm.

7.3.2 3D medical imaging

As already mentioned in chapter 2, PET images have a lower resolution than CT images but they contain functional information for the study and diagnosis of neurodegenerative diseases. The resolution of these images can be improved by deconvolving the tomographic reconstructions.

To this purpose, an evaluation of the 3D PSF, which describes the blurring of the tomographic apparatus, is required. The images used are provided by the Medical Physics Department of the San Martino Genoa Hospital and were acquired with a CT/PET. The technical characteristics provided by the manufacturer, indicate the following values for the FWHM.

- Axial FWHM at 1 cm from the center: 4.6 mm.
- Axial FWHM at 10 cm from the center: 5.8 mm.
- Transaxial FWHM at 1 cm from the center: 4.6 mm.
- Transaxial FWHM at 10 cm from the center: 5.8 mm.

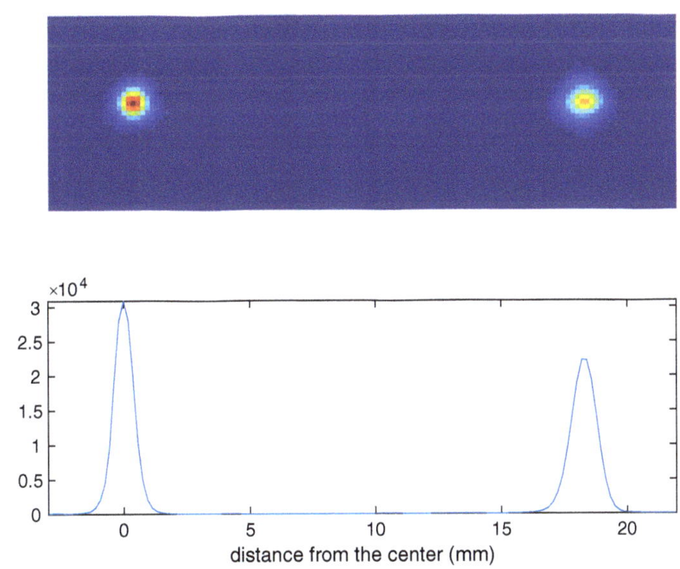

Figure 7.17. Reconstruction of a pair of linear sources and the corresponding intensity profile as a function of the distance from the center (PET rotation axis).

An image of a pair of linear sources is acquired to check the response of the tomographic reconstruction. These sources are metallic rods with a radioactive core. The 3D reconstruction voxel size is 5.33 × 5.33 × 5 mm. The diameter of the radioactive part is 2.8 mm, therefore smaller than the voxel size. The zoom of the reconstruction of these line sources and the corresponding intensity profiles are shown in figure 7.17. The figure suggests that the PSF may be modeled by a Gaussian, even if it exhibits a weak space-variant behavior. An accurate deconvolution should take into account this variability (see section 8.5); however, in this case, the PSF is considered space-invariant with a FWHM of 5.2 mm.

The acquisition of line sources provides only an axial shape of the PSF. In order to deconvolve the PET real data, we need a 3D Gaussian PSF. In consideration of the technical characteristics of the apparatus, a PSF with the same FWHM in the three directions is chosen.

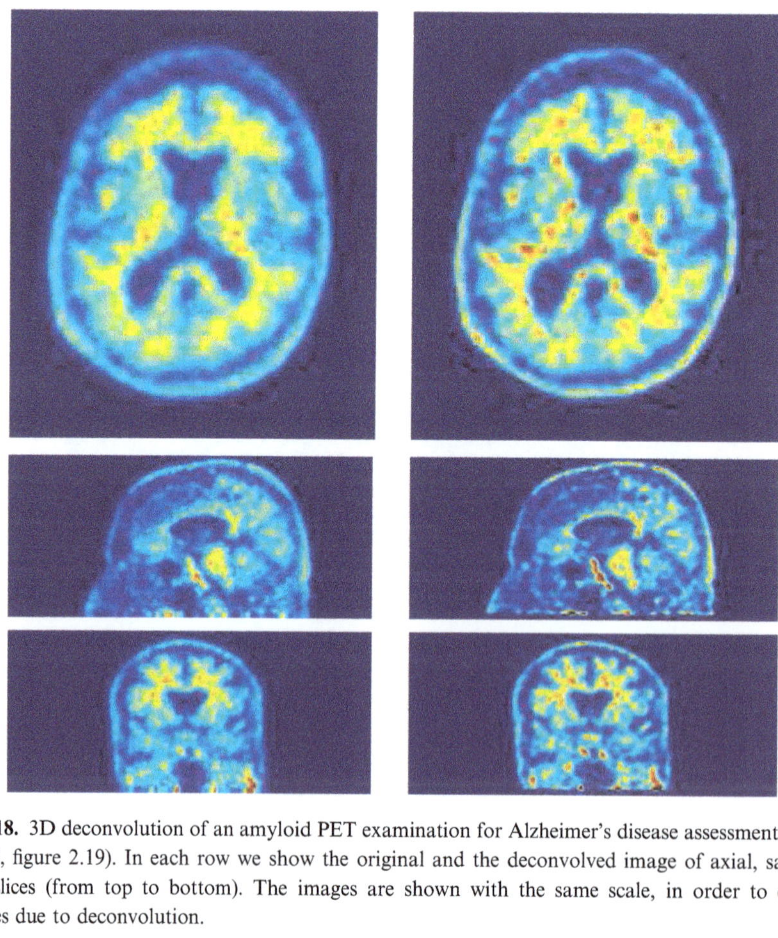

Figure 7.18. 3D deconvolution of an amyloid PET examination for Alzheimer's disease assessment (shown in chapter 2, figure 2.19). In each row we show the original and the deconvolved image of axial, sagittal and coronal slices (from top to bottom). The images are shown with the same scale, in order to clarify the differences due to deconvolution.

In figure 7.18 we show the improvement of the tomographic reconstruction obtained after a few iterations of the SGP algorithm.

7.3.3 Astronomical imaging

As an example in astronomy we consider the images of a volcanic eruption on Io, the inner one of the four Jupiter moons discovered in 1619 by Galileo Galilei and called by him 'Astri Medicei', in honor of Cosimo II de' Medici (see also the beginning of section 2.3).

Raw images are kindly provided to us by Imke De Pater, Professor of Astronomy at Berkeley University, California (USA). The images capture the eruption of one of Io volcanoes, Pillan, occurred on 14 August 2014 and observed with near-infrared camera NIRC2, coupled to the adaptive optics system on the 10 m W M Keck II telescope on Mauna Kea, Hawaii. The images are acquired using different filters corresponding to J, H, K, L and M bands (for this astronomical nomenclature, see the table at the beginning of section 2.3.1). For each band, images of the star HD 161903 are also acquired for PSF estimation. Details of acquisition and

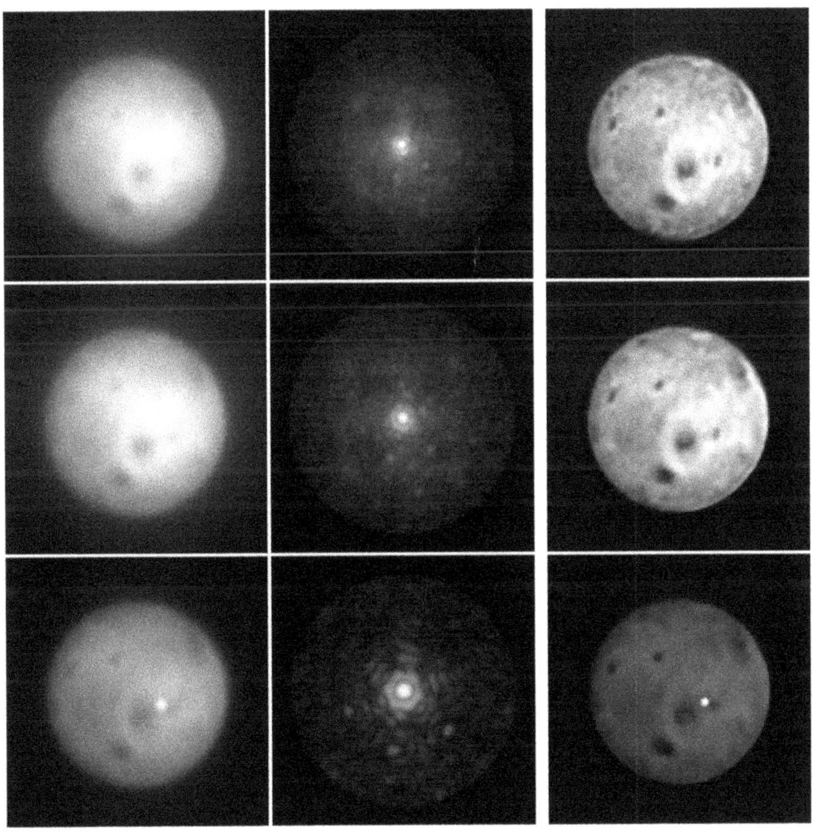

Figure 7.19. From top to down, J, H and K images of the Pillan eruption (left panels; linear scale), the corresponding PSFs (central panels; log scale) and the reconstructions (left panels; linear scale).

astronomical considerations can be found in [12] and [13]. As is known, Io is a highly volcanic moon of Jupiter and astronomers are interested in studying its temporal evolution.

As stressed in chapter 2, raw images are affected by various physical phenomena and, before their processing, it must be ensured that all the conditions required for the deconvolution are respected. Moreover, the following procedure is applied for obtaining a valid PSF:

- suppression of 'hot pixels', i.e. pixels with values greater than a threshold value computed on the eight neighbor's pixels;
- subtraction of an estimated background value;

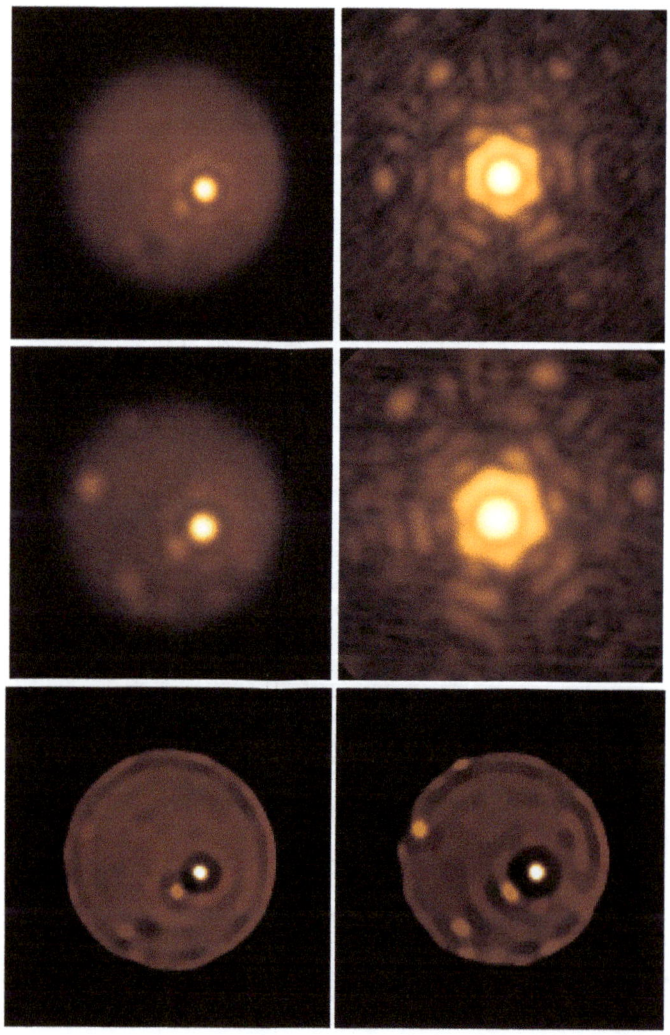

Figure 7.20. From the top down, L and M images of the Pillan eruption (left panels; linear scale), the corresponding PSFs (right panels; log scale); last row: the reconstructions of L (left) and M (right) images.

- zeroing of the negative values;
- application of a circular mask;
- normalization of the result to unit sum.

The J, H and K band images with the corresponding PSFs (central parts of the Keck PSF) and their reconstructions are shown in figure 7.19. The reconstructions are obtained using SGP with HS regularization after a search of the parameters β, δ based on inspection of the corresponding reconstructions. The results look quite satisfactory. We believe that the Pillan eruption appears as a gray small spot in the J band, close to the darker one in the Pillan region, which becomes weakly bright in the H band and finally very bright in the K band.

Deconvolution of Io images in the L and M bands, leads to the reconstructions shown in figure 7.20. These reconstructions show that other hot sources besides Pillan are present on the surface of Io; however, these reconstructions are not satisfactory because, as discussed at the beginning of section 5.1, the reconstruction of bright sources over an unknown background is affected by ringing artifacts. In such a case the use of the two component methods discussed in section 8.4 is required. For this reason the problem of the deconvolution of the L and M band images of the Pillan eruption is reconsidered in that section.

References

[1] Beck A and Teboulle M 2009 A fast iterative shrinkage-thresholding algorithm for linear inverse problems *SIAM J. Imaging Sci.* **2** 183–202

[2] Bertero M and Boccacci P 1998 *Introduction to Inverse Problems in Imaging* (Bristol: Institute of Physics)

[3] Biggs D S C and Andrews M 1997 Acceleration of iterative image restoration algorithms *Appl. Opt.* **36** 1766–75

[4] Bonettini S, Benfenati A and Ruggiero V 2016 Scaling techniques for ϵ-subgradient methods *SIAM J. Optim.* **26** 1741–72

[5] Bonettini S, Loris I, Porta F and Prato M 2016 Variable metric inexact line-search based methods for nonsmooth optimization *SIAM J. Optim.* **26** 891–921

[6] Bonettini S, Porta F and Ruggiero V 2016 A variable metric forward-backward method with extrapolation *SIAM J. Sci. Comput.* **38** A2558–84

[7] Bonettini S and Ruggiero V 2011 An alternating extragradient method for total variation based image restoration from Poisson data *Inverse Probl.* **27** 095001

[8] Bonettini S, Zanella R and Zanni L 2009 A scaled gradient projection method for constrained image deblurring *Inverse Probl.* **25** 015002

[9] Brännlund U, Kiwiel K C and Lindberg P O 1995 A descent proximal level bundle method for convex nondifferentiable optimization *Oper. Res. Lett.* **17** 121–6

[10] Chambolle A and Pock H 2011 Diagonal preconditioning for first order primal-dual algorithms in convex optimization, *Int. Conf. on Computer Vision, Barcelona* pp 1762–9

[11] Dai Y H and Fletcher R 2006 New algorithms for singly linearly constrained quadratic programming problems subject to lower and upper bounds *Math. Program.* **106** 403–21

[12] de Pater I 2015 Time evolution of Io's volcanoes Pele and Pillan from 1996-2015, as derived from GALILEO, NIMS, Keck, Gemini, IRTF and LBTI observations, *AAS/Division for*

Planetary Sciences Meeting Abstracts #47, volume 47 of AAS/Division for Planetary Sciences Meeting Abstracts p 409.06

[13] de Pater I, Laver C, Davies A G, de Kleer K, Williams D A, Howell R R, Rathbun J A and Spencer J R 2016 Io: Eruptions at Pillan, and the time evolution of Pele and Pillan from 1996 to 2015 *Icarus* **264** 198–212

[14] Figueiredo M A T and Bioucas-Dias J M 2010 Restoration of Poissonian images using alternating direction optimization *IEEE Trans. Image Process* **19** 3133–45

[15] Goffin J L and Kiwiel K C 1999 Convergence of a simple subgradient level method *Math. Program.* **85** 207–11

[16] Landi G and Loli Piccolomini E 2008 A projected Newton-CG method for nonnegative astronomical image deblurring *Numer. Algebra.* **48** 279–300

[17] Porta F, Zanella R, Zanghirati G and Zanni L 2015 Limited-memory scaled gradient projection methods for real-time image deconvolution in microscopy *Commun. Nonlinear. Sci. Numer. Simul.* **21** 112–27

[18] Willett R and Nowak R 2003 Platelets: a multiscale approach for recovering edges and surface in photon limited medical imaging *IEEE Trans. Med. Imaging.* **22** 332–50

[19] Setzer S, Steidl G and Teuber T 2010 Deblurring Poissonian images by split Bregman techniques *J. Vis. Commun. Image Represent* **21** 193–9

[20] Staglianó A, Boccacci P and Bertero M 2011 Analysis of an approximate model for Poisson data reconstruction and a related discrepancy principle *Inverse Probl.* **27** 125003

IOP Publishing

Inverse Imaging with Poisson Data
From cells to galaxies
Mario Bertero, Patrizia Boccacci and Valeria Ruggiero

Chapter 8

Specific topics in image deblurring

In this chapter we discuss specific problems which arise mainly in the case of image deconvolution, i.e. the imaging matrix has a Toeplitz or cyclic structure and is given in terms of a PSF. As we discussed in chapter 3 this model applies to problems of image reconstruction in microscopy and astronomy. The first specific problem we consider concerns super-resolution, namely the possibility of going beyond the diffraction limit by means of suitable data processing. The second concerns the correction of boundary artifacts, a problem arising when the object to be imaged is partially contained in the image domain. The third problem is the so-called *blind deconvolution* which is relevant when not only the object but also the PSF is not known or only partially known. It is important mainly in astronomy but it can arise also in microscopy. The fourth problem is specific to astronomical imaging and consists of the design of specific methods for dealing with images containing the superposition of very bright point objects (such as brilliant stars) to smooth and diffuse objects, i.e. images with a very high dynamic range. Finally, the fifth problem originates both in microscopy and astronomy if the model of a space-invariant PSF is not satisfactory; the problem is treatable if the space-variant PSF is slowly varying over the image domain and can be locally approximated by a space-invariant one.

In this chapter H is a circulant block-matrix with circulant blocks; therefore, it can be expressed in terms of a PSF h (with periodic boundary conditions). Its action on an object x is given by $Hx = h * x$, where $*$ denotes cyclic convolution. The PSF h can be an array depending on two indices (2D case) or a cube depending on three indices (3D case).

8.1 Super-resolution by data inversion

Nowadays the term super-resolution is used in very different contexts with different meanings. We mention, for instance, the so-called *super-resolution image reconstruction* which is a very active research area and consists in obtaining a high-resolution (HR) image by combining multiple low-resolution (LR) images obtained, for

doi:10.1088/2053-2563/aae109ch8

instance, by sub-pixel misalignment. In other words the purpose is to obtain an improvement in the quality of the image beyond the limit due to the size of the pixels of the detector, for instance a CCD camera or a CMOS (complementary metal-oxide–semiconductor) image sensor. The resolution enhancement is achieved via software since one has to solve a problem of image reconstruction which, as far as we know, is formulated in the framework of a least-squares approach and therefore is beyond the scope of this book. A review, with a comprehensive list of references, is given in [74].

However, the term super-resolution was first coined in the advent of digital images and data processing for indicating the possibility of going beyond the diffraction limit by a suitable processing of the detected image. In this section we discuss super-resolution in this sense. We recall that the diffraction limit, also called the Rayleigh or Abbe limit in the case of a circular pupil, is discussed in section 2.1.2.1 as concerns microscopy and in section 2.3.3 as concerns astronomy. In the case of microscopy, we also recall that nowadays the new microscopes based on STED methodology (see section 2.1.3) are super-resolving precisely in this sense. They achieve super-resolution by means of specific hardware. It is possible that the super-resolving numerical methods developed in recent years can allow us to go beyond STED resolution by processing STED images.

Historically, the possibility of improving the diffraction limit is first discussed in a paper by Toraldo di Francia [86] where the possible relationship between super-gain antennas and optical resolving power is discussed. The problem is reconsidered in a paper by Wolter [94] and in the papers of Harris [55] and McCutchen [68]. The main argument for justifying the possibility of super-resolution is the following. Let us consider an optical system described by a convolution operator such as a microscope or telescope. As discussed in section 3.1.1 the PSF of this system is band-limited and consequently also the image of any given object. Therefore, if we assume absence of noise and we divide the FT of the image by the OTF, we obtain the FT of the object over the band of the system (see equation (3.11)). Now, if the object has a bounded support, i.e. it is zero outside a bounded domain, its FT is analytic; since it is known over the band of the optical instrument, thanks to existence and uniqueness of analytic continuation, we can conclude that it is possible to determine everywhere the FT of the object, hence the object itself with unlimited resolution.

The previous argument neglects noise and, unfortunately, the dream of unlimited resolution does not take into account that analytic continuation is a classic example of an ill-posed problem, as pointed out in [92]. The limitations due to noise were already considered in [80, 87, 88]; in particular, in the last paper the analysis is based on a simple model and on the use of the well-known prolate spheroidal wave functions (PSWF) [83].

Indeed, in [88] a simple 1D model of an optical instrument is considered. The band $\mathcal{B}$ of the instrument is the interval $[-\omega_{\max}, \omega_{\max}]$ so that the corresponding amplitude PSF is a sinc function; if we further assume that the object has a bounded support contained in the interval $[-a, a]$, then the image formation is described by the following forward operator

$$(\hat{L}x)(r) = \int_{-a}^{a} \frac{\sin[\omega_{\max}(r - r')]}{\pi(r - r')} x(r')dr', \; |r| \leqslant a; \tag{8.1}$$

the amplitude image (also called coherent image) is given by $y = \hat{L}x$ and therefore it is assumed that it is observed on the same interval of the object. By means of the change of variable $t = r/a$, $\hat{L}$ can be reduced to the standard form

$$(\hat{L}x)(t) = \int_{-1}^{1} \frac{\sin[c(t - s)]}{\pi(t - s)} x(s)ds, \; |t| \leqslant 1, \tag{8.2}$$

where, with an abuse of notation, we write $x(s)$ instead of $x(as)$; moreover, if $\rho = \pi/\omega_{\max}$ denotes the sampling distance, or the resolution limit achievable by detecting the signal in the range of the operator $\hat{L}$, the parameter c is given by

$$c = a\,\omega_{\max} = \pi\frac{a}{\rho}; \tag{8.3}$$

sometimes c is called the *space-band-width product* and, except for the factor $\pi/2$, coincides with the number of sampling points (resolution elements) contained within the support of the object, i.e. $2a/\rho$.

The operator $\hat{L}$ is self-adjoint, positive definite and compact in $L^2(-1,1)$, the space of the square-integrable functions on the interval $[-1,1]$; it is also injective since the condition $\hat{L}x = 0$ implies that the FT of x is zero on the band $\mathcal{B}$ and therefore is zero everywhere. It follows that there exists a countable set of positive eigenvalues $\{\lambda_k\}_{k=0}^{\infty}$ (see section 3.6.4, proposition 3.2); the corresponding eigenfunctions, which form an orthonormal basis in $L^2(-1,1)$, are related to the PSWF $\psi_k(c, t)$ [83] by

$$v_k(s) = \lambda_k^{-1}\psi_k(c, s), \;\; \hat{L}v_k = \lambda_k v_k; \;\; k = 0, 1, 2, \ldots. \tag{8.4}$$

The eigenfunctions v_k may be regarded as elements of 'information' which retain their identity under the forward imaging transformation, except for a scaling in magnitude by the value λ_k. The larger λ_k, the more efficient the transmission of the corresponding information element, but, if we remember that the eigenvalues of a compact operator tend to zero and we order the λ_k in decreasing magnitude, there will come a point at a sufficiently large value of k where the information is transmitted so weakly that it cannot be distinguished from the noise affecting the data.

The eigenvalue 'spectrum' λ_k of the operator $\hat{L}$ has a very typical behavior when the value of c is significantly larger than 1: the eigenvalues have values near unity up to $k \simeq 2c/\pi = \frac{2a}{\rho}$ and then they drop very quickly to values near zero. Therefore, since a large value of c implies that a is large with respect to the resolution limit ρ, this behavior implies that in such a case one can recover at most a number of information elements equal to the number of resolution elements contained in $2a$, the size of the object support. In other words, super-resolution is not achieved.

The previous result does not exclude the possibility of super-resolution when the support of the object is not much larger than the resolution distance. As a consequence of this remark, the possibility of super-resolution is reconsidered in [24] (and in [20] in the case of incoherent illumination) on the basis of two important remarks:

- if we consider a localized object with a size of the order of the resolution limit, then the corresponding value of c is not very large and the eigenvalues of $\hat{L}$ do not drop to zero so quickly as in the situation previously considered;
- if we assume that the image is detected for any value of t and not only in the interval $[-1, 1]$, then we must consider the integral operator

$$(Lx)(t) = \int_{-1}^{1} \frac{\sin[c(t-s)]}{\pi(t-s)} x(s)ds, \quad t \in \mathbf{R}, \tag{8.5}$$

which is a compact and injective operator from $L^2(-1, 1)$ into $L^2(\mathbf{R})$ so that the analysis must be performed in terms of its singular values (see section 3.6.4).

Since the adjoint of the operator L is given by

$$(L^*y)(s) = \int_{-\infty}^{\infty} \frac{\sin[c(s-t)]}{\pi(s-t)} y(t)dt, \quad |s| \leqslant 1, \tag{8.6}$$

by taking into account the projection properties of the sinc-kernel, it is easy to show that $L^*L = \hat{L}$. Then, noting that the PSWF are also an orthonormal basis in the subspace of the square-integrable band-limited functions with a band interior to the interval $[-c, c]$, if we introduce the functions

$$u_k(t) = \psi(c, t); \; k = 0, 1, 2, \ldots, \tag{8.7}$$

we obtain the shifted eigenvalue problem of the operator L

$$Lv_k = \sqrt{\lambda_k} u_k, \; L^*u_k = \sqrt{\lambda_k} u_k. \tag{8.8}$$

By taking into account the slower decay of the singular values $\sqrt{\lambda_k}$ with respect to that of the eigenvalues λ_k of $\hat{L}$ (we recall that $\lambda_k \leqslant 1$) it is possible to estimate, as a function of the noise level and of the size of the support of the object, the amount of super-resolution which is measured by the number of singular values greater than the noise level.

The idea is further developed in a series of papers [19, 20, 23] (for a review, see [22]) demonstrating the theoretical possibility of super-resolution by data inversion in specific situations. Roughly speaking, the basic condition is that the size of the support of the object to be super-resolved is not much greater than the resolution limit of the optical instrument.

Unfortunately, a similar analysis cannot be easily extended to the case of Poisson data; indeed, an analysis in terms of singular values and singular functions is strictly related to linear data inversion while the methods for Poisson data inversions are highly nonlinear. While the least-squares functional can be decomposed into the sum of the squares of the contributions of the different components, such a decomposition is not possible in the case of the data-fidelity function for Poisson data. However, the lesson arising from the previous analysis is that a constraint on the support of the object can be useful also in the Poisson case, whenever an estimate of this support can be derived from the observed data. In other words, the constraint on the support of the object to be imaged can be added to the other constraints discussed in section 4.1.

Before discussing this point, we point out that it has been claimed by other authors that other constraints may provide super-resolution. One is non-negativity, which is implemented, for instance, in the maximum entropy method and is also a constraint required in the problems discussed in this book. In this context, it has been proved that maximum entropy provides super-resolution in the case of the so-called nearly black objects, i.e. objects that are zero in the vast majority of the pixels [46]. Moreover, non-negativity works well when combined with the sparsity of the object, as shown in [45]. If we remember that the minimizers of the Poisson data-fidelity function are non-negative and, in the case $b = 0$, satisfy the flux condition (see section 4.1), one can conjecture that they can have some super-resolving property (see, for instance, [13]).

Now, the EM–RL method, in the case $b = 0$, implements in a quite natural way both the constraint of non-negativity and that on the total flux of the image. Moreover, it can also implement, with no additional computational cost, the constraint on the domain of the object if a suitable initialization of the iterations is used. Indeed, if the initial guess is zero in one pixel, then as easily follows by induction from the structure of the iteration, all the subsequent iterates will be zero in that pixel. The same remark applies to the OSEM method, both in the case of emission tomography and in the case of astronomical imaging [12]. Therefore, these methods provide algorithms implementing all the relevant constraints and they appear quite suitable for estimating the amount of super-resolution achievable in practical situations.

An application to astronomical imaging, including the case of Fizeau interferometry (see section 2.3.4.1), is discussed in [4], freely downloadable from https://www.aanda.org/articles/aa/abs/2005/08/aa0366-04/aa0366-04.html.

The purpose is to resolve unresolved binary stars or small unresolved star clusters. An example is shown in figure 1 of that paper. To this purpose a three-step approach is proposed:

- **Step 1**: Deconvolve the image with one of the available deconvolution methods. Identify unresolved groups of stars (their images appear as small clouds). If their angular separation is much larger than the Rayleigh limit, associate a mask to each cloud by setting 1 in the pixels where the values of the reconstructed cloud are greater than a given percentage of their maximum value (for instance 10% or even less). In this way a mask is associated to each cloud.
- **Step 2**: Deconvolve again the given image using now the previous mask as an additional constraint, i.e. require that the object is zero outside the masks (as already remarked, in the case of EM–RL or OSEM, it is sufficient to use the masks for initializing the algorithms). If the support constraint is sufficient for

resolving the point sources, the result of this second step is a (small) number of point sources inside the domains identified in the previous step.

- **Step 3**: As follows from numerical simulations, the first two steps provide the correct locations of the stars but not their magnitudes. Therefore, the detected image is locally modeled by a set of stars with given positions but unknown intensities. These are determined by a least-squares approach which compares the images of these sets of stars with the detected image.

Since the results of [4] demonstrate that, in principle, an improvement of resolution by a factor from 4 to 6 is possible, such an improvement in resolution may not be compatible with the image sampling implemented in the detector. A possible solution of this problem is indicated in that paper.

In order to exemplify the improvement in resolution provided by the image deconvolution methods based on the assumption of Poisson data, with the constraints of non-negativity and flux conservation without the additional constraint on the support of the object, we show two examples of super-resolution.

The first example is obtained by zooming-in on the central part of the images shown in the second row of figure 7.16; these partial images are shown in figure 8.1. From these enlarged versions it is evident that, after deconvolution, the filaments are much narrower and it is possible to identify crossing points and small filaments which are not visible in the original image.

We point out that this is not a localized object. To estimate the improvement in resolution we compare the modulus of the FT of the images before and after deconvolution. From this comparison we can estimate an increase in band-width by a factor between 2 and 3; hence this is also an improvement in resolution, or, if you prefer, the amount of super-resolution achievable by means of deconvolution methods based on the Poisson data approach.

The second example is the deconvolution of the simulated image of a binary star with an angular separation between the two stars which is about ten times smaller than the Rayleigh limit. The result is shown in figure 8.2. The binary is resolved even if an extremely large number of iterations is required.

More precisely, the binary image is simulated starting from an object which consists of two pixels, with intensity $1.e + 09$, separated by 10 pixels. We assume that the pixel size is 5 mas and that the object is convolved with an ideal PSF (Airy

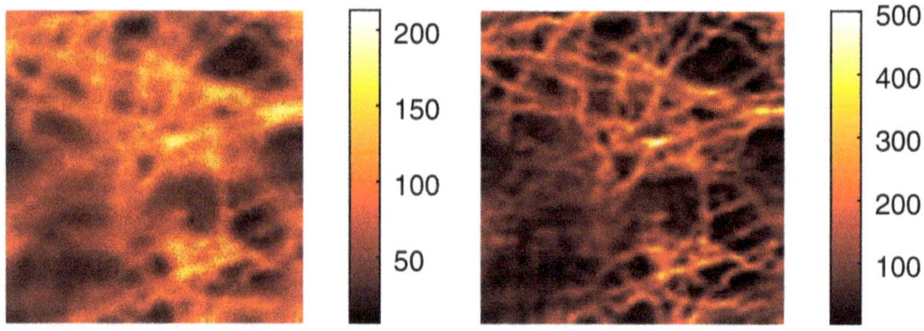

Figure 8.1. Comparison of a sub-domain of a confocal image before (left) and after (right) deconvolution.

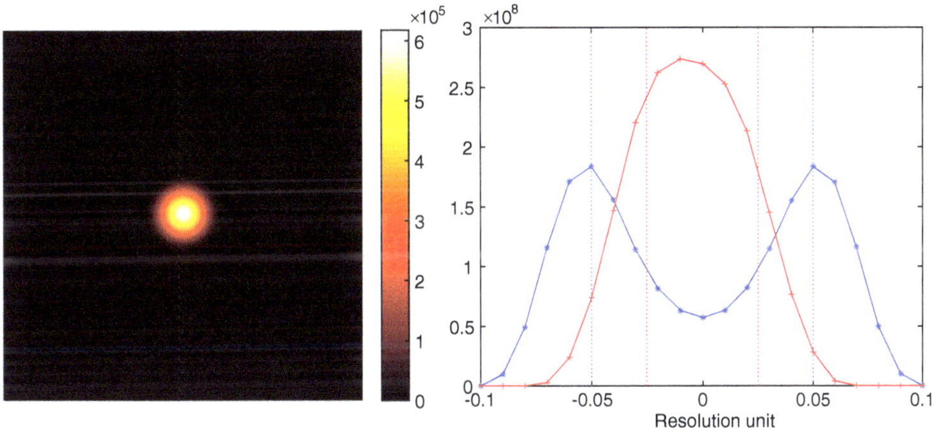

Figure 8.2. Reconstruction of a binary star showing an improvement in resolution by a factor of 10. Left panel: the image of the binary with an angular separation of 50 mas (see the text). Right panel: the profile of the reconstruction (blue line). In the same panel we show the profile of the reconstruction of a binary with an angular separation of 25 mas (red line). The binary is not resolved.

pattern), corresponding to a telescope with a diameter of 1 m at a wavelength of 2.2 μm. Atmospheric turbulence is neglected, so that we assume a diffraction-limited observation. The resulting image is corrupted with Poisson noise. The Rayleigh distance, in this case, is 550 mas and therefore a considerable over-sampling of the image is considered. Moreover, the angular separation of the two stars is 50 mas. The reconstruction is performed by pushing to convergence the reconstruction algorithm (SGP without regularization). If the distance between the two point sources is reduced by a factor of 2 (5 pixels) the reconstruction algorithm is unable to separate the two sources (red lines in figure 8.2).

8.1.1 Application to confocal microscopy

The case of non-uniform illumination of the object is considered in [23], having in mind, for instance, the illumination by a Gaussian beam. However, precisely in that period the group of Fred Brakenhoff published the first convincing images taken by a confocal microscope and was able to answer to biological questions [32], proving the importance of confocal microscopy in biology. As described in section 2.1.2, in confocal microscopy the illumination of the sample is very localized and this is a typical situation for getting super-resolution. Therefore, in [21] (see also [15]), thanks to theoretical results demonstrated in previous works, the authors propose to replace, in the confocal plane (see figure 2.2), the pinhole of the standard confocal microscope with an array of detectors and to process the data by an image reconstruction method for improving the quality of the image. The resolution achievable with such a system is discussed in [82] in the case of a simple data processing.

A mathematical model of the approach proposed in [21] can be formulated in a simple way by assuming that the difference between excitation and emission wavelength can be neglected and introducing the PSF $h(\vec{r}) = |h_{\mathrm{ill}}(\vec{r})|^2 = |h_{\mathrm{det}}(\vec{r})|^2$) (see equation (2.8)); then, in the case of magnification one, the image in the confocal plane, for a given scanning position $\vec{s}$ of the object, is given by

$$y(\vec{r}, \vec{s}) = \int h(\vec{r} - \vec{r}\,')h(\vec{r}\,')x(\vec{r}\,' + \vec{s})d\vec{r}\,'. \tag{8.9}$$

If we neglect the finite-size of the pinhole, we can obtain the confocal image by setting $\vec{r} = 0$ in the previous equation. Then, by means of a change of variable, if h is a symmetric function of $\vec{r}$ we obtain the image provided by a confocal microscope

$$y_{\mathrm{con}}(\vec{s}) = y(0, \vec{s}) = \int h^2(\vec{s} - \vec{r}\,')x(\vec{r}\,')d\vec{r}\,', \tag{8.10}$$

showing that, under the previous simplifying assumptions, the band-width of a confocal microscope is twice the band-width of a wide-field one. However the improvement of resolution is not by a factor of 2 because of the behavior of the OTF over the band.

Then, the idea described in [21] consists in solving the integral equation (8.9), with a given $y(\vec{r}, \vec{s})$, for obtaining an approximate solution $\tilde{x}(\vec{r}, \vec{s})$; the value $\tilde{x}(0, \vec{s})$ is used as an improvement of $y(0, \vec{s})$ in the scanning procedure. Concerning the solution of the integral equation, the proposed method is based on truncated singular function expansion; the purpose is to 'fill' the band of the confocal microscope for obtaining a confocal resolution which is twice the resolution of a wide-field microscope; this is what is called super-resolution in confocal scanning microscopy. Both 2D [17] and 3D [18] super-resolution is demonstrated theoretically. Due to technological difficulties in the realization of the idea, a version of this confocal microscope obtained by replacing the pinhole in the confocal plane by a suitable optical mask able to perform data inversion is proposed in [16]. The feasibility of super-resolution is soon demonstrated [53] and the idea is also extended to 3D imaging [3]. An account of this kind of confocal microscope, which provides the expected gain in resolution and contrast with respect to standard confocal microscopy, is given in [39], chapter 10.

The idea of replacing the pinhole with an array of detectors, in particular a CCD camera, is re-proposed in [69], where the optical scheme of the microscope is given and a much simpler method of data processing, coinciding with that proposed in [82], is described. Using the notation previously introduced, the authors define the following *effective* image:

$$y_{\mathrm{eff}}(\vec{s}) = \int y\left(\vec{r}, \vec{s} - \frac{\vec{r}}{2}\right)d\vec{r}. \tag{8.11}$$

This operation, in the discrete case, is called *pixel reassignment* (for a discussion see for instance [35]). Then, from the expression (8.9) of the confocal image, by means of a change of variable and using again the symmetry of $h(\vec{r})$ one can write equation (8.11) in the following form

$$y_{\text{eff}}(\vec{s}) = \int h_{\text{eff}}(\vec{s} - \vec{r})x(\vec{r})d\vec{r}, \tag{8.12}$$

where

$$h_{\text{eff}}(\vec{s}) = \int h\left(\vec{s} + \frac{\vec{r}}{2}\right)h\left(\vec{s} - \frac{\vec{r}}{2}\right)d\vec{r} = 4\int h(2\vec{s} - 2\vec{r})h(2\vec{r})d\vec{r}. \tag{8.13}$$

By computing the FT of h_{eff} we obtain

$$\hat{h}_{\text{eff}}(\vec{\omega}) = \left[\hat{h}\left(\frac{\vec{\omega}}{2}\right)\right]^2, \tag{8.14}$$

and therefore h_{eff} has the same band-width of the confocal microscope, as is obvious.

To understand the improvement provided by this approach, which is called by the authors of [69] *image scanning microscopy* (ISM), in figure 8.3 we compare the 2D OTFs of wide-field and confocal microscope with that of ISM. It follows that the OTF of ISM is slightly larger than the confocal OTF so that a deconvolution approach based on equation (8.11) should be more effective than a direct deconvolution of the confocal image (8.10). Moreover, one should take into account that: (i) the confocal OTF plotted in the figure corresponds to a point-like pinhole while that corresponding to a finite-size pinhole takes smaller values; (ii) the SNR of the image provided by ISM is higher than the SNR of a confocal image. For the OTF achievable with the approach described at the beginning of this section see [17].

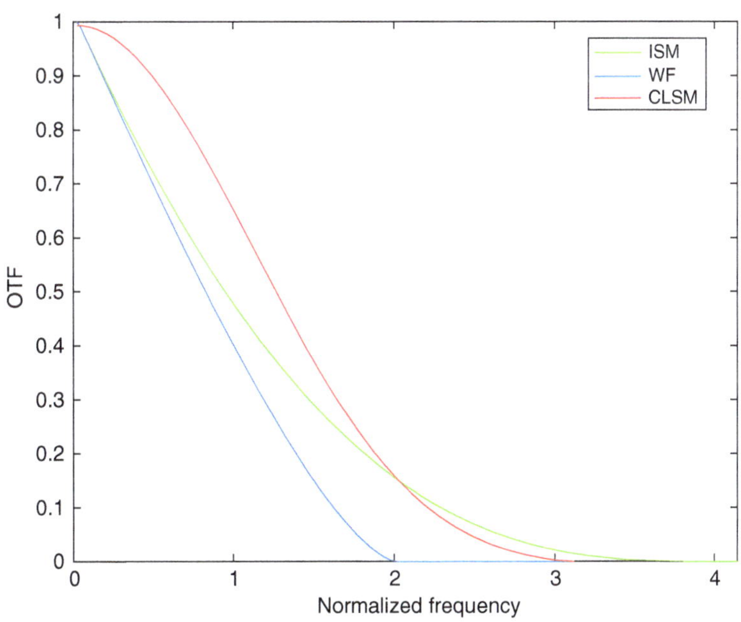

Figure 8.3. Comparison of the OTF of wide-field (WF), confocal (CLSM) and image scanning microscopy (ISM) as a function of normalized frequency, i.e. the true frequency multiplied by the factor NA/λ.

Due to the limited frame-rate of the camera, the implementation discussed in [69] represents more a proof-of-principle than an effective solution of confocal imaging with detector array. More recently new detector devices, such as the airy-scan (a hexagonal-shaped bundle of optical fibers connected to a one-dimensional array of GaAsP photo multiplier tubes (PMT) [58]) or the single photon avalanche diode (SPAD) array [35] have provided effective implementations. It is interesting that computationally expensive image reconstructions, such as those obtained by means of nonlinear deconvolution, have been implemented on the resulting scanned images, i.e. the images generated by each element of the detector array at the end of the beam scanning across the sample [35], but, so far, no method working directly on the micro-images, generated point-by-point, has been explored. The authors remark that the multiple-image deconvolution, described in this book for Fizeau interferometry, could be applied to the set of different images generated by each point of the detector.

8.2 Boundary artifacts correction

In astronomy and microscopy, if the observed object is in the center of the image domain and is surrounded by a signal-free region whose angular size is greater than the main width of the PSF, then FFT-based methods for image deconvolution can be successfully and efficiently used. However, as a consequence of the limited field of view (FoV) of the optical instrument, it may happen that an extended object is not completely contained within the image domain; in other words the boundaries of the image do not correspond to a signal-free region. In such a case the use of FFT implicitly assumes a periodic continuation of the image outside the original domain; as a consequence, discontinuities appear at the boundaries and, in the deconvolved image, these discontinuities generate Gibbs oscillations (sometimes called *ripples*), which can propagate inside the image domain and degrade completely the quality of the reconstruction.

Two main approaches are frequently used for overcoming this difficulty. The first, *apodization*, consists in multiplying the image by a suitable mask which is equal to 1 in the central region and smoothly tends to 0 at the boundary, so that also the masked image smoothly tends to zero at the boundary and its periodic continuation is free of discontinuities. The physical interpretation is that the effect of the mask consists in reducing the effect of all the sources, both internal and external to the FoV, which are close to the boundary.

The second approach consists of a suitable continuation of the image outside the FoV, ensuring continuity of the extended image. This is the approach based on the so-called *boundary conditions* (BC). Different kinds of BC are proposed in the literature. Reflexive (or symmetric) BC are proposed in [2], Neumann's BC in [71] and anti-reflexive BC in [81]. These conditions remove the discontinuities at the boundary. For instance, reflexive BC preserve the continuity of the image while the anti-reflexive ones preserve the continuity of the image and its gradient. In general, the use of BC provides a reduction of the ripples which, however, depends on the particular image. As a general comment, we must point out that:

- the use of some specific BC is equivalent to extend the image outside the FoV by means of some symmetry rule; for instance, in the case of reflexive BC, the data outside the FoV are obtained by reflection of the data inside the FoV, a procedure which can hardly produce the correct (and unknown) values of the image and of the target outside the FoV, i.e. the extension of the image provided by these approaches is somehow artificial;
- all the BC approaches have been implemented and tested in the framework of least-squares methods (LSM) even if some of them can be applied also to EM–RL;
- as far as we know, efficient implementation is possible only in the case of symmetric PSF, a condition which is approximately satisfied in microscopy but is hardly satisfied in astronomy.

The basic point is that, as a consequence of the extent of the PSF, the image of an object close to the boundary is not completely contained in the FoV, so that some information is missed; on the other hand, an object outside the FoV, but close to the boundary, can significantly contribute to the values of the image in the pixels which are close to the boundary, so that the image contains some additional information coming from external sources. Such a situation is illustrated in figure 8.4 where we

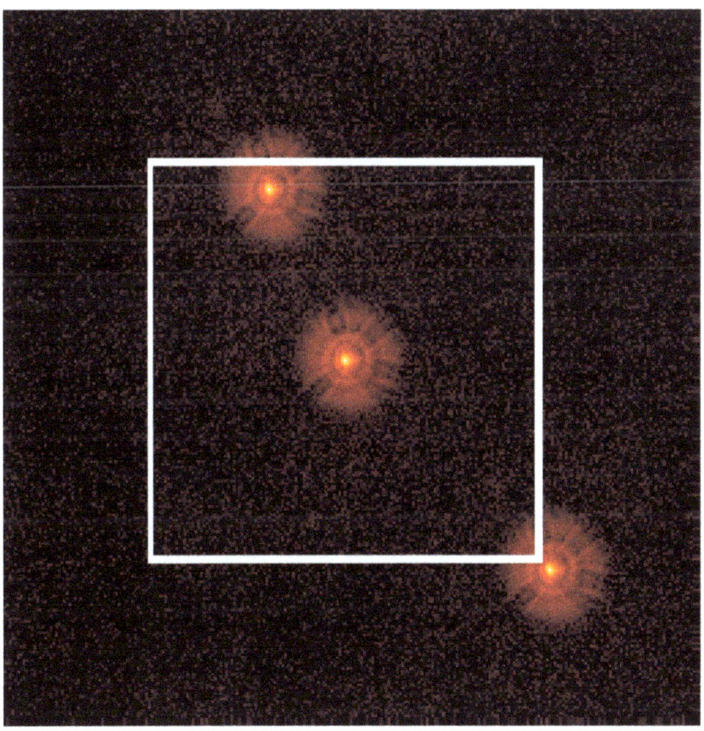

Figure 8.4. Example of an image with boundary effects. The white square indicates the region of the FoV. The image of the central star is completely within the FoV while that of the upper star is partially external and that of the lower star is partially internal.

show three stars, one with its image (the PSF of the telescope) completely internal to the FoV, one external to the FoV but sufficiently close to the boundary so that a part of its image lies inside the FoV and finally one internal to the FoV but sufficiently close to the boundary so that a part of its image lies outside the boundary of the FoV.

As a consequence of this remark, one should formulate the problem by introducing external pixel values as unknown parameters to be estimated by the deconvolution method. The main difficulties generated by this approach are evident:

- the problem is under-determined because the number of unknown values is greater than the number of data;
- the computational efficiency provided by FFT is lost because the problem cannot be formulated as a standard deconvolution problem.

However, these difficulties can be overcome. For the physical motivations mentioned above we prefer a data-driven approach which was proposed years ago and implemented in the *lucy* task of STSDAS [85, 93], the software developed for the deconvolution of HST images; unfortunately, a complete description of the implementation details is not available in the refereed literature. The approach, derived from ideas developed in [76] in the case of medical imaging, is based on the use of a weight array compensating for the flat field and masking the bad pixels, as well as the outer parts of the image. A slightly modified version of this approach is independently proposed in [14] (extended to interferometric imaging in [5]) as a modification of the EM–RL method; in [77] and [95] it is extended to SGP for applications to astronomy and microscopy, respectively.

Here, we present the approach in a context more general than that considered in the original papers. For simplicity, we consider the 2D case and we assume that both the image and the source object have the same dimension so that both y and x are $N \times N$ arrays, denoted by S. The extension to the 3D case is trivial.

According to the previous discussion we intend to reconstruct the object on a broader array, denoted by $\tilde{S}$ and centered on S, let us say 2N × 2N; this choice may provide a too broad array and is only justified by the use of the most efficient FFT in the reconstruction algorithms, but it can be replaced by another suitable choice. We denote by $\tilde{x}$ the object defined on $\tilde{S}$; moreover, since we are considering deconvolution problems, we need an imaging matrix $\tilde{H}$ defined in terms of a PSF $\tilde{h}$, which is also an array defined on $\tilde{S}$, i.e. $\tilde{H}\tilde{x} = \tilde{h} * \tilde{x}$. If the PSF is known only on S, then the extension to $\tilde{S}$ can be performed in several ways: for instance, by zero padding, if the PSF is sufficiently localized at the center of S, or by means of a good model, if available. In any case $\tilde{h}$ must be normalized to 1 on $\tilde{S}$ (if the extension is by zero padding it is obvious that no further normalization is required since the original PSF h is normalized to 1 on S). In conclusion the imaging matrix $\tilde{H}$ is given by

$$(\tilde{H}\tilde{x})_{i,\,k} = \sum_{j,l=0}^{2N-1} \tilde{h}_{i-j,\,k-l}\tilde{x}_{j,\,l};\ \ i, k = 0, 1, \dots, 2N - 1 \tag{8.15}$$

and therefore it transforms an object defined on $\tilde{S}$ into an image also defined on $\tilde{S}$; similarly, its transposed $\tilde{H}^T$ transforms an image z defined on $\tilde{S}$ into an object also defined on $\tilde{S}$

$$(\tilde{H}^T z)_{j,\,l} = \sum_{i,k=0}^{2N-1} \tilde{h}_{i-j,\,k-l} z_{i,\,k};\ j,\, l = 0,\, 1,\, \dots,\, 2N-1. \tag{8.16}$$

We also extend the background b to the broader array and we denote the extension by $\tilde{b}$. We recall that in microscopy and astronomy we always have $b > 0$. If b is a constant array, the extension is obvious, otherwise one must search for an extension which does not introduce discontinuities.

The very simple idea, which can be derived from [14], consists of: (a) introduce in $\tilde{S}$ an array M_S which is 1 on S and zero elsewhere, and (b) extend y by zero padding to $\tilde{y}$, defined on $\tilde{S}$.

Taking into account all the previous definitions, the data-fidelity function can be written as follows

$$f_0(\tilde{x};\tilde{y}) = \sum_{i,k=0}^{2N-1} \left\{ \tilde{y}_{i,\,k} \ln \frac{\tilde{y}_{i,\,k}}{(\tilde{H}\tilde{x}+\tilde{b})_{i,\,k}} + (\tilde{H}\tilde{x}+\tilde{b})_{i,\,k} - \tilde{y}_{i,\,k} \right\}, \tag{8.17}$$

and its gradient is given by

$$\nabla_{\tilde{x}} f_0(\tilde{x};\tilde{y}) = \tilde{v} - \tilde{H}^T \frac{\tilde{y}}{\tilde{H}\tilde{x}+\tilde{b}}, \tag{8.18}$$

where the quotient is intended component-wise, $\tilde{v}$ is an $2N \times 2N$ array given by

$$\tilde{v}_{j,\,l} = \sum_{i,k=0}^{N-1} \tilde{h}_{i-j,\,k-l};\ j,\, l = 0,\, 1,\, \dots,\, 2N-1 \tag{8.19}$$

and $\tilde{H}^T$ is defined in equation (8.16). Note that the denominator in equation (8.17) is automatically restricted to S by the numerator $\tilde{y}$.

The point is to compute efficiently the gradient, i.e. $\tilde{v}$ and $\tilde{H}^T$. They can be computed efficiently by means of FFT if we remark that

$$\tilde{v} = \tilde{H}^T M_S,\ \tilde{H}^T \frac{y}{\tilde{H}\tilde{x}+\tilde{b}} = \tilde{H}^T \frac{\tilde{y}}{\tilde{H}\tilde{x}+\tilde{b}}. \tag{8.20}$$

It is clear that, with respect to a standard deconvolution, the computational cost is increased from $4N^2 \log_2 N$ to $16N^2 \log_2(2N)$.

The crucial quantity is the array $\tilde{v}$ which indicates the *extent of influence* of the PSF. To clarify this point we show an example of $\tilde{v}$ in figure 8.5. Its values are approximately 1 inside S (exactly 1 in the central pixel) and they start to decrease in points close to the boundary becoming very small, possibly zero, in pixels distant from S. Such behavior is coherent with the fact that distant objects, external to S, contribute very little to the image inside S.

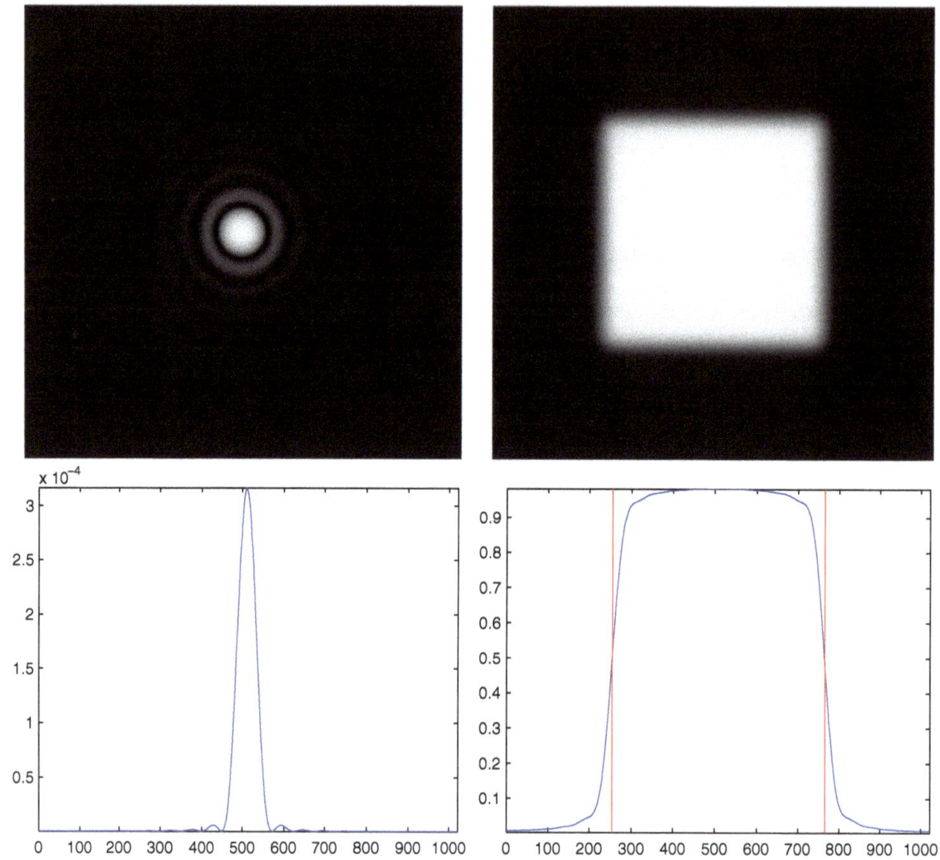

Figure 8.5. Upper panels: plot of the PSF, log scale (left panel) and plot of the $\tilde{v}$ function, linear scale (right panel). Lower panels: cut of the PSF (left panel) and of the $\tilde{v}$ function (right panel), both in linear scale.

We must take into account this behavior in deriving a modified EM–RL, able to correct for boundary artifacts, in a way similar to that used in chapter 5 for deriving standard EM–RL. Indeed from the first KKT condition for the data-fidelity function (8.17), if $\tilde{x}^*$ is a minimizer, we obtain the relationship

$$\tilde{v}\tilde{x}^* = \tilde{x}^* \tilde{H}^T \frac{\tilde{y}}{\tilde{H}\tilde{x}^* + \tilde{b}}, \tag{8.21}$$

which can be transformed into a fixed point equation if we divide by $\tilde{v}$. But this division can imply a division by very small numbers and possibly by zero if the PSF has been extended by zero padding. It is obvious that such an annoying feature is due to the fact that the image defined on S does not contain sufficient information on parts of the object which are too far from the boundary of S. Therefore, the point is to determine a region, let us say R with $S \subset R \subset \tilde{S}$, such that the pixels of R provide a non-negligible contribution to the image in S. One can proceed as follows:

- introduce a threshold σ, for instance $\sigma = 10^{-2}$;
- define R as the set of the pixels j, l where $\tilde{v}_{j,l} \geqslant \sigma$.

The final step is to consider a constrained minimization of $f_0(\tilde{x};\tilde{y})$ by constraining $\tilde{x}$ to be zero outside R.

In the case of EM–RL such a constraint can be introduced in a very simple way by defining the following 'window'

$$(w_R)_{j,\,l} = \begin{cases} \dfrac{1}{\tilde{v}_{j,\,l}}, & j,\, l \in R, \\ 0, & \text{elsewhere}. \end{cases} \tag{8.22}$$

Then, the modified EM–RL method, taking into account boundary artifacts correction is given by

$$\tilde{x}^{(k+1)} = w_R \tilde{x}^k \tilde{H}^T \frac{\tilde{y}}{\tilde{H}\tilde{x}^{(k)} + \tilde{b}}, \tag{8.23}$$

which is just the algorithm proposed in [14].

The very simple approach described above can be applied not only to EM–RL and OSEM, as proposed in [5, 14] but also to other iterative methods, such as, for instance, SGP for the minimization of the data-fidelity function without or with differentiable regularization [77].

In the case of OSEM with balanced subsets, boundary-artifact correction is given by algorithm 5.1 where step 2 is replaced by

$$\tilde{x}^{(k,\,l)} = w_R^{(l)} \tilde{x}^{(k,\,l-1)} \left(\tilde{H}^{(l)}\right)^T \frac{\tilde{y}^{(l)}}{\tilde{H}^{(l)}\tilde{x}^{(k,\,l-1)} + \tilde{b}}, \tag{8.24}$$

where $w_R^{(l)}$ is the window function associated with the imaging matrix $\tilde{H}^{(l)}$. As remarked in section 5.2 this version of OSEM can be used in the deconvolution of the images of a Fizeau interferometer.

In the case of SGP, it is sufficient to modify the scaling of the gradient which now has the form

$$(D_k)_{i,\,i} = \max\left\{\frac{1}{L_k},\, \min\left\{L_k,\, \frac{1}{x_i^{(k)}(w_R)_i}\right\}\right\}, \quad i = 1, \ldots, N \tag{8.25}$$

with $L_k \geqslant 1$ given as in equation (6.19). The software implementing this version of SGP is described in [77].

In figure 8.6 we demonstrate the accuracy that can be obtained by means of this approach. A blurred image of the Crab nebula, perturbed by Poisson noise, is partitioned into four quadrants, as shown in the left panel. Three of them are deconvolved using the algorithm with boundary correction, while the fourth one is deconvolved using standard EM–RL; the full object is obtained as a mosaic of the four reconstructions (500 iterations in all cases). The reconstructions obtained with equation (8.23) are essentially free of ripples while these are clearly visible in the reconstruction provided by equation (5.1). For more details we refer to [14].

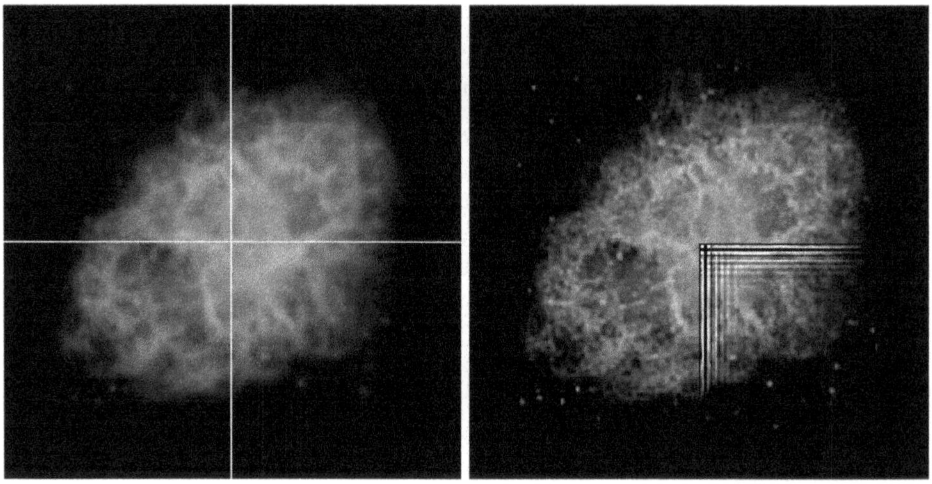

Figure 8.6. Left panel: the blurred image of the Crab Nebula with the indication of the partition into four quadrants. Right panel: the mosaic obtained by deconvolving three quadrants with the algorithm of equation (8.23) and the fourth one with standard EM–RL (reproduced from: Bertero *et al* 2009 *Inverse Problems* **25** 123006).

8.3 Blind deconvolution

Blind deconvolution is the problem of image deblurring when the PSF is not known or only approximately known, in the sense that a rough approximation from observation or a model, containing a limited number of unknown parameters, is available; in these cases the term *myopic deconvolution* is sometimes used. For simplicity, the problem is in general considered by assuming a space-invariant model for the PSF.

The naive formulation of the problem is to solve the equation $y = h * x$ where both the PSF h and the object x are unknown and must be estimated from the recorded image y. It is obvious that the problem is extremely under-determined and that there exists an infinite set of solutions. One of them is the trivial one, i.e. $x = y$ and $h = \delta$, where δ is the delta-array, which is 1 at the center and zero elsewhere. Moreover, if the pair $\{\tilde{h}, \tilde{x}\}$ is a solution and R is an injective linear operation commuting with the cyclic convolution, then the pair $\{R\tilde{h}, R^{-1}\tilde{x}\}$ is also a solution.

Blind deconvolution is the subject of many publications in the literature and the different approaches concern specific classes of images and PSFs. For instance, approaches applicable to natural images may not be suitable in microscopy or astronomy; approaches developed for motion blur are not applicable to other classes of blur, and so on. The obvious reason is that, since the problem is extremely ill-posed, specific kinds of prior knowledge must be introduced, for both the object and the PSF, in order to reduce the class of possible solutions. The interested reader can find a description of several approaches and methods in [33]; as concerns natural images, a recent paper [67] contains a critical analysis of proposed methods as well as several relevant references.

Since this book is devoted to Poisson data, we restrict the discussion to the methods of blind deconvolution proposed in this framework. To this purpose we provide a formulation of the problem along the lines of chapter 4. Since now the unknown quantities are h, x, the likelihood is given by

$$\mathcal{L}_y(h, x) = \prod_{i \in \mathcal{I}} e^{-(h*x)_i} \frac{(h * x)_i^{y_i}}{y_i!}; \quad h, x \geqslant 0, \tag{8.26}$$

so that its negative logarithm, except for an irrelevant constant, leads to the data-fidelity function

$$f_0^B(h, x; y) = \mathrm{KL}(y; h * x + b). \tag{8.27}$$

Thanks to the symmetry of the convolution product, the data-fidelity function is convex with respect to each one of the two blocks of variables h and x, with the other fixed, but is not convex with respect to the totality of the variables.

The Bayesian approach is obtained by considering h, x as realizations of multi-valued r.v.s, respectively H and X (here, and only here, H denotes a r.v. and not the imaging matrix) so that, the conditional probability of H, X for a given value y of Y is given by

$$\mathrm{P}_{H,X}(h, x|y) = \frac{\mathrm{P}_Y(y|h, x)\mathrm{P}_H(h)\mathrm{P}_X(x)}{\mathrm{P}_Y(y)}, \tag{8.28}$$

where $\mathrm{P}_H(h)$, $\mathrm{P}_X(x)$ are the priors of the PSF and object respectively.

Finally, by assuming Gibb's priors and proceeding as in chapter 4, we obtain that the MAP estimates of h, x can be obtained by minimizing, with respect to both sets of variables h, x the following objective function

$$f_{\beta,\gamma}^B(h, x; y) = f_0^B(h, x; y) + \beta f_1(h) + \gamma f_2(x), \tag{8.29}$$

where f_1, f_2 are the regularization functions of the PSF and object, respectively; β, γ are the corresponding regularization parameters. In the case of convex regularization functions, the objective function is convex with respect to each one of the two blocks of variables, with the other fixed, but is not convex with respect to the totality of the variables. Of course, we can also consider different constraints for the two blocks of variables, possibly defining convex sets as we discuss in a moment; however, in any case we must solve a constrained and non-convex minimization problem, so that we expect the existence of local minima, stationary points, etc, an additional difficulty to be managed in the case of blind problems.

Let C_h, C_x be the convex sets defined by the constraints on the two blocks of variables; then the problem is a particular case of the classic optimization problem

$$\min_{h \in C_h, x \in C_x} f(h, x), \tag{8.30}$$

where f is a non-convex, continuously differentiable function which is convex with respect to each block of variables. Since in the case of blind deconvolution we have a natural grouping of the variables into two blocks, we consider only this case and not the more general case of an arbitrary number of blocks. However, we point out that the general case is relevant in Fizeau interferometry where multiple images of the same target, corresponding to different PSFs, are available. In this case, if L is the number of observed images hence the number of unknown PSFs, then the number of blocks is $L + 1$.

A classic approach to the solution of the previous problem is the so-called *alternating optimization method,*, also known as *nonlinear block Gauss–Seidel* or *block coordinate descent method* ([25], chapter 2); it consists in solving problem (8.30) by successively minimizing f with respect to each block of variables, over the corresponding constraint set, by keeping the other fixed. The iteration can be schematized as follows:

- give an initial point $h^{(0)} \in C_h$, $x^{(0)} \in C_x$;
- for $k = 0, 1, \ldots$

$$\begin{aligned} h^{(k+1)} &= \arg\min_{h \in C_h} f(h, x^{(k)}) \\ x^{(k+1)} &= \arg\min_{x \in C_x} f(h^{(k+1)}, x); \end{aligned} \tag{8.31}$$

therefore, each iteration implies the solution of two convex minimization problems.

As concerns the convergence of the method, it is known that the limit points of the sequence generated by the method are stationary for problem (8.30) when the minimum problems in equation (8.31) have a unique solution (see, for instance, [25]). However, in the case of two blocks, which is the case of blind deconvolution, the most remarkable result is given in [52] where it is shown that, in the case of two blocks, every limit point of the sequence is a stationary point of equation (8.30) without specific convexity assumptions. Therefore, the result is of interest when the function f, restricted to each block, is not strictly convex.

The result has a theoretical relevance but it is hardly usable in practice because, at each iteration, one must solve two convex minimization problems. To this purpose, in the case of non-negativity as a unique constraint and $\beta = \gamma = 0$, one can use EM–RL or one of its accelerated versions for each one of the sub-problems, while, if at least one of the regularization parameters is greater than zero, one can use SGM for minimizing the corresponding objective function. In all cases one can use, for instance, SGP if additional constraints are introduced. The obvious difficulty is the excessive computational cost.

For possible applications the result proved in [31] on *inexact block coordinate descent* approach is very interesting. An inexact minimization is obtained if the approximate solutions in problem (8.31) are computed by means of a finite number of iterations of an algorithm converging to the minimizer. According to [31] the limit points of the resulting sequence $\{h^{(k)}, x^{(k)}\}_{k=0}^{\infty}$ are stationary points of problem (8.30), for any possible choice of the number of iterations used. The basic condition is that each iteration consists of a gradient projection step based on an Armijo line-search along the feasible direction with variable step-size. The result holds true for an arbitrary number of blocks, without particular convexity assumption on the objective function, and therefore holds true for the problem of blind deconvolution. This result provides a framework for a critical analysis of some of the proposed blind deconvolution methods.

Inspired by the alternating approach proposed in [8], the early methods are based on a single EM–RL iteration without reference to minimization of an objective function. In these methods the alternation is given by

- initialize with $h^{(0)}$ such that $\|h^{(0)}\|_1 = 1$ and $x^{(0)}$ such that $\|x^{(0)}\|_1 = \|y\|_1$;
- for $k = 0, 1, \ldots$

$$\begin{aligned} h^{(k+1)} &= \frac{h^{(k)}}{\|y\|_1}[x^{(k)}]^T * \frac{y}{h^{(k)} * x^{(k)}} \\ x^{(k+1)} &= x^{(k)}[h^{(k+1)}]^T * \frac{y}{h^{(k+1)} * x^{(k)}}. \end{aligned} \tag{8.32}$$

The iteration takes into account that, if $b = 0$, then $\|h^{(k)}\|_1 = \|h^{(0)}\|_1 = 1$ and $\|x^{(k)}\|_1 = \|x^{(0)}\|_1 = \|y\|_1$.

This algorithm is proposed, independently, in [57] for microscopy, in [61, 91] for astronomy and in [48] for a more general context. It may be interesting to remark that equation (8.32) is a particular case of inexact alternating minimization, with only one iteration in the two approximate minimizations, when the objective function is $f_0^B(h, x; y)$ and only non-negativity is required. However, the iterative method, i.e. EM–RL, does not satisfy the conditions required in [31]. Indeed, for a general descent method to be convergent, the strongest Armijo-like decreasing conditions have to be verified [72].

A similar algorithm is proposed in [27], but using the accelerated version of EM–RL proposed by the authors in [26]. In this case different numbers of iterations are used for the two approximate minimizations. Also, in this case no convergence proof is available. Finally, a method for Fizeau interferometry is proposed in [43] where the OSEM algorithm is used for the object while EM–RL is used for each one of the PSFs. In this problem the number of blocks is greater than 2.

The problem of blind deconvolution is strictly related to the problem of *non-negative matrix factorization*, a matrix decomposition technique with a variety of applications ranging from signal and image processing to document classification and bio-informatics [11, 65]. In this context, an interesting result is that the algorithm (8.32) is also proposed in [64, 65] for non-negative matrix factorization (in [65] the result is applied to PET imaging); in [65] the algorithm is derived by the method of surrogate functions (majorization–minimization method, see section 6.7) and, in this way, monotone convergence of the objective function is proved. It may be useful to recall the basic steps of the derivation of the result since the method of surrogate functions has been recently applied to regularized non-negative matrix factorization and blind deconvolution [40, 63].

In view of an extension to regularized blind deconvolution, we report here the derivation of the Lee and Seung algorithm. Let Y be a matrix with non-negative entries (possibly a vector) and let us wish to estimate a factorization of Y in the form

$Y = HX$, with H, X non-negative matrices. Then the following objective function is considered

$$\mathrm{KL}(Y;\, HX) = \sum_{i,j}\left\{ Y_{i,j} \ln \frac{Y_{i,j}}{(HX)_{i,j}} + (HX)_{i,j} - Y_{i,j} \right\}, \tag{8.33}$$

which is clearly a generalization of the data-fidelity function $f_0^B(h, x; y)$ of equation (8.27).

As discussed in section 6.7 a surrogate function of $\mathrm{KL}(Y;\, HX)$ with respect to X, taking into account also the terms depending only on Y, is given by

$$\begin{aligned} G(X, Z) = &\sum_{i,j}\{ Y_{i,j} \ln Y_{i,j} - Y_{i,j} + (HX)_{i,j}\} \\ &- \sum_{i,j} \frac{Y_{i,j}}{(HZ)_{i,j}} \sum_k H_{i,k} Z_{k,j} \ln\left[\frac{X_{k,j}}{Z_{k,j}} (HZ)_{i,j} \right] \end{aligned} \tag{8.34}$$

and a similar surrogate of $\mathrm{KL}(Y;\, HX)$ with respect to H is obtained by exchanging the role of X and H; in both cases one gets

$$G(X, Z) \geqslant \mathrm{KL}(Y;\, HX), \quad G(X, X) = \mathrm{KL}(Y;\, HX); \tag{8.35}$$

$$G(H, Z) \geqslant \mathrm{KL}(Y;\, HX), \; G(H, H) = \mathrm{KL}(Y;\, HX). \tag{8.36}$$

Then, assuming that the matrix H is $M \times P$ and the matrix X is $P \times N$, the algorithm is as follows:

- Initialize with H, X strictly positive.
- Given $X^{(k)}$, $H^{(k)}$, minimize $G(H, X^{(k)})$ with respect to H and take the minimizer as $H^{(k+1)}$; the result is

$$H^{(k+1)} = \frac{H^{(k)}}{1_{M\times N}(X^{(k)})^T} \circ \frac{Y}{H^{(k)}X^{(k)}} (X^{(k)})^T. \tag{8.37}$$

- Minimize $G(H^{(k+1)}, X)$ with respect to X and take the minimizer as $X^{(k+1)}$; the result is

$$X^{(k+1)} = \frac{X^{(k)}}{(H^{(k+1)})^T 1_{M\times N}} \circ (H^{(k+1)})^T \frac{Y}{H^{(k+1)}X^{(k)}}. \tag{8.38}$$

In the previous equations $\circ$ denotes the element-wise product of two matrices for avoiding confusion with the standard matrix product; moreover $\mathbf{1}_{M\times N}$ denotes the $M \times N$ matrix with all elements equal to 1.

The iteration defined by equations (8.37) and (8.38) defines the Lee and Seung algorithm. In [65], the authors prove, using the properties of the surrogate functions, that both steps produce a decrease of the objective function (8.33); this monotonicity proves convergence of the values of the objective function since the KL divergence is non-negative. However, convergence of the iterates is not proved.

The previously mentioned approaches do not consider the introduction of regularization terms nor the introduction of constraints besides non-negativity. Indeed, only the constraints which are assured by the EM–RL method are considered, namely non-negativity and value of the $\|.\|_1$ norm of the PSF (normalization) and of the object (flux conservation), the latter coinciding with the flux of the image. However, it is important to discuss other possible constraints on PSF and object since additional constraints reduce the set of possible solutions.

An example is the additional constraint provided by the knowledge of the Strehl ratio (SR) of the optical instrument, whose relevance is stressed in [44]. We recall that the SR is the ratio between the maximum value of an aberrated PSF versus that of a perfect PSF (the diffraction-limited one in astronomy, for instance); therefore it is a measure of the aberration of an optical system. In astronomical imaging the aberration is mainly due to atmospheric turbulence. Modern telescopes are equipped with AO systems (see section 2.3.4) which are able to partially compensate for the atmospheric blur, attempting to reach as far as possible the diffraction limit; in such a case SR is a measure of the achieved compensation. During the observation of an astronomical object the SR can be estimated by the astronomers and provided with a few percent error (about 4%–5%).

The SR constraint implies an upper bound on the PSF in addition to the lower bound of non-negativity and therefore, with such a constraint, the trivial solution provided by the δ-array as PSF is not allowed. In [78] the efficacy of this box constraint, combined with the normalization of the PSF, is tested by applying the result proved in [31] to the minimization of $f_0^B(h, x; y) = \mathrm{KL}(y; h*x + b)$; more precisely, in both steps of equation (8.31) the exact minimization is replaced by a finite number of SGP iterations, since this method satisfies the conditions required for convergence. We remark that SGP can be efficiently implemented because the projection on a convex set defined by box plus equality constraint, as required for the PSF, can be easily computed [36]. As concerns the object, only the non-negativity constraint is considered.

For providing an example to the reader we show the results of a numerical simulation discussed in [78]. In this simulation four PSFs, shown in figure 8.7, first column, are considered. The three PSFs with SR = 0.67, 0.40 and 0.17 are those used in [44] and can be generated by means of the software package CAOS (code for adaptive optics systems) [34]. The parameters corresponding to these PSFs are given in [44]. We only specify that they correspond to a telescope with an effective diameter of 8.22 m and an observation wavelength $\lambda = 1.65$ μm (H-band). The fourth PSF is a simulation of HST PSF before COSTAR (corrective optics space telescope axial replacement) correction, since it is frequently used in the testing of deconvolution methods. Of course, in such a case the SR must be computed by comparison with the ideal PSF, which is that of a telescope with a diameter of about 2.4 m, assuming an observation wavelength of 0.55 μm.

Since regularization terms on the object are not introduced we expect that the approach can work mainly on point-wise objects, i.e. images of star clusters in

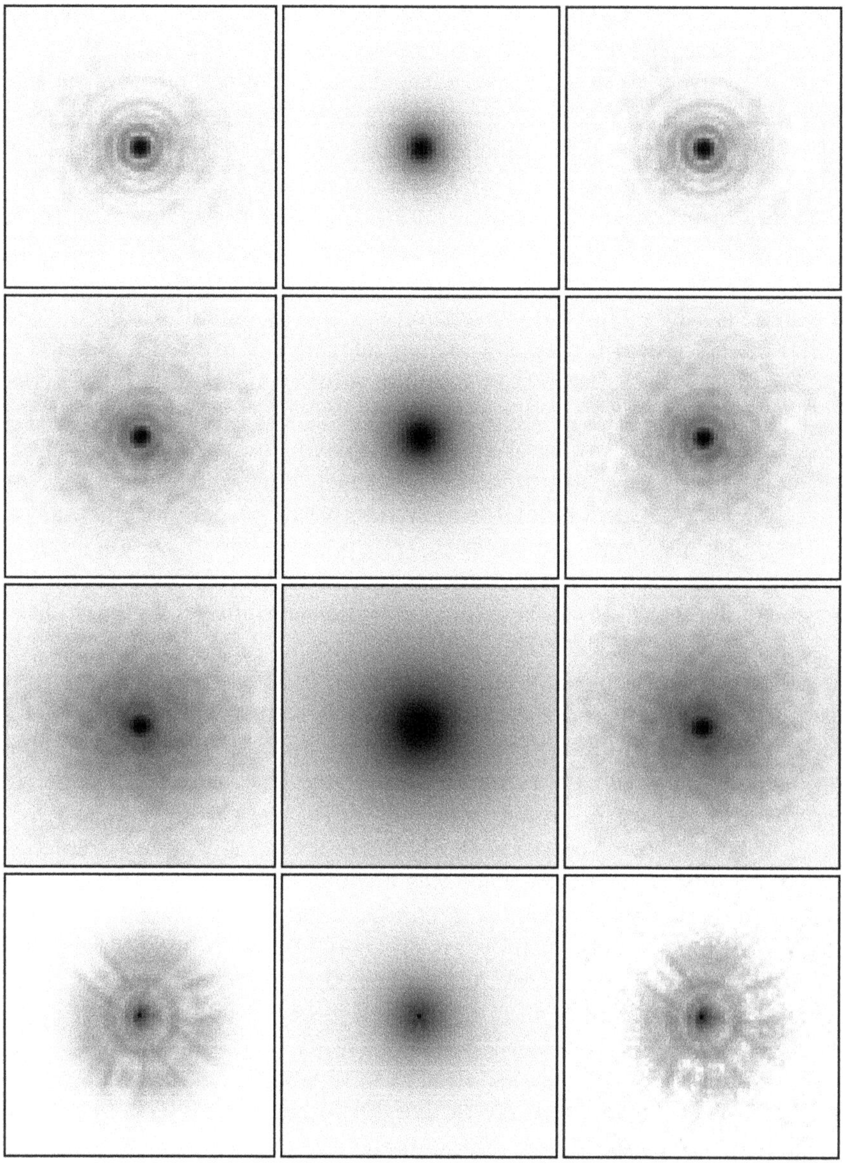

Figure 8.7. First column: the PSFs used for image generation; second column: the PSFs used for initializing the blind algorithm; third column: the PSFs reconstructed by the blind algorithm. First row: AO-corrected PSF with SR = 0.67; second row: AO-corrected PSF with SR = 0.40; third row: AO-corrected PSF with SR = 0.17; fourth row: HST PSF before COSTAR correction (reproduced from [78]).

astronomy. In figure 8.8 we show images of three stellar objects used in this simulation: in the upper panels the images of a binary and of an open star cluster based on an image of the Pleiades, both obtained by convolving the point-wise objects with the PSF with SR = 0.64; in the lower panel a simulation of a star cluster,

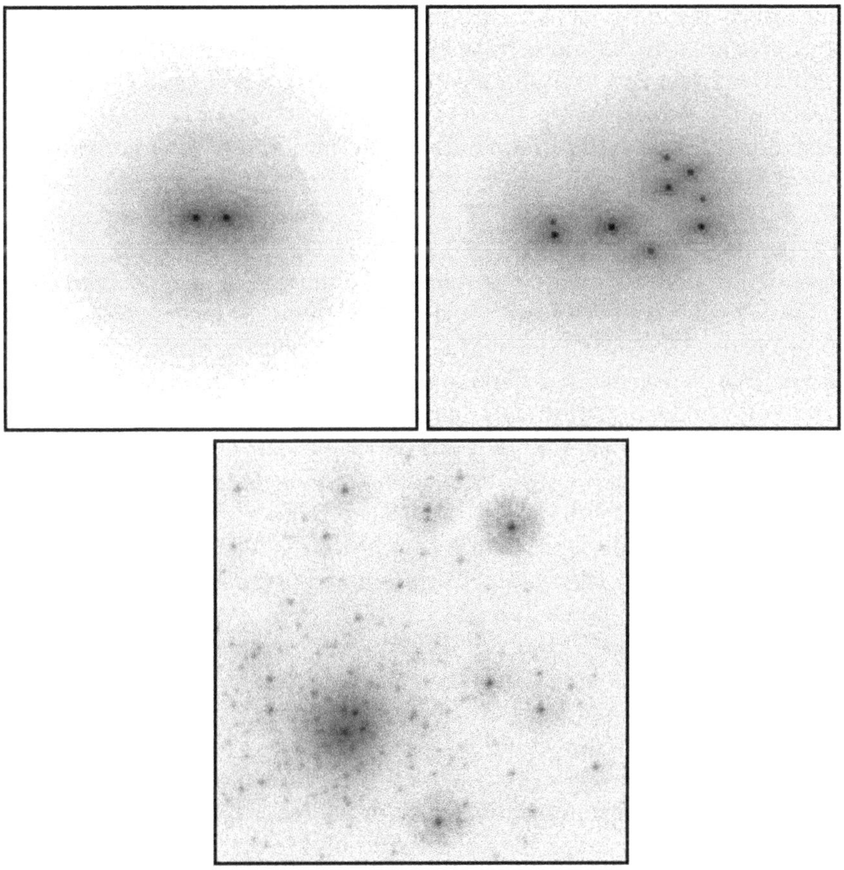

Figure 8.8. Images of the binary and of the star cluster (upper panels) in the case of a PSF with SR = 0.67. In the lower panel the HST image of the star cluster (reproduced from [78]).

consisting of 470 light sources, as observed by the Hubble space telescope (HST) before COSTAR correction, an image frequently used for numerical simulations. For this case only, we do not use an AO-corrected PSF but the aberrated HST PSF, which corresponds to SR = 0.09. These data can be obtained via anonymous ftp from ftp://ftp.stsci.edu/software/stsdas/testdata/restore/sims/star_cluster/. The images are obtained by convolving the corresponding point-wise objects with the aberrated PSFs, by adding background and perturbing with Poisson and Gaussian noise (for RON simulation); in the deconvolution the latter is compensated by means of the Poisson approximation proposed in [84].

As concerns initialization of the SGP-based inexact block coordinate descent method, a constant array is used for the object while a PSF satisfying all the constraints (non-negativity, SR upper bound and normalization) is used for the PSF. This initial choice is very important. To this purpose we point out a typical feature of the PSF of a telescope: it is a band-limited function and, if the telescope consists of a circular mirror, the band, i.e. the support of its Fourier transform, is a disc with a radius proportional to the ratio between the diameter D of the telescope and the observation wavelength λ. It is not easy to insert this property as a constraint on the PSF because the projection on the resulting set of constraints (including SR and normalization) is not easily computable. For this reason this constraint is not considered in this simulation. However, one can try to force the estimated PSF to have this property using an initial PSF which has the correct band-limiting property and satisfies the other constraints.

The ideal PSF of the telescope is not suitable as an initial guess because it does not satisfy the SR constraint. However, one can consider, as suggested for instance in [26], the autocorrelation of the ideal PSF, which has the same band. In the described simulations, which assume in general a telescope of the 8 m class and observations in H-band, the autocorrelation is a good choice if SR $\geqslant$ 0.46 (remark that the maximum value of the ideal PSF depends on the ratio D/λ, where D is the diameter of the principal mirror and λ the observation wavelength). For lower values of SR one can take the autocorrelation of the autocorrelation and so on, until the SR constraint is satisfied. This is the choice considered in [78] and, quite surprisingly, it seems that the algorithm, in spite of its high nonlinearity, preserves the band-limiting property satisfied by the initial guess.

For all these examples 50 inner SGP iterations on the object and 1 inner SGP iteration on the PSF are used. This choice can be justified by the features of the deconvolution method used for inexact minimization, since we need a sufficiently large number of SGP inner iterations for obtaining a nearly point-wise object. Moreover, with a few numerical experiments in the case of the binary, one can verify that this choice is a good compromise which provides a sufficiently fast convergence for all cases. In a first instance 300 outer iterations are performed. We do not show the results obtained for object reconstruction because, except for a few artifacts, the result consists essentially of single pixels located in the correct position with correct intensity. More details on the blind deconvolution, on the values used for SR and on artifact discussion are given in [78]. As concerns PSF reconstruction, the results are shown in the last column of figure 8.7 and they are quite satisfactory since characteristic features of the PSFs, not contained in the initializations (shown in the second column) are reconstructed by the blind method.

Additional examples of astronomical targets, obtained from HST images, are considered in [78]: the CRAB nebula NGC1952, the spiral galaxy NGC 6946 and the planetary nebula NGC 7027. These objects are shown in the first column of figure 8.9 while their images obtained by convolving with the PSF with SR = 0.67 are shown in the second column, including background addition and noise perturbation. An application of the blind method previously described is attempted without adding object regularization to $f_0^B(h, x; y)$. Presumably, because of the lack of regularization, a difficult and crucial point, already remarked in [27], is the choice of the number of inner iterations for the two blocks. Satisfactory numbers can be obtained for each object and PSF by several attempts; moreover, outer iteration must not be pushed to convergence because a sort of semi-convergence seems to

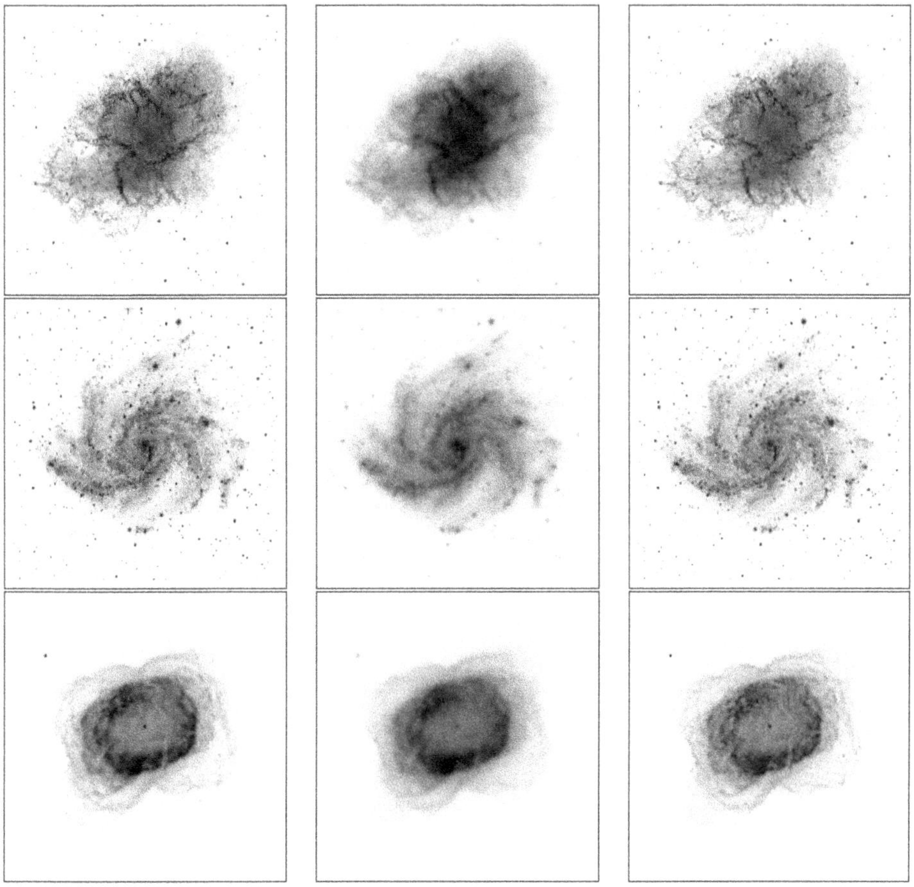

Figure 8.9. First column: the objects used for image generation; second column: the blurred images in the case of the PSF with SR = 0.67; third column the reconstructed objects obtained with the blind algorithm. First row: the Crab nebula NGC1952; second row: the spiral galaxy NGC 6946; third row: the planetary nebula NGC 7027 (reproduced from [78]).

apply to this case, i.e. regularization is obtained by early stopping of outer iterations. Some results are shown in the last column of figure 8.9.

It is obvious that the uncertainty in the selection of the numbers of inner iterations is a difficulty in the application of the inexact block coordinate descent method and further research is required to clarify this point.

A different approach based on surrogate functions is proposed in [40, 63] for constrained and regularized blind deconvolution. The approach is presented in the framework of non-negative matrix factorization; therefore, using the notation previously introduced, the functional to be minimized is given by

$$F(H, X; Y) = \mathrm{KL}(Y; HX) + \frac{\mu}{2}\|H\|_F^2 + \lambda\|X\|_1 + \frac{\nu}{2}\|X\|_F^2 \tag{8.39}$$

where $\mathrm{KL}(Y; HX)$ is defined in equation (8.33) and $\|.\|_F$ denotes the Frobenius norm. Therefore, in this approach, the matrix H (the PSF) is regularized by penalization of its ℓ_2-norm while X (the object) is regularized by penalization of its ℓ_1 and ℓ_2 norms. As concerns the constraints, in [63] normalization of the PSF is considered besides non-negativity of both H and X while in [40] only non-negativity is considered. We give the results only in this second case.

Surrogate functions for $F(H, X; Y)$ can be easily obtained from the surrogate (8.34) by observing that the regularization terms are separable. Remark that the approach can be extended to the case of HS regularization of the object, using the separable surrogate for this penalty introduced in [41]. As already explained, the surrogate approach assures monotonic decrease of the objective function. The algorithm is as follows:

- Initialize with $H^{(0)}$, $X^{(0)}$ strictly positive.
- Given $H^{(k)}$, $X^{(k)}$ compute

$$A^{(k)} = H^{(k)} \circ \frac{Y}{H^{(k)}X^{(k)}} (X^{(k)})^T, \tag{8.40}$$

$$B^{(k)} = 1_{M\times N} (X^{(k)})^T, \tag{8.41}$$

$$H^{(k+1)} = \frac{2A^{(k)}}{B^{(k)} + \sqrt{B^{(k)} \circ B^{(k)} + 4\mu A^{(k)}}}. \tag{8.42}$$

- Next compute

$$C^{(k+1)} = X^{(k)} \circ (H^{(k+1)})^T \frac{Y}{H^{(k+1)}X^{(k)}} \tag{8.43}$$

$$D^{(k+1)} = \lambda 1_{N\times P} + (H^{(k+1)})^T 1_{M\times P} \tag{8.44}$$

$$X^{(k+1)} = \frac{2C^{(k+1)}}{D^{(k+1)} + \sqrt{D^{(k+1)} \circ D^{(k+1)} + 4\nu C^{(k+1)}}}. \tag{8.45}$$

For $\lambda = \mu = \nu = 0$ the algorithm of Lee and Seung is re-obtained.

In [63] numerical experiments on the application of the method to blind deconvolution are reported showing that it is possible to obtain satisfactory results while in [40] the application of the method to dynamic PET and to joint estimation of activity and attenuation in time-of-flight PET is discussed.

Convergence of the iterates is not proved; however, the following results can be obtained [40, 63].

- The values of the objective function decrease monotonically as a consequence of the properties of the surrogate functions.
- The sequence of the values of the cost function is convergent.
- An asymptotic regularity property holds for the sequence of iterates: $\forall\, i, j, l$

$$\lim_{k\to\infty} \left(H_{i,j}^{(k+1)} - H_{i,j}^{(k)}\right) = 0,$$

$$\lim_{k\to\infty} \left(X_{j,l}^{(k+1)} - X_{j,l}^{(k)}\right) = 0.$$

- As a consequence of Ostrowski theorem [73], theorem 26.1 (see section A.5), the set of limit points of the iterates $(H^{(k)}, X^{(k)})$ is compact and connected.

Details on the previous results can be found in [62]. The last result implies convergence of the iterates if it is possible to prove that the number of limit points is finite. Indeed in such a case, the set of limit points can contain only one point and the sequence of iterates is convergent; moreover, its limit is a stationary point of the objective functions satisfying KKT conditions. Unfortunately, a complete convergence proof is not available. However, convergence results can be obtained for other approaches. One example is the method of inexact block coordinate descent method previously described. Other examples are provided by *block coordinate proximal gradient* and FB methods [6, 29, 30].

8.4 Images with point and smooth sources

A specific problem arising in astronomy is the deconvolution of images where very bright and localized sources (stars or localized and bright objects), are superimposed on weak diffuse structures which deserve scientific interest and are above the (approximately constant) background due to sky emission. The latter can be estimated and inserted in the imaging model, as discussed in section 3.1.1, while the image of the diffuse structures must be improved by the deconvolution method. In general the reconstructions obtained by standard deconvolution methods are affected by significant ringing artifacts around the bright sources (see figure 5.1), even if the methods take into account background emission. These artifacts destroy the quality of the reconstructed image and perturb the correct value of the intensity of a bright source since a part of it is contained in the rings around the peak corresponding to the star reconstruction.

A useful approach to circumvent this difficulty is provided by a model of the object x which assumes that it is the sum of two components: a point-wise one, x_P, and a smooth one, x_S, so that $x = x_S + x_P$.

To our knowledge this model was first proposed in [75] as a way of obtaining more accurate estimates of star magnitudes in the framework of the application of

EM–RL to the deconvolution of HST images before COSTAR correction. The same model was proposed independently in the framework of a least-squares approach, by assuming different regularization penalties for the two components [37, 51]. Therefore, the model leads to an under-determined problem because, as in the case of blind deconvolution, one has to estimate two arrays, x_S, x_P from one image y.

In the case of Poisson data, if $f_0(x; y)$ is the data-fidelity function defined in equation (4.7), then the model requires the constrained minimization, with respect to the two blocks x_S, x_P, of the function

$$f_{\beta,\gamma}(x_S, x_P; y) = f_0(x_S + x_P; y) + \beta f_1(x_S) + \gamma f_2(x_P), \tag{8.46}$$

with different regularization functions for the two components. The difference, with respect to blind deconvolution, is that the objective function (8.46) is not only convex with respect to the two blocks separately but also with respect to the totality of the variables. Therefore, one can use block coordinate descent methods or also methods of global minimization.

Since the component x_P is sparse, one can take $f_2(x_P) = \|x_P\|_1$, as suggested in [37, 51]. However, with this regularization, the method seems to be unable to identify accurately the positions of the bright spots. For this reason, in the numerical experiments considered in [51], the case of known star positions is considered. The same assumption is used in an application of the model to the deconvolution of multiple interferometric images [59]. In the latter, regularization of the smooth component is obtained by means of T0 regularization and the proposed algorithms is given in terms of EM–RL and SGM; a more refined algorithm is proposed in [10]. In conclusion, the strong *prior* assumption that the position of the stars (or generic bright localized sources) is known seems to be required for a successful application of the method. Such a requirement is not unreasonable in astronomy since methods for accurate astrometry are available; moreover the stars are unresolved so that they can be represented by a delta-function in the object model.

The assumption that the positions of the point sources are known requires a reformulation of the model. Indeed, in such a case x_P is zero everywhere except in the pixels corresponding to the known positions of the sources and only the values of x_P in these pixels are unknown. In this approach, let us assume that the image contains Q sources, corresponding to the pixels with index values $\{j_1, l_1\}$, $\{j_2, l_2\}$, ..., $\{j_Q, l_Q\}$; then the point-wise component can be represented as follows

$$x_P = \sum_{q=1}^{Q} c_q \delta_{\{j_q,\, l_q\}} \tag{8.47}$$

where the c_q are unknown parameters corresponding to source intensities and $\delta_{\{j_q,\, l_q\}}$, is an array which is 1 in the pixel $\{j_q, l_q\}$, and 0 elsewhere. With this model the computation of $H\,x$ leads to the result

$$(Hx)_{i,\,k} = (Hx_S)_{i,\,k} + \sum_{i=1}^{Q} c_q h_{i-j_q,\, k-l_q}. \tag{8.48}$$

The physical interpretation of this model is obvious: the computed image is the sum of a smooth image and of Q PSFs centered on the positions of the Q point sources with coefficients given by their intensities.

The prior information about the positions of the point sources greatly reduces the number of unknowns which is now $N^2 + Q$, i.e. the array x_S and the column vector c of length Q, with components $c_1, \ldots, c_Q$. If we introduce the arrays $(h_q)_{i,k} = h_{i-j_q, k-l_q}$ and the row vector of arrays $h^{(Q)}$ whose components are the arrays h_q, the notation can be greatly simplified by introducing

$$\tilde{H} = (H\, h^{(Q)}), \;\; \tilde{x} = \begin{pmatrix} x_S \\ c \end{pmatrix} \tag{8.49}$$

so that

$$\tilde{H}\tilde{x} = Hx_S + h^{(Q)}c, \;\; \tilde{H}^T z = \begin{pmatrix} H^T z \\ (h^{(Q)})^T z \end{pmatrix}, \tag{8.50}$$

where $(h^{(Q)})^T$ is the column vector whose components are given by h_q^T.

Since no additional regularization is required on the component x_P which is reduced to a vector of length Q, let us introduce the following objective function of the problem

$$f_\beta(\tilde{x};y) = f_0(\tilde{x};y) + \beta f_1(x_S) \tag{8.51}$$

If we consider a convex and differentiable regularization function $f_1(x_S)$, then the objective function is convex and differentiable with respect to the set of variables $\tilde{x}$; moreover, gradient methods for the constrained minimization can be used. As concerns the constraints, a standard one is non-negativity. If it is useful to introduce also a constraint on the flux of the computed image, then, as easily follows from equation (8.48), by taking into account the normalization of the PSF, this constraint is given by the equality condition

$$\sum_{j,l=0}^{N-1} (x_S)_{j,l} + \sum_{q=1}^{Q} c_q + \sum_{i,k=0}^{N-1} b_{i,k} = \sum_{i,k=0}^{N-1} y_{i,k}. \tag{8.52}$$

If gradient methods are used for the minimization of the objective function, then, by taking into account the normalization of the PSF, the gradient of $f_0(\tilde{x};y)$ is given by

$$\nabla_{\tilde{x}} f_0(\tilde{x};y) = \tilde{1} - \tilde{H}^T \frac{y}{\tilde{H}\tilde{x}+b} \tag{8.53}$$

and, similarly the Hessian is given by

$$\nabla^2_{\tilde{x}} f_0(\tilde{x};y) = \tilde{H}^T \mathrm{diag}\left(\frac{y}{(\tilde{H}\tilde{x}+b)^2}\right)\tilde{H}. \tag{8.54}$$

Starting from this expression of the Hessian of $f_0(\tilde{x};y)$, in [10] it is proved that, when f_1 is one of the Tikhonov regularizers or the HS regularizer, then the minimizer of the objective function (8.51) is unique.

In order to minimize the objective function (8.51) several methods discussed in the previous chapters are available. We focus on SGP, discussed in section 6.3.1 and, for the convenience of the reader, we give some implementation details.

As follows from equation (8.53), the computation of the gradient implies the computation of $\tilde{H}\tilde{x}$ and the computation of $\tilde{H}^T z$. The first implies the computation of $H\, x_S$, which can be computed efficiently by means of FFT, and the sum of Q scalar-array products; moreover, the computation of $\tilde{H}^T z$ implies the computation of $H^T\, z$, again computable by FFT, and Q additional scalar-array products.

As concerns the scaling, we use the rule suggested by SGM, section 5.4. By taking into account that in the case of deconvolution $v = 1$, where, as usual, 1 is the array with all entries equal to 1, and that the scaling applies only to x_S, we obtain that, for the gradient of f_β, the decomposition (5.20) provides a function $\tilde{V}$ given by

$$\tilde{V}(\tilde{x}) = 1 + \beta\begin{pmatrix} V_1(x_S) \\ 0 \end{pmatrix} \tag{8.55}$$

$V_1(x_S)$ deriving from the decomposition of $f_1(x_S)$. In conclusion, as follows from equation (6.20), at iteration $k + 1$ the scaling will be given by

$$D_k = \max\left\{\frac{1}{L_k}, \min\left\{L_k, \frac{\tilde{V}(\tilde{x}^{(k)})}{\tilde{x}^{(k)}}\right\}\right\}. \tag{8.56}$$

To demonstrate the improvement, with respect to standard deconvolution, provided by the approach with two components described above, we reconsider the reconstruction of the images of the Pillan eruption on Io discussed in section 7.3.3, in particular the images in the L and M bands, shown in figure 7.20. These reconstructions are not satisfactory, even if they show the existence of a number of bright spots, in general not detectable in the raw images. They can be considered essentially non-resolved by the deconvolution method. Indeed, the most frequent volcanic structures on Io are lakes of lava with a size of about 30 km, which approximately coincides with the pixel size. The positions of these structures appearing in the L and M images can be approximately derived from the first deconvolution by computing the centroids of the bright spots, in particular of those appearing in the reconstruction of the M-band images. Subsequently, this information can be used in the approach outlined above.

In order to verify the feasibility of this approach, we first describe the results of a numerical experiment attempting to simulate an Io-like object providing images with SNR values comparable to those of the real ones.

To this end we generate a disc with the same diameter of Io in the Keck image and a smoothly variable brightness, including limb darkening; we superimpose on the disc very bright sources and we convolve the result with a PSF modeling the Keck

PSF in the M-band. Finally, the result is perturbed with Poisson noise. The values are adjusted in such a way that, if we compute the ratio between mean value and standard deviation on domains of uniformity in the simulated image, it coincides with the value computed on the real image. The simulated object and its image are shown in figure 8.10.

In the first step we use a number of iterations of SGP with HS regularization in order to sharpen the images of the nine hot spots. We compute the centroids of their reconstructions and we find that they coincide with their positions in the original model, except in one case where we find a shift of one pixel. These centroids are used to produce the map of the bright spots (second step) to be used in the third step.

In the third step, taking into account the existence of a sharp edge due to the limb, we decide to test the three regularizers HS, MRF (equation (5.13)), and MIS (equation (5.62)). The mean value of the modulus of the gradient on the blurred and noisy image is about $\delta_{\text{mean}} = 266$. Therefore, for the three regularizers, we consider a grid of the two parameters δ, β consisting of 11×9 points, with δ varying from 10 to

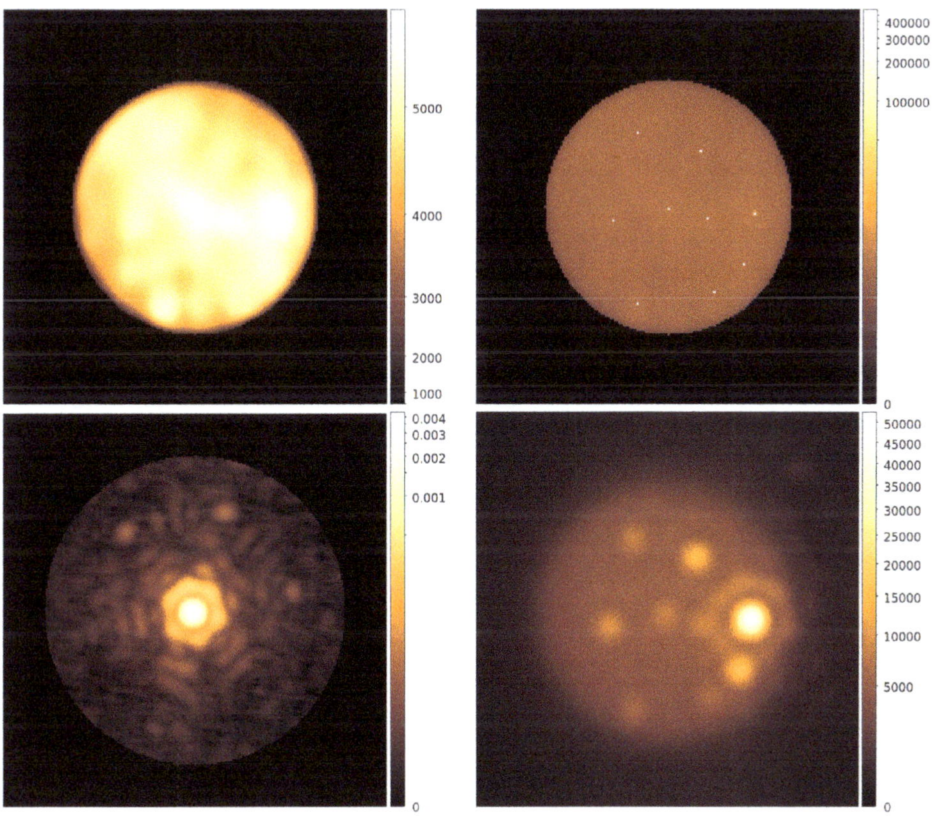

Figure 8.10. Simulated image of an Io-like object at the M-band. First row—left: model of the surface of the planet (quadratic scale). Right: the model after the addition of nine hot spots (log scale with saturation of the hot spots). Second row—left: the model of the PSF (log scale). Right: the blurred and noisy image (sqrt scale).

10^3 (11 values) and β varying from 10^{-5} to 10^{-1} (9 values). For each regularizer we consider SGP both with and without flux constraint. For each one of these six cases we find only one minimum of the rms error, computed as the ℓ_2 norm of the difference between the reconstructed and the original surface of the Io-like object. The results are reported in table 8.1.

We point out that, for the six tested algorithms, we do not find significant differences between the values of the minimum rms errors. Moreover, for all of them, the best value of β is the same. We only find variations in the best values of δ, even if we find the same value for the same regularization without and with the flux constraint. In this example it seems that the flux constraint does not have a significant effect.

It is important to note that, for a given value of β, the dependence of the relative rms error on δ is not strong so that, if we take the same value for all algorithms (for instance 50) we have variations on the relative rms error of at most 1%. We also note that the values of δ are much smaller than the average value of the modulus of the gradient computed on the blurred and noisy image; indeed they are of the order of the average modulus of the gradient computed inside the surface of the model without the bright spots and without the contribution of the limb (about 35).

Finally, the last step is a non-regularized SGP, also pushed to convergence, applied to the simulated image, using as a background each one of the six surfaces of the Io-like object reconstructed in the previous step. Also in this case we consider the algorithm without and with flux constraint. Therefore, we obtain 12 reconstructions of the nine bright spots of the model.

In table 8.2 we give the average relative error of the intensities of the nine bright spots for each one of the 12 reconstructions. The result is promising since, in all cases, we have an average error of about 3%. Of course we must take into account that we have used the same PSF for convolving and deconvolving the Io-like object (inverse crime); moreover, we have considered only one noise realization. Therefore, before stating that this is a way for photometry, one should do several numerical experiments with different noise realizations and with slightly different PSFs for the deconvolution.

In conclusion, all methods provide similar reconstruction errors with similar efficiency. A difference between the different methods consists of the presence of

Table 8.1. For the three regularizers MRF, MIS and HS, without and with flux constraint, the values of δ, β, corresponding to the minimum of the relative rms error, are given, together with the corresponding number of iterations and the value of the relative rms error.

	No flux				Flux			
REG TYPE	β	δ	IT	RMSE	β	δ	IT	RMSE
MRF	10^{-2}	10	694	6.72%	10^{-2}	10	1387	6.69%
MIST-REG	10^{-2}	50	247	6.83%	10^{-2}	20	521	6.81%
HS	10^{-2}	80	272	6.69%	10^{-2}	80	246	6.69%

Table 8.2. Average relative error on the intensities of the nine reconstructed bright spots for the 12 reconstructions considered in this section. In the first column we give the results obtained without the flux constraint in step 4 and in the second those obtained with this constraint. Each row corresponds to the indicated algorithm used in the third step.

REG type	No flux		Flux	
Third step	IT	Err mean	IT	Err mean
MRF (no flux)	465	3.28%	367	3.00%
MRF (flux)	783	3.41%	283	3.06%
MIS (no flux)	922	3.81%	328	3.45%
MIS (flux)	796	3.97%	289	3.64%
HS (no flux)	370	3.70%	834	2.98%
HS (flux)	519	3.52%	2031	2.38%

point-wise artifacts which can be different in the different reconstructions. This can be a criterion for the user, i.e. select the reconstruction with minimal number of artifacts. For this reason in figure 8.11 we show the reconstruction obtained with MIS regularization, with no flux constraint at the third step and flux constraint at the fourth one. It corresponds to the third line, last column of table 8.2.

Since the result of the simulation is promising, the same approach is applied to the images of the Pillan eruption in the L and M bands. Therefore, the first step is a standard deconvolution. The raw images, the corresponding PSFs and the reconstructions obtained by means of SGP with HS regularization are shown in figure 7.20. We note that the values of the real images are rescaled for calibration so that the mean value of the modulus of the gradient of the observed image is 2.7×10^{-15}. By taking into account the effect of this scaling on the parameters of an edge-preserving regularizer we must search for values of δ of the order of 10^{-15} while values of β must be searched around 10^{-2}, the value obtained in the previous simulation. After some attempts we choose $\delta = 10^{-15}$ and $\beta = 5 \times 10^{-3}$.

The reconstructions obtained in the first step allow one to identify eight bright spots which are used for producing the map to be used in the two-component deconvolution. In figure 8.12 we show the identified hot spots.

With this map, in the third step, we consider the three edge-preserving regularizations and the two possible combinations corresponding to the use or no use of the flux constraint at the third and fourth step. The result of this step is the reconstruction of Io surface without the bright spots. Finally, this image is used as a background in the fourth step which is based on SGP without regularization. Also in this case we consider both the algorithm with and that without flux constraint. The final result is obtained by adding the result of the fourth step to the surface image obtained at the third one. Again, for each image we obtain 12 reconstructions. We find that the combination providing a reconstruction with reduced number of point-wise artifacts is obtained if we do not use the flux constraint at the third step but we use it at the fourth one. The result obtained for the smooth surface of Io is shown in figure 8.13 while the final results obtained in the fourth step (point-wise plus smooth

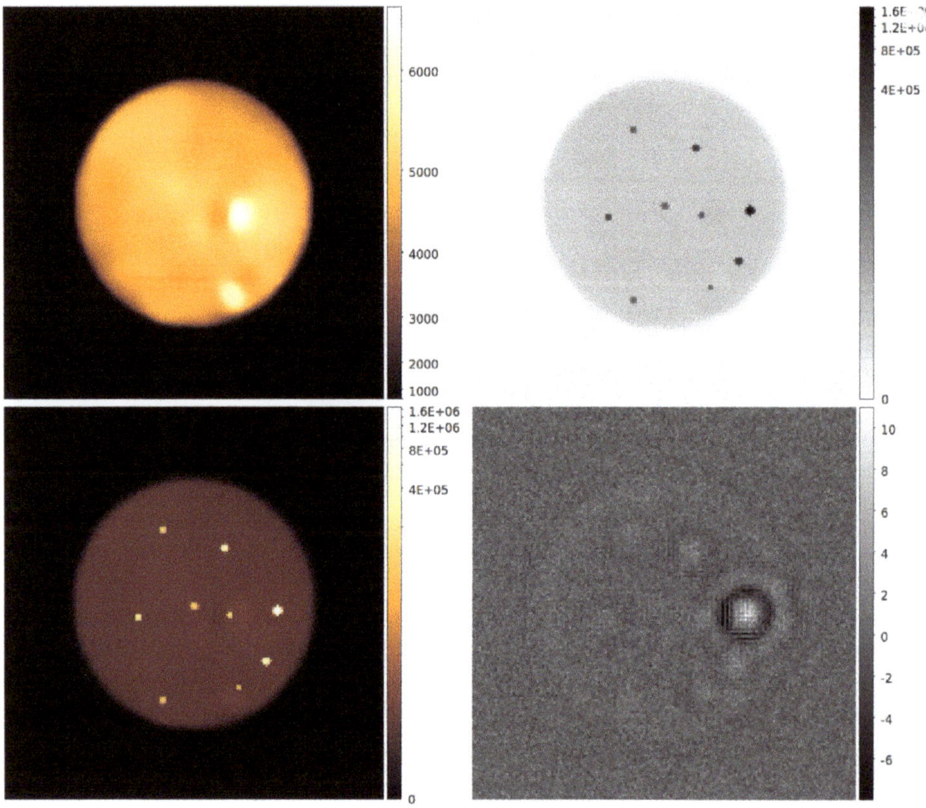

Figure 8.11. Reconstruction of the simulated image shown in figure 8.10. First row—left: reconstruction of the surface as obtained at the 3rd step (quadratic scale). Right: the mask (2nd step) shown in reverse b/w scale. For visualization purposes the limb of Io is identified with a dashed disc. Second row—left : complete reconstruction (surface plus hot spots, represented in log scale). Right: plot of the normalized residual of equation (4.45), showing moderate artifacts, concentrated in the region of the brightest source.

component) is shown in figure 8.14. The reconstruction of the limb is not perfectly circular and this effect is presumably due to the presence of hot spots close to the limb.

By comparison with a map of Io's surface shown in figure 1 of [38], it is possible to recognize (the numbers are those indicated in figure 8.12) Loki (hot spot nr. 3), Pele (hot spot nr. 2) and Pillan (hot spot nr. 1) in our reconstructed images together with five other hot spots to be identified. As concerns photometry, we remark that from our reconstruction we obtain a ratio between the fluxes of Pillan and Pele of about 6, while the value obtained in [38] is about 5; the two values are compatible within the errors estimated in that paper.

8.5 Images with space-variant blur

The model of an imaging system based on the convolution of a PSF h with the source object x is satisfactory if the image of a point source, i.e. the PSF, is

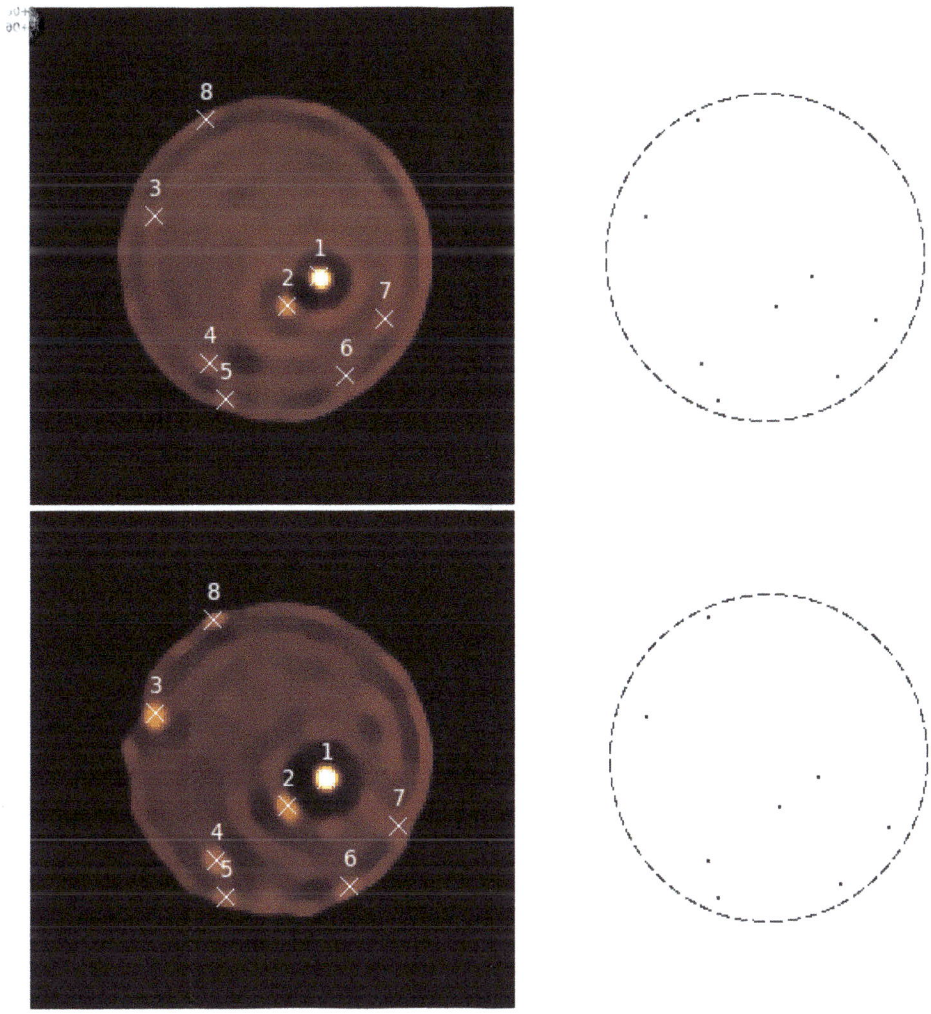

Figure 8.12. Identification of the hot spots in the L (upper-left) and M (lower-left) band images and the corresponding maps in the right panels.

independent of the position of the source. In such a case one says that the system is space-invariant or also that the PSF is space-invariant.

However, it may happen, both in microscopy and in astronomy (even if for different reasons), that the image of a point source depends on its position. In such a case one says that the system is space-variant or also that the system is described by a space-variant blur. It follows that the convolution model of equation (3.2) must be replaced by the following one

$$(Lx)(\vec{r}) = \int_S h(\vec{r}, \vec{r}\,')x(\vec{r}\,')d\vec{r}\,', \ \vec{r} \in S, \tag{8.57}$$

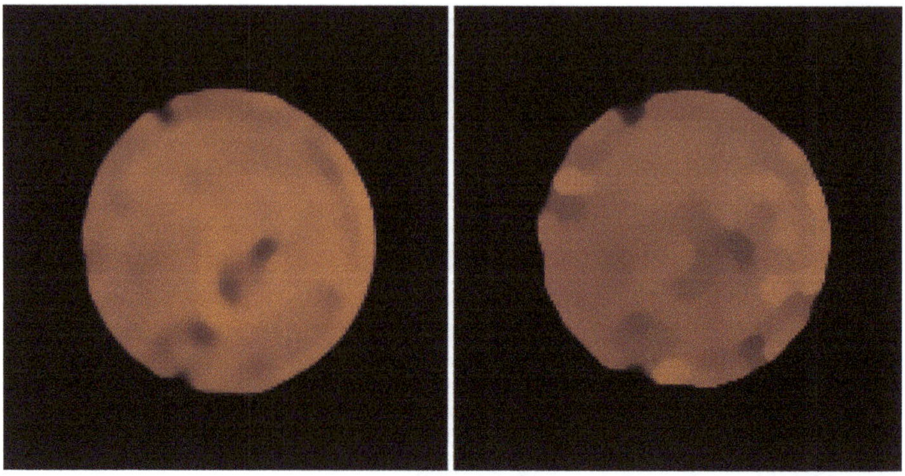

Figure 8.13. Reconstruction of the smooth component of the L (left) and M (right) band images.

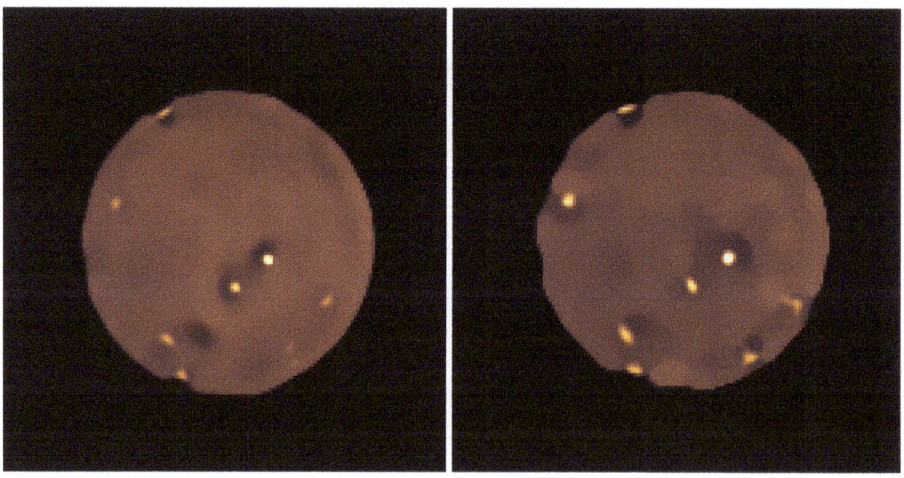

Figure 8.14. The final reconstructions of the L (left) and M (right) band images.

i.e. the linear operator is a general integral operator. The integral kernel $h(\vec{r}, \vec{r}')$ has an interesting and important meaning: for a given value of $\vec{r}'$ it is the image of a point source (a delta-function) located in this point. It is called the *space-variant PSF* of the optical system.

If the image of a point is strongly varying over the FoV of the image, then the problem is numerically intractable for large images; indeed, if the image has size N^2, then one needs to store a matrix with N^4 entries and one has both storage and efficiency problems in matrix–vector computations. The problem can be tractable in the case of a slowly and smoothly changing space-variant PSF if it can be locally approximated by a space-invariant one. However, before describing possible approaches, we must briefly describe the different origin of space-variance in microscopy and astronomy. *We must also point out that in all cases of a space-variant blur a computable approximation of the space-variant PSF should be available.*

In microscopy the problem arises when one acquires 3D images of a thick sample by means of optical sectioning (see section 2.1.2). Indeed, when light crosses a medium with varying refraction index, the image of a point source located in the focal plane depends mainly on the position of this plane and not on its distance from the optical axis of the instrument. Therefore, in such a case, the blur depends essentially on the depth and the expression depth-varying, or depth-variant, blur is used.

In astronomy the space-variance of the blur is mainly due to the AO system (see section 2.3.4). We recall that an AO system includes a deformable mirror, which compensates for the time-evolving effects of the atmospheric turbulence on the optical wavefront of the science target. The compensation is calculated by a real-time control system on the basis of measurements of the disturbances performed on a guide source, for instance a natural star. The goal of such an AO system, also known as single-conjugate adaptive optics (SCAO), is to make the guide star wavefront flat. However, the science target is usually not coincident with the guide star: the light beams from the science target and from the guide star cross different volumes of atmosphere and are affected by different wavefront aberrations because of the stratified structure of the atmospheric turbulence. A perfect correction of the guide star wavefront is not perfect for the science object. As a consequence of this mismatch, the PSF in different points of the FoV is typically elongated towards the guide star and therefore space-variant since the elongation increases with the angular distance from the guide star. More complex AO systems, such as multi-conjugate AO (MCAO) [9], based on the use of multiple deformable mirrors and of multiple guide stars provide a remarkable uniformity with respect to SCAO systems, but some residual PSF variation is still possible. Therefore, space-variance correction can be a relevant topic especially for the future giant telescopes.

In the case of a slowly and smoothly varying local PSF two approaches have been proposed that we call the *sectioning* and the *interpolation* approach. In both cases one assumes that it is possible to decompose the image domain into non-overlapping patches where the PSF can be considered approximately space-invariant so that the imaging operator is locally described by a convolution product. In the case of 3D microscopic images the patches can be obtained by decomposing the image volume into a finite number of planar layers, orthogonal to the optical axis of the microscope. In the case of an astronomical image, corresponding for instance to an $N \times N$ array, the patches can be non-overlapping arrays with a smaller dimension, covering the full image.

Both approaches and both applications have a common feature which, for simplicity, is presented in the case of 2D astronomical images. As already remarked, the first point is that a computable model, exact or approximate, of the space-variant PSF $h(\vec{r}, \vec{r}')$, is available. The domain S of the image y, for instance a square, is divided into D non-overlapping sub-domains S_i such that $\cup_{i=1}^{D} S_i = S$ and the space-variant PSF is computed in the centers $\vec{r}_i'$ of the sub-domains S_i, i.e. $h(\vec{r}, \vec{r}_i')$; then, the function $h_i(\vec{u}) = h(\vec{u} + \vec{r}'_i, \vec{r}'_i)$ is the PSF associated with the sub-domain S_i, in the sense that $h_i(\vec{r} - \vec{r}')$ approximates $h(\vec{r}, \vec{r}')$ within S_i

$$\int h(\vec{r}, \vec{r}')x(\vec{r}')d\vec{r}' \approx \int h_i(\vec{r} - \vec{r}')x(\vec{r}')d\vec{r}', \ \forall \vec{r} \in S_i. \tag{8.58}$$

The subdivision is performed in such a way that the difference between the PSFs associated to adjacent sub-domains is not too large, for instance 15%, if measured with a suitable metric distance (for instance, that provided by the ℓ_1 or ℓ_2 norm). In conclusion a mosaic of sub-domains of the full image and of associated space-invariant PSFs is produced.

To visualize this approach for the reader, in figure 8.15 we show a mosaic of 5×5 computed PSFs for the wide-field camera of HST before COSTAR correction. It can be downloaded from http://www.stsci.edu/ftp/software/tables/testdata/restore/sims/star_cluster/. It is evident that the variation of the PSF over the image domain is very slow.

In the case of 3D images in microscopy, since the PSF is depth-variant, the sub-domains are suitable layers and the space-invariant PSFs are obtained by computing the depth-variant one in the central points of these layers.

As we said, the previous modeling is common both to sectioning and to interpolation approaches. The first approach proposed for treating this image reconstruction problem is the sectioning one: in [89] the authors propose to apply to space-variant deconvolution the sectioned method they propose in [90] for dealing with the case of a space-variant noise. The idea is to consider partially overlapping sub-domains (or sectors or patches) of the image domain, containing the non-overlapping sub-domains discussed above. The size of the overlap is dictated by the extent of the PSF. The PSF associated to one sector is used in a deconvolution method for estimating the source object within that sector. Finally, non-overlapping sub-sectors are extracted from the overlapping ones and the corresponding reconstructed images are reassembled so that the final reconstructed image is the mosaic of these local reconstructions. The drawback of this approach is mainly due to the possible existence of boundary artifacts (see section 8.2) at the boundaries of the different patches and discontinuities because of the use of different PSFs.

The approach is applied to different situations with different choices of the sub-domains and models of the space-variant PSF. The application to astronomical images was motivated by the aberration of the HST mirror before COSTAR correction. The method of [89] is proposed in [1] with EM–RL as deconvolution method for the different sectors. Implementation on a massively parallel computer is considered in [28, 47], while a method based on local SVD is proposed in [49]. An approach for boundary effect correction is proposed in [7], where local deconvolution is also performed on suitable overlapping blocks and only the reconstructions on non-overlapping central sub-blocks are kept to form the complete image as a mosaic. A refinement of this approach is proposed in [60] where deconvolution methods with boundary effect correction are used. The method is implemented in

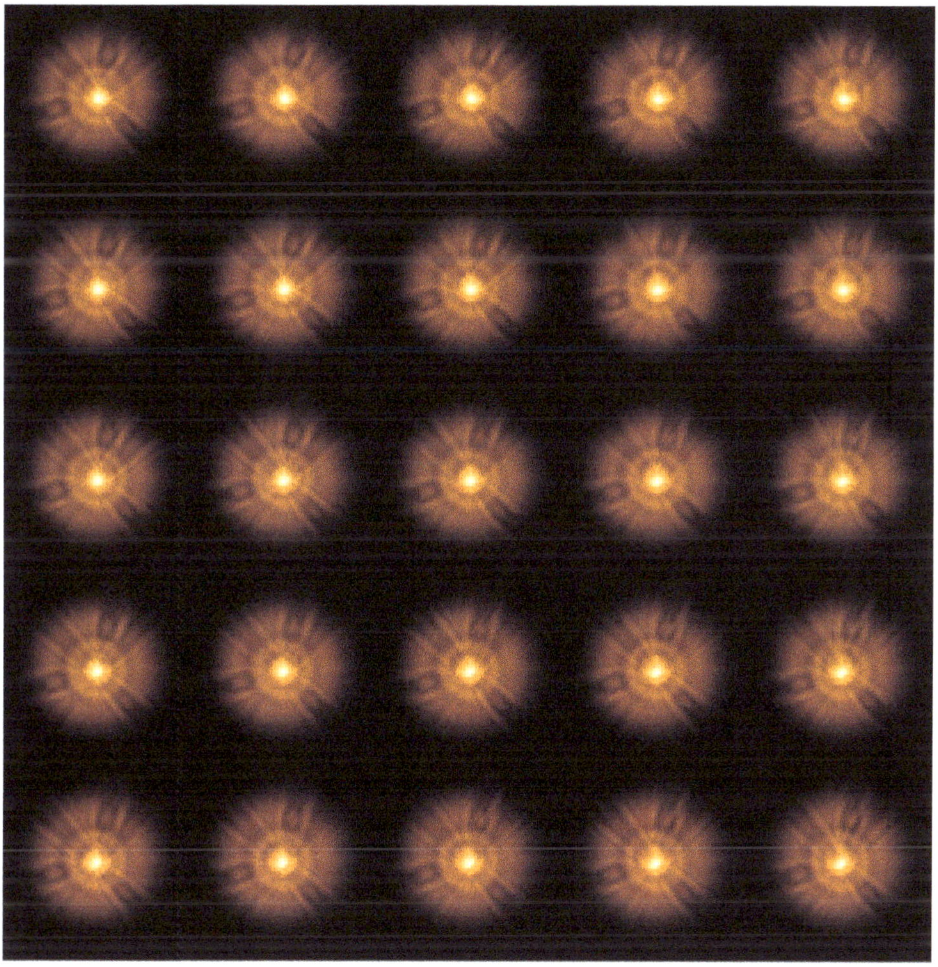

Figure 8.15. An example of a mosaic of samples of the space-variant PSF of the wide-field camera of HST before COSTAR correction.

the software *PATCH* which can be freely downloaded from the section *Software* of the URL http://www.airyproject.eu; it is described in https://doi.org/10.1117/2.1201508.006107 and is written in IDL because this is the language more frequently used in astronomy.

The interpolation method is first introduced in [70] while a different interpolation approach is proposed in [50, 56]. In the first case one interpolates the images provided by the different PSFs while, in the second case, one interpolates the PSFs.

To be more precise, we consider first the case of 2D square images and we assume that the local PSFs are given by the approach described above; then, if ϕ is a continuous weight function approximately equal to 1 on a square with the size of the sub-domains S_i and tending to zero outside, the model proposed in [70] for the computed image is given by

$$y(\vec{r}) = \sum_{i=1}^{D} \phi(\vec{r} - \vec{r}_i)(h_i * x)(\vec{r}) = \int_S \tilde{h}(\vec{r}, \vec{r}\,')x(\vec{r}\,')d\vec{r}\,', \tag{8.59}$$

where h_i is the PSF associated with the ith sub-square and the approximation of the space-variant PSF is given by

$$\tilde{h}(\vec{r}, \vec{r}\,') = \sum_{i=1}^{D} \phi(\vec{r} - \vec{r}_i)h_i(\vec{r} - \vec{r}\,'). \tag{8.60}$$

This equation defines a linear integral operator and the transposed (adjoint) operator is also a linear integral operator with a kernel given by

$$\tilde{h}^T(\vec{r}, \vec{r}\,') = \tilde{h}(\vec{r}\,', \vec{r}) = \sum_{i=1}^{D} \phi(\vec{r}\,' - \vec{r}_i)h_i(\vec{r}\,' - \vec{r}). \tag{8.61}$$

The other interpolation approach proposed in [50, 56] is given by

$$y(\vec{r}) = \sum_{i=1}^{D} \big(h_i * (\phi_i x)\big)(\vec{r}) = \int_S \hat{h}(\vec{r}, \vec{r}\,')x(\vec{r}\,')d\vec{r}\,', \tag{8.62}$$

where, for simplicity, we set $\phi_i(\vec{r}\,')x(\vec{r}\,') = \phi(\vec{r}\,' - \vec{r}_i)x(\vec{r}\,')$ and the approximation of the space-variant PSF is given by

$$\hat{h}(\vec{r}, \vec{r}\,') = \sum_{i=1}^{D} h_i(\vec{r} - \vec{r}\,')\phi(\vec{r}\,' - \vec{r}_i). \tag{8.63}$$

Both interpolation models provide an approximation of the space-variant PSF in a form which makes numerically tractable the deconvolution problem; therefore, they can be used in conjunction with any deconvolution method. They were proposed and compared mainly in the framework of the least-squares approach. For instance, a comparison of the two approaches in the case of fluorescence microscopy is discussed in [54] while a comparison for application to astronomy is discussed in [42]. The conclusion of both papers is that the approximation provided by equation (8.63), which in a sense seems to be more natural since it interpolates the set of given PSFs, provides more accurate results than the approximation given in equation (8.60). Even if this result is obtained in the framework of a least-squares approach to image deconvolution, we believe that it holds true also in the case of deconvolution methods derived in the framework of Poisson data. In particular, the interpolation method proposed in [79] for 3D fluorescence microscopy, with particular interpolation kernels, is based on the use of the EM–RL method.

In [42], the approximation (8.63) is recommended not only on the basis of the numerical results but also of the following remarks:

- PSF interpolation (8.63) preserves PSF symmetry while the other does not;
- the approximation error implicit in the PSF interpolation (8.63) tends to zero when the PSF grid is indefinitely refined;
- positivity and normalization are preserved.

The two approximations provide comparable results when the supports of the space-invariant PSFs are small with respect to the distance between adjacent sample locations.

Regarding the computational burden, all matrix–vector products can be computed efficiently by means of FFT; moreover, parallel computation can be easily implemented.

An important point is the choice of the interpolation kernel ϕ. In [54], since the kernel depends only on the depth variable, a triangular function extending outside the extent of the layer is used. However, as suggested in [42, 50] one can select a function ϕ which minimizes the approximation errors measured, for instance, by the L^2 distance

$$\int_S |h(\vec{r}, \vec{r}\,') - \sum_{i=1}^{D} h_i(\vec{r} - \vec{r}\,')\phi(\vec{r}\,' - \vec{r}_i)|^2 d\vec{r}\, d\vec{r}\,'. \tag{8.64}$$

In a more refined model one can introduce different interpolation functions ϕ_i for the different patches and minimize the previous least-squares functional with respect to both ϕ_i, h_i.

References

[1] Adorf H M 1994 Towards HST restoration with a space-variant PSF, cosmic rays and other missing data *The Restoration of HST Images and Spectra–II* ed R J Hanish and R L White 72–8

[2] Aghdasi F and Ward R K 1996 Reduction of boundary artifacts in image restoration *IEEE Trans. Image Process* **5** 611–8

[3] Akduman U, Brand J, Grochmalicki G, Hester G and Pike E R 1998 Super-resolving masks for incoherent high-numerical-aperture scanning microscopy in three dimensions *J. Opt. Soc. Am.* A **15** 2275–87

[4] Anconelli B, Bertero M, Boccacci P and Carbillet M 2005 Restoration of interferometric images—IV. An algorithm for super-resolution of stellar systems *Astron. Astrophys.* **431** 747–55

[5] Anconelli B, Bertero M, Boccacci P, Carbillet M and Lantéri H 2006 Reduction of boundary effects in multiple image deconvolution with an application to LBT-LINC NIRVANA *Astron. Astrophys.* **448** 1217–24

[6] Attouch H, Bolte J, Redond P and Soubeyran A 2010 Proximal alternating minimization and projection methods for nonconvex problems: an approach based on the Kurdyka-Lojasiewicz inequality *Math. Oper. Res.* **35** 438–57

[7] Aubailly M, Roggemann M C and Schulz T J 2007 Approach for reconstructing anisoplanatic adaptive optics images *Appl. Opt.* **46** 6055–63

[8] Ayers G R and Dainty J C 1988 Iterative blind deconvolution method and its applications *Opt. Lett.* **13** 547–9

[9] Beckers J M 1988 Increasing the size of the isoplanatic patch with multiconjugate adaptive optic, *ESO Conf. and Workshop Proc.* **2** 693

[10] Benfenati A and Ruggiero V 2015 Inexact Bregman iteration for deconvolution of superimposed extended and point sources *Commun. Nonlinear Sci. Numer. Simul.* **20** 882–96

[11] Berry M W, Browne M, Langville A N, Pauca V P and Plemmons R J 2007 Algorithms and applications for approximate nonnegative matrix factorization *Comput. Stat. Data Anal.* **52** 156–73

[12] Bertero M and Boccacci P 2000 Application of the OS-EM method to the restoration of LBT images *Astron. Astrophys. Suppl. Ser.* **144** 181–6

[13] Bertero M and Boccacci P 2003 Super-resolution in computational imaging *Micron* **34** 265–73

[14] Bertero M and Boccacci P 2005 A simple method for the reduction of boundary effects in the Richardson-Lucy approach to image deconvolution *Astron. Astrophys.* **437** 369–74

[15] Bertero M, Boccacci P, Brakenhoff G J, Malfanti F and van der Voort H T M 1990 Three-dimensional image restoration and super-resolution in fluorescence confocal microscopy *J. Microsc.* **157** 3–20

[16] Bertero M, Boccacci P, Davis R E, Malfanti F, Pike E R and Walker J G 1992 Super-resolution in confocal scanning microscopy: IV. Theory of data inversion by the use of optical masks *Inverse Probl.* **8** 1–23

[17] Bertero M, Boccacci P, Davis R E and Pike E R 1991 Super-resolution in confocal scanning microscopy: III. The case of circular pupils *Inverse Probl.* **7** 655–74

[18] Bertero M, Boccacci P, Malfanti F and Pike E R 1994 Super-resolution in confocal scanning microscopy: V. Axial super-resolution in the incoherent case *Inverse Probl.* **10** 1059–77

[19] Bertero M, Boccacci P and Pike E R 1982 Resolution in diffraction-limited imaging, a singular value analysis II. The case of incoherent illumination *Opt. Acta* **29** 1599–611

[20] Bertero M, Brianzi P, Parker P and Pike E R 1984 Resolution in diffraction-limited imaging, a singular value analysis III. The effect of sampling and truncation of the data *Opt. Acta* **31** 181–201

[21] Bertero M, Brianzi P and Pike E R 1987 Super-resolution in confocal scanning microscopy *Inverse Probl.* **3** 195–212

[22] Bertero M and De Mol C 1996 Super-resolution by data inversion *Progress in Optics* vol 36 ed E Wolf (Amsterdam: Elsevier) ch 3

[23] Bertero M, De Mol E R, Pike C and Walker J G 1984 Resolution in diffraction-limited imaging, a singular value analysis IV. The case of uncertain localization and non-uniform illumination of the object *Opt. Acta* **31** 923–46

[24] Bertero M and Pike E R 1982 Resolution in diffraction-limited imaging, a singular value analysis I. The case of coherent illumination *Opt. Acta* **29** 727–46

[25] Bertsekas D P 1999 *Nonlinear Programming* 2nd edn (Belmont, MA: Athena Scientific)

[26] Biggs D S C and Andrews M 1997 Acceleration of iterative image restoration algorithms *Appl. Opt.* **36** 1766–75

[27] Biggs D S C and Andrews M 1998 Asymmetric iterative blind deconvolution of multi-frame images *Proc. SPIE* **3461** 328–38

[28] Boden A F, Redding D C, Hanisch R J and Mo J 1996 Massively parallel spatially variant maximum-likelihood restoration of Hubble Space Telescope imagery *J. Opt. Soc. Am.* **13** 1537–45

[29] Bolte J, Combettes P L and Pesquet J C 2010 Alternating proximal algorithm for blind image recovery, *2010 IEEE Int. Conf. on Image Processing* pp 1673–6

[30] Bolte J, Sabach S and Teboulle M 2014 Proximal alternating linearized minimizatio for nonconvex and nonsmooth problems *Math. Program.* **146** 459–94

[31] Bonettini S 2011 Inexact block coordinate descent methods with application to the nonnegative matrix factorization *IMA J. Numer. Anal.* **31** 1431–52

[32] Brakenhoff G J, van der Voort H T M, van Spronsen E A, Linnemans W A M and Nanninga N 1985 Three-dimensional chromatin distribution in neuroblastoma nuclei shown by confocal scanning laser microscopy *Nature* **317** 748–9

[33] Campisi P and Egiazarian K (ed) 2007 *Blind Image Deconvolution: Theory and Applications* (Boca Raton, FL: CRC Press)

[34] Carbillet M, Femenía B, Riccardi A and Fini L 2004 Modeling astronomical adaptive optics —I. The software package CAOS *Mon. Not. R. Astron. Soc.* **356** 1263–75

[35] Castello M *et al* 2018 Image scanning microscopy with single-photon detector array, bioRxiv

[36] Dai Y H and Fletcher R 2006 New algorithms for singly linearly constrained quadratic programming problems subject to lower and upper bounds *Math. Program.* **106** 403–21

[37] De Mol C and Defrise M 2004 Inverse imaging with mixed penalties, *Proc. URSI Symp. on Electromagnetic Theory, Pisa* pp 798–800

[38] de Pater I, Laver C, Davies A G, de Kleer K, Williams D A, Howell R R, Rathbun J A and Spencer J R 2016 Io: Eruptions at Pillan, and the time evolution of Pele and Pillan from 1996 to 2015 *Icarus* **264** 198–212

[39] de Villers G and Pike E R 2017 *The Limits of Resolution* (Boca Raton, FL: CRC Press)

[40] Defrise M and De Mol C 2018 *On blind imaging, NMF and PET, The Radon Transform—The First 100 years and Beyond* ed R Ramlau and O Scherzer (Berlin: De Gruyter & Co.)

[41] Defrise M, Vanhove C and Liu X 2011 An algorithm for total variation regularization in high-dimensional linear problems *Inverse Probl.* **27** 065002

[42] Denis L, Thiébaut E and Soulez F 2011 Fast model of space-variant blurring and its application to deconvolution in Astronomy, *Proc. 18th IEEE Int. Conf. on Image Proc.* pp 2817–20

[43] Desiderá G, Anconelli B, Bertero M, Boccacci P and Carbillet M 2006 Application of iterative blind deconvolution to the reconstruction of LBT LINC-NIRVANA images *Astron. Astrophys.* **402** 727–34

[44] Desiderá G and Carbillet M 2009 Strehl-constrained iterative blind deconvolution for post-adaptive-optics data *Astron. Astrophys.* **507** 1759–62

[45] Donoho D L 1992 Superresolution via sparsity constraints *SIAM J. Math. Anal.* **23** 1309–31

[46] Donoho D L, Johnstone I M, Hock J C and Stern A S 1992 Maximum entropy and the nearly black object *J. R. Stat. Soc.* B **54** 41–81

[47] Faisal M, Lanterman A D, Snyder D L and White R L 1995 Implementation of a modified Richardson-Lucy method for image restoration on a massively parallel computer to compensate for space-variant point-spread-function of a charged-coupled-device camera *J. Opt. Soc. Am.* A **12** 2593–603

[48] Fish D A, Brinicombe A M and Pike E R 1995 Blind deconvolution by means of the Richardson—Lucy algorithm *J. Opt. Soc. Am.* A **13** 58–65

[49] Fish D A, Grochmalicki J and Pike E R 1996 A scanning singular value decomposition method for restoration of images with space variant blur *J. Opt. Soc. Am.* A **13** 464–9

[50] Gilad E and von Hardenberg J 2006 A fast algorithm for convolution integrals with space and time variant kernels *J. Comput. Phys.* **216** 326–36

[51] Giovannelli J-F and Coulais A 2005 Positive deconvolution for superimposed extended source and point sources *Astron. Astrophys.* **439** 401–12

[52] Grippo L and Sciandrone M 2000 On the convergence of the block nonlinear Gauss-Seidel method under convex constraints *Oper. Res. Lett.* **26** 127–36

[53] Grochmalicki J, Pike E R and Walker J G 1993 Experimental confirmation of super-resolution in incoherent scanning microscopy *Pure Appl. Opt.* **2** 565–8

[54] Hadi S B and Blanc-Féraud L 2012 Modeling and removing depth variant blur in 3D fluorence microscopy, *Proc. IEEE Int. Conf. ICASSP 2012* pp 689–92

[55] Harris J L 1964 Diffraction and resolving power *J. Opt. Soc. Am.* **54** 931–6

[56] Hirsch M, Sra S, Schölkopf B and Armeling S 2010 Efficient filter flow for space-variant multiframe blind deconvolution *Proc. IEEE CVPR* **2010** 607–14

[57] Holmes T J 1992 Blind deconvolution of quantum-limited incoherent imagery: maximum-likelihood approach *J. Opt. Soc. Am.* A **9** 1052–61

[58] Huff J 2015 The airyscan detector from ZEISS: confocal imaging with improved signal-to-noise ratio and super-resolution *Nat. Methods* **12** 1205

[59] La Camera A, Antoniucci S, Bertero M, Boccacci P, Lorenzetti D, Nisini B and Arcidiacono C 2014 Reconstruction of high dynamic range images: Simulations of LBT observations of a stellar jet, a pathfinder study for future AO-assisted giant telescopes *Pub. Astron. Soc. Pacific* **126** 180–93

[60] La Camera A, Schreiber L, Diolaiti E, Boccacci P, Bertero M, Bellazzini M and Ciliegi P 2015 A method for space-variant deblurring with application to adaptive optics imaging in astronomy *Astron. Astrophys.* **579** A1

[61] Lantéri H, Aime C, Beaumont H and Gaucherel P 1994 Blind deconvolution using the Richardson-Lucy algorithm *Proc. SPIE* **2312** 182–92

[62] Lecharlier L 2014 Blind inverse imaging with positivity constraints *PhD thesis* Université Libre de Bruxelles https://orbi.uliege.be/bitstream/2268/220716/1/TheseLoic_ResumeIntro.pdf

[63] Lecharlier L and De Mol C 2014 Regularized blind deconvolution with Poisson data, *J. Phys: Conf. Ser.* **464** 012003

[64] Lee D D and Seung H S 1999 Learning the parts of objects by non-negative matrix factorization *Nature* **401** 788–91

[65] Lee D D and Seung H S 2001 Algorithms for non-negative matrix factorization *Advances in Neural Information Processing Systems 13, Proc. NIPS 2000* (Cambridge, MA: MIT Press) pp 556–62

[66] Lee S and Wright S J 2008 Implementing algorithms for signal and image reconstruction on graphical processing units *Tech. Rep.* (Computer Sciences Department, University of Wisconsin-Madison) www.researchgate.net/publication/236736825_Implementing_Algorithms_for_Signal_and_Image_Reconstruction_on_Graphical_Processing_Units

[67] Levin A, Weiss Y, Durand F and Freeman W T 2009 Understanding and evaluating blind deconvolution algorithm, *IEEE Conf. Computer Vision and Pattern Recognition* pp 1964–71

[68] McCutchen C W 1967 Superresolution in microscopy and the Abbe resolution limit *J. Opt. Soc. Am.* **57** 1190–2

[69] Müller C B and Enderlein J 2010 Image scanning microscopy *Phys. Rev. Lett.* **104** 198101

[70] Nagy J G and O'Leary D P 1998 Restoring images degraded by spatially-variant blur *SIAM J. Sci. Comput.* **19** 1063–82

[71] Ng M K, Chan R H and Tang W-C 1999 A fast algorithm for deblurring models with Neumann boundary conditions *SIAM J. Sci. Comput.* **21** 851–66

[72] Nocedal J and Wright S J 1999 *Numerical Optimization* (New York: Springer)

[73] Ostrowski A M 1973 *Solution of Equations in Euclidean and Banach Spaces* (New York: Academic)

[74] Park S C P, Park M K and Kang M G 2003 Super-resolution image reconstruction: a technical overview *IEEE Sign. Process. Mag.* **20** 21–36

[75] Pirzkal N, Hook R N and Lucy L B 2000 GIRA—Two-channel photometric restoration vol 216 *Astronomical Data Analysis Software and Systems IX, ASP Conf. Proc.* vol 216 ed N Manset, C Veillet and D Crabtree (Astronomical Society of the Pacific) pp 655–8
[76] Politte D G and Snyder D L 1991 Correction for accidental coincidences and attenuation in maximum-likelihood image reconstruction for positron-emission tomography *IEEE Trans. Med. Imaging* **10** 82–9
[77] Prato M, Cavicchioli R, Zanni L, Boccacci P and Bertero M 2012 Efficient deconvolution methods for asfronomical imaging: algorithms and IDL-GPU codes *Astron. Astrophys.* **539** A133
[78] Prato M, La Camera A, Bonettini S and Bertero M 2013 A convergent blind deconvolution method for post-adaptive-optics astronomical imaging *Inverse Probl.* **29** 065017
[79] Preza C and Conchello J-A 2004 Depth-variant maximum-likelihood restoration for three-dimensional fluorescence microscopy *J. Opt. Soc. Am.* A **21** 1593–601
[80] Rushforth C K and Harris R W 1968 Restoration, resolution, and noise *J. Opt. Soc. Am.* **58** 539–45
[81] Serra Capizzano S 2003 A note on anti-reflective boundary conditions and fast deblurring models *SIAM J. Sci. Comput.* **25** 1307–25
[82] Sheppard C J R 1988 Super-resolution in confocal imaging *Optik* **80** 53–4
[83] Slepian D and Pollak H O 1961 Prolate spheroidal wave functions, Fourier analysis and uncertainty-i *Bell Labs Tech. J.* **40** 43–63
[84] Snyder D L, Helstrom C W, Lanterman A D, Faisal M and White R L 1994 Compensation for read-out noise in HST image restoration *The Restoration of HST Images and Spectra II* ed R J Hanish and R L White (Baltimore, MD: The Space Telescope Science Institute) pp 139–54
[85] Stobie E B, Hanish R J and White R L 1994 Implementation of the Richardson-Lucy algorithm *STDAS—ASP Conf. Ser.* **61** pp 296–9
[86] Toraldo di Francia G 1952 Super-gain antennas and optical resolving power *Supp. Nuovo Cimento* **9** 426–38
[87] Toraldo di Francia G 1955 Resolving power and information *J. Opt. Soc. Am.* **45** 497–501
[88] Toraldo di Francia G 1969 Degrees of freedom of an image *J. Opt. Soc. Am.* **59** 799–804
[89] Trussell H J and Hunt B R 1978 Image restoration of space variant blurs by sectioned methods *IEEE Trans. Acoust. Speech Signal Process.* **26** 608–9
[90] Trussell H J and Hunt B R 1978 Sectioned methods for image restoration *IEEE Trans. Acoust. Speech Signal Process.* **26** 157–64
[91] Tsumuraya F, Miura N and Baba N 1994 Iterative blind deconvolution method using Lucy's algorithm *Astron. Astrophys.* **282** 699–708
[92] Viano G A 1976 On the extrapolation of optical image data *J. Math. Phys.* **17** 1160–5
[93] White R L 1993 Improvements to the Richardson-Lucy method, Newslett *STScI's Image Restoration Project* **1** 11–9
[94] Wolter H 1961 On basic analogies and principal differences between optical and electronic information *Progress in Optics* vol I ed E Wolf (Amsterdam: North-Holland) ch V
[95] Zanella R, Zanghirati G, Cavicchioli R, Zanni L, Boccacci P, Bertero M and Vicidomini G 2013 Towards real-time image deconvolution: application to confocal and STED microscopy *Sci. Rep.* **3** 2523

Mario Bertero, Patrizia Boccacci and Valeria Ruggiero

Chapter 9

Towards a regularization theory

As follows from the previous treatment, most research on Poisson data inversion deals with discrete settings along the lines proposed in the paper of Shepp and Vardi. Currently a rich variety of methods is available and only the future will indicate those that will survive either for their accuracy and efficacy in specific applications or for their pedagogical value.

Unfortunately, a generally accepted formulation of regularization methods in infinite dimension is lacking and this refers to the difficulty of devising completely satisfactory models in an infinite-dimensional setting. In this final chapter we provide a concise account of the different approaches proposed up to now. The interested reader is referred to the cited papers to study in more depth the methods and results. We subdivide the proposed methods into two classes: deterministic and statistical.

9.1 Deterministic regularization approaches

We qualify 'deterministic' the approaches that mimic Tikhonov regularization theory by converting the data-fidelity functional, i.e. the least-squares functional, into a generalized KL divergence derived from the discrete data-fidelity function (4.7) by a limiting process. In these cases, the datum y is treated as a function perturbed by an error that possibly approaches zero. As discussed in section 3.4, this assumption is not consistent with the peculiarities of Poisson data since the latter are more and more accurate, in the sense of relative error, if the number of detected photons tends to infinity. Anyway, the mathematical models are correctly managed in the literature: definition of regularized solutions, convergence properties as the error approaches zero, etc. The questionable point is that their physical significance is not clear. For a critical comment on the limiting process leading to the data-fidelity function considered in the deterministic approaches we refer to [7], Remark 2.7.

doi:10.1088/2053-2563/aae109ch9

9.1.1 Infinite-dimensional extension of EM–RL

The first analysis of Poisson data inversion in an infinite-dimensional setting is due to Mülthei and Shorr [13, 14] and Mülthei [12]; they focus on the extension of the EM–RL method for solving the first kind Fredholm integral equations

$$y(\vec{r}) = \int_S h(\vec{r}, \vec{r}')x(\vec{r}')d\vec{r}' \doteq (Ax)(\vec{r}), \tag{9.1}$$

with strictly positive datum y and kernel h. Here S is both the object and image domain. The datum y and the unknown x are generically qualified as densities, presumably densities of photons, i.e. number of photons per unit area. By taking the logarithm of the discrete likelihood function defined in equation (4.2) and replacing sums by integrals, the continuous version of the maximum-likelihood approach becomes equivalent to maximize the functional

$$\Lambda(x; y) = \int_S y(\vec{r})\ln(Ax)(\vec{r})d\vec{r}, \tag{9.2}$$

under the constraint

$$\int_S (Ax)(\vec{r})d\vec{r} = \int_S y(\vec{r})d\vec{r}. \tag{9.3}$$

Then, the continuous version of the EM–RL algorithm becomes

$$\begin{aligned} x^{(k+1)} &= G(x^{(k)}) \\ G(x)(\vec{r}) &= \frac{x(\vec{r})}{a(\vec{r})}\int_S \frac{h(\vec{r}', \vec{r})y(\vec{r}')}{F(\vec{r}')}d\vec{r}' \\ a(\vec{r}) &= \int_S h(\vec{r}', \vec{r})d\vec{r}',\ F(\vec{r}') = \int_S h(\vec{r}', \vec{r}'')x(\vec{r}'')d\vec{r}''. \end{aligned} \tag{9.4}$$

The convergence of the iterates is not proved in the above-mentioned papers even if, in [14], a proof does appear in the discrete case (see section 5.5.1). The results that these papers offer in the infinite-dimensional case can be summarized as follows:

- if the sequence $x^{(k)}$ converges in the L^1 norm, then its limit point is a maximizer of $\Lambda(x; y)$;
- the values $\Lambda(x^{(k)}, y)$ increase for increasing k;
- the KL divergence between the iterates and the limit point decreases as k increases.

The convergence of the iterates (9.4) is reconsidered in [17]. The starting point is the minimization of the following functional

$$\Phi_0(x; y) = \int_S \left\{y(\vec{r})\ln\frac{y(\vec{r})}{(Ax)(\vec{r})} + (Ax)(\vec{r}) - y(\vec{r})\right\}d\vec{r} := \mathrm{KL}(y, Ax), \tag{9.5}$$

a continuous analog of the data-fidelity function (4.7). Here A is the linear integral operator defined in equation (9.1). In [17] the authors first investigate convergence of the continuous EM–RL algorithm in the case of exact data $\bar{y}$: for a given $\bar{y}$ there exists a non-negative solution $\bar{x}$ of the integral equation (9.1). In such a case $\Phi(\bar{x}; \bar{y}) = 0$. They assume that both $\bar{y}$ and the kernel h are strictly positive and the latter satisfies the additional conditions

$$\int_S h(\vec{r}, \vec{r}')d\vec{r} = 1,\ 0 < m < h(\vec{r}, \vec{r}') < M. \tag{9.6}$$

where m, M are positive constants.

The following results are proved [17, 18]

- if the initial point $x^{(0)}$ is such that $\Phi_0(x^{(0)}; \bar{y}) < \infty$, then

$$\lim_{k\to\infty} \|Ax^{(k)} - \bar{y}\|_p = 0, \tag{9.7}$$

  for any $p \in [1, \infty)$, where $\|.\|_p$ denotes the L^p-norm;
- if, in addition, both $x^{(0)}$ and $\bar{x}$ are bounded, then iterates $x^{(k)}$ have a subsequence that converges to a solution of the equation in the weak topology of L^p for any $p \in [1, \infty)$. If, in addition, the solution is unique, then the whole sequence converges to the solution $\bar{x}$ in the same topologies.

Next, the authors consider the case of noisy data, denoted as y^δ and assumed to satisfy the condition

$$\int_S y^\delta(\vec{r})d\vec{r} = \int_S \bar{y}(\vec{r})d\vec{r} = 1,\ \|y^\delta - \bar{y}\|_1 \leqslant \delta,\ \delta > 0, \tag{9.8}$$

which forces them to approach the exact data $\bar{y}$ when $\delta \to 0$.

If we denote by $x^{(k),\delta}$ the iterates that come out of equation (9.4) when y is replaced by y^δ, then convergence of the sequence obviously fails to hold. However, the continuous EM–RL method is possibly semi-convergent, in the sense discussed in [5], chapter 6 or in section 4.1: the distance between the iterates and the solution has an initial decay and then increases, so that the method can provide a regularized solution via a suitable stopping rule.

To prove the last mentioned property of the EM–RL, additional bounds are prescribed on y^δ, i.e. it is assumed that there exist positive constants m_1, M_1 such that

$$m_1 \leqslant y^\delta \leqslant M_1. \tag{9.9}$$

Thanks to bounds (9.6) and (9.9) a stopping rule is proposed based on the iteration index defined by

$$k_*(\delta) = \min\left\{k|\Phi(x^{(k),\delta}; y^\delta) \leqslant \eta\delta \max\left(\ln\frac{M_1}{m}, \ln\frac{M}{m_1}\right)\right\} \tag{9.10}$$

for some $\eta > 1$. Indeed, the following results are proved [17, 18]:

- the stopping index $k_*(\delta)$ is finite and satisfies $k_*(\delta) = O(\delta^{-1})$; moreover,

$$\lim_{\delta\to 0} \|Ax^{k_*(\delta),\delta} - \bar{y}\|_p = 0 \tag{9.11}$$

  for any $p \in [1, \infty)$;

- if the initial guess $x^{(0)}$ is bounded and equation (9.1) has a bounded solution $\bar{x}$, then the sequence $x^{k_*(\delta),\delta}$ has a subsequence which, when $\delta \to 0$, converges to a solution of the equation in the weak topology of L^p for any $p \in [1, \infty)$; if the solution is unique, then the whole sequence converges to $\bar{x}$ in the same topology.

In principle, the stopping rule based on the iteration index (9.10) might work for solving a discrete problem but we do not know any relevant concrete example.

9.1.2 Regularization of the data-fidelity function

A regularized problem is investigated in [16]. Here the data-fidelity function involves the following KL divergence

$$\mathrm{KL}(y, x) = \int_S y(\vec{r})\ln\frac{y(\vec{r})}{x(\vec{r})}d\vec{r}, \tag{9.12}$$

and the regularization function is also a KL divergence. In other words, the regularized functional is given by

$$\Psi(x, y^\delta) = \mathrm{KL}(y^\delta, Ax) + \beta\mathrm{KL}(x, x^*), \tag{9.13}$$

where y^δ is again a noisy datum satisfying condition (9.8) and x^* is a prior estimate of the solution. Under the assumptions that the perturbed datum y^δ, the operator A and the solution x are positive and satisfy quite natural boundedness conditions, existence, uniqueness and convergence for the regularized approximations are derived together with stability results and convergence rates.

The regularization of the data-fidelity functional (9.5) is investigated in a series of papers by Bardsley and coworkers [1–4, 11]. In these papers, the functional under minimization has the following form

$$\Phi_\beta(x; y) = \int_S \{(Ax)(\vec{r}) - y(\vec{r})\ln(Ax)(\vec{r}))\}d\vec{r} + \beta R(x), \tag{9.14}$$

where $R(x)$ is the regularization term and terms depending only on y are neglected in the data-fidelity functions. In all papers it is assumed that A is a linear bounded operator with the additional property $Ax \geqslant 0$ for any $x \geqslant 0$.

The papers focus on existence of the solution and on possible regularization schemes. The case of Tikhonov regularization

$$R(x) = \frac{1}{2}\|x\|_{L^2}^2, \tag{9.15}$$

is investigated in [3] while the regularization by diffusion, proposed in [8], is considered in [2]. Other considered regularizations include L^1 regularization [11] and the TV-smoothed (also called HS regularization in this book) given by

$$R(x) = \int_S \sqrt{|\nabla x(\vec{r})|^2 + \delta^2}\,d\vec{r} \tag{9.16}$$

in the case of a differentiable x (see [4]). A theoretical framework for the regularization of Poisson likelihood estimation is proposed in [1]. In all these papers the behavior of the minimizers is considered when the error on the data function y, and possibly also the error on the operator A, tends to zero with respect to some metric.

The case of TV regularization is investigated in [19] where the EM–TV method, discussed in section 6.4.1, is also proposed. The authors start from a semi-discrete model where the data are discrete while the unknown is a function x of a continuous space variable. They assume that a function y exists such that its integral over a sub-domain S_i of the detector domain S is the number of photons detected in S_i

$$\int_{S_i} \bar{y}(\vec{r})d\vec{r} = y_i. \tag{9.17}$$

Moreover, y_i is Poisson distributed with an expected value given by

$$(Ax)_i = \int_{S_i} (Ax)(\vec{r})d\vec{r}, \tag{9.18}$$

where x is the unknown function and A is a linear bounded operator.

They consider the semi-discrete data-fidelity function

$$\sum_i \{(Ax)_i - y_i \ln(Ax)_i\}, \tag{9.19}$$

and recast it as follows

$$\Phi_0(x; y) = \int_S \{(Ax)(\vec{r}) - y(\vec{r})\ln(Ax)(\vec{r})\} d\mu. \tag{9.20}$$

Here μ is the measure defined by $d\mu = \sum_i \chi_i d\vec{r}$, χ_i is the characteristic function of S_i and $d\vec{r}$ is the density of the standard Lebesgue measure. Finally, the following functional

$$\Phi_\beta(x; y) = \Phi_0(x; \bar{y}) + \beta \|x\|_{BV(S)} \tag{9.21}$$

is minimized, where $\Phi_0(x; y)$ is defined in the previous equation and $\|.\|_{BV(S)}$ denotes the norm in the space of functions of bounded variation over S. The latter is given by

$$\|x\|_{BV(S)} = \sup_{\phi \in C_0^\infty; \|\phi\|_\infty \leqslant 1} \int_S x \operatorname{div} \phi \, d\mathbf{r}, \tag{9.22}$$

and coincides with

$$\|x\|_{BV(S)} = \int_S |\nabla x| d\mathbf{r}. \tag{9.23}$$

if x is sufficiently regular.

If the operator A is compact and preserves non-negativity, i.e. $(Ax) \geqslant 0$ for any $x \geqslant 0$, the minimization of $\Phi_\beta(x; y)$ in $BV(S)$, is well-posed for given y and A; more precisely the following results hold.

- $\Phi_\beta(x; y)$ is lower semi-continuous in $BV(S)$.
- If the operator A does not annihilate the constant functions, then $\Phi_\beta(x; y)$ is BV-coercive, i.e. $\Phi_\beta(x; y) \to \infty$ whenever $\|x\|_{BV(S)} \to \infty$.
- The functional has a minimizer; the proof follows from the direct method of the calculus of variations, thanks to the lower semi-continuity and coercivity of the functional.
- If A is injective and $y > 0$, then the functional is strictly convex and therefore the minimizer is unique.
- If $\{y^{(k)}\}$ is a sequence converging to y in the following sense

$$\lim_{k\to\infty} \mathrm{KL}(y^{(k)}, y) = 0, \tag{9.24}$$

  and $\{x^{(k)}\}$ is the sequence of the corresponding (unique) minimizers of the functionals $\Phi_\beta(x; y^{(k)})$, then this sequence has convergent subsequences and each convergent subsequence converges in the L^1 norm to a minimizer of $\Phi_\beta(x; y)$.

The proof of the last point requires some additional assumptions, namely that $\ln y$ and $\ln(Ax)$ belong to L_μ^∞ and finally that the sequence $\{y^{(k)}\}$ is uniformly bounded in the L_μ^1 norm.

9.2 Statistical approaches

Statistical approaches have been recently proposed. The basic assumption is that randomly distributed mathematical points describe the photons hitting the detector. Although infinite precision in the determination of the position of a photon conflicts with Heisenberg's uncertainty principle (in case the incident radiation is monochromatic, photons have a well-defined energy and momentum), photons are certainly well localized because each element of the detector has a small size. Since, as far as we know, a mathematical model taking into account all the features of the physical phenomenon is not available, we believe that approximate models which have been proposed by mathematicians deserve careful consideration and may provide a first insight into the problem. The advantage of the statistical approaches that assume perfect localization of photons is that they can take into account the correct statistics of the photon counting process.

9.2.1 The approach of Hohage and Werner

The basic assumption in [7] (see also [6, 22]) is the following:

- the positions of the detected photons are known for any given observation;
- these positions are a realization of a Poisson process.

The notion of a Poisson process, a random point process with specific properties, is fundamental here. An excellent reference on this topic is the book of Kingman [9] (see also [21]). In appendix A.8 we provide the basic definitions for the convenience of the reader.

If the realization of the Poisson process underlying detection consists of n points $\vec{r}_1, \ldots, \vec{r}_n$ (n random) and y stands for the unknown intensity function, then, as derived in [7] (see appendix A.9), the negative logarithm of the relevant likelihood functional is given by

$$\mathcal{S}(y; \vec{r}_1, \ldots, \vec{r}_n) = \int_S y(\vec{r})d\vec{r} - \sum_{k=1}^{n} \ln y(\vec{r}_k) \tag{9.25}$$

and it can be written in an alternative form in terms of the random measure given in equation (A.35)

$$\mathcal{S}(y; \mathcal{G}) = \int_S y(\vec{r})d\vec{r} - \int_S \ln y(\vec{r})d\mathcal{G}. \tag{9.26}$$

Function y must be non-negative. However, because of the logarithm, the functional does not have a minimum without extra *a priori* constraints on the solution or the addition of a suitable penalty term.

In this context, one assumes that there exists a source described by a Poisson process with an unknown intensity function x and that there exists a relationship between the source intensity x and the image intensity y. In [7] a general model based on a nonlinear operator F, satisfying suitable conditions, is considered, i.e. the relationship between source intensity and image intensity is written in the form $y = F(x)$. Here, for simplicity, we assume that the relationship is linear affine, i.e. $y = Ax + b$ where b is a known and positive background function and A is either the integral operator defined in equation (9.1) or any linear compact operator in $L^1(S)$. Then estimating x, when a realization of the Poisson process $\mathcal{G}$ with intensity y is given, amounts to minimizing the following functional

$$\mathcal{S}_0(x; \mathcal{G}) = \int_S (Ax + b)(\vec{r})d\vec{r} - \int_S \ln(Ax + b)(\vec{r})d\mathcal{G}. \tag{9.27}$$

on a set of non-negative functions. This functional is precisely the *data-fidelity function* of our problem.

Minimizing functional (9.27) is an ill-posed task since minimizers need not exist in general. A regularization approach involving the convergence of the solution to the 'true' one when 'error' on the data tends to zero needs understanding in what sense 'error' tends to zero. Since in the present situation data consist of a realization of a Poisson process, the common practice for improving the quality of an image obtained by photon counting is to accumulate photons as much as possible as already remarked in section 3.4. In astronomy this can be obtained by increasing the number of frames; in emission tomography it should be necessary to increase the dose administered to the patient and, of course, this is not a safe practice. Similarly, in confocal microscopy, one should increase the intensity of the laser beam but, again, such a procedure can kill the observed living cell.

In a theoretical approach one can ignore practical limitations on the number of photons under detection and investigate what happens when the statement 'number of photons tending to infinity' replaces 'error on the data tending to zero'. The point is to formulate a model where this statement has a precise meaning.

The first step is to introduce an observation time τ (which could have the meaning of an administered dose in emission tomography or laser intensity in fluorescence microscopy), already discussed at the end of section 3.4. The statement 'number of photons tending to infinity' can be replaced by 'observation time tending to infinity' since more and more photons are accumulated if one increases the observation time (see also section 3.4). The model introduced in [7] consists of a family of Poisson processes $\{\mathcal{G}_\tau\}_{\tau \geqslant 0}$ such that the intensity function of $\mathcal{G}_\tau$ is given by $y_\tau = \tau\, y$. Obviously one assume that the Poisson process associated with the random source can be described in a similar way so that in this context $y = Ax + b$ and x is the intensity function under estimation.

Next, for each τ, the point process defined either by $\tilde{\mathcal{G}}_\tau = \frac{1}{\tau}\mathcal{G}_\tau$ or by

$$\tilde{\mathcal{G}}_\tau = \frac{1}{\tau}\sum_{\tau_j \leqslant \tau} \delta_{\vec{r}_k,\, \tau_k}, \tag{9.28}$$

can be introduced [7] and called *temporally normalized Poisson process*. If we remark that, for any sub-domain S_{sub} of S

$$\mathrm{E}\left[\int_{S_{\text{sub}}} d\tilde{\mathcal{G}}_\tau\right] = \int_{S_{\text{sub}}} y d\vec{r}, \ \mathrm{VAR}\left[\int_{S_{\text{sub}}} d\tilde{\mathcal{G}}_\tau\right] = \frac{1}{\tau}\int_{S_{\text{sub}}} y d\vec{r}, \tag{9.29}$$

we observe that the realizations of $\tilde{\mathcal{G}}_\tau$, when τ increases, tend to accumulate in the regions where the density function y is larger. Thus patterns are provided that mimic this function under vanishing fluctuations when τ approaches infinity.

The data-fidelity function can be easily manipulated and written in terms of the temporally normalized Poisson process as follows

$$\mathcal{S}_0(x; \tilde{\mathcal{G}}_\tau) = \int_S (Ax + b)(\vec{r})d\vec{r} - \int_S \ln(Ax + b)(\vec{r})d\tilde{\mathcal{G}}_\tau. \tag{9.30}$$

Since the minimization of this data-fidelity function is ill-posed, a regularization similar to those considered in the discrete case, is also considered in [7], i.e. the problem is converted into the minimization of the following functional

$$\mathcal{S}_\beta(x; \tilde{\mathcal{G}}_\tau) = \int_S (Ax + b)(\vec{r})d\vec{r} - \int_S \ln(Ax + b)(\vec{r})d\tilde{\mathcal{G}}_\tau + \beta R(x) \tag{9.31}$$

where as usual $R(x)$ is a suitable convex regularization function.

In [7], the existence of minimizers of the regularized functional, for any $\beta > 0$ is proved for a more general class of imaging operators, including nonlinear ones, and for a class of convex regularizers $R(x)$ including TV. Existence can be easily proved when A is the integral operator of equation (9.1) with a bounded kernel. The minimizer (in general, not unique) depends also on the realization of $\mathcal{G}_\tau$; we omit this dependence and we use the notation $x^*_{\beta,\tau}$ for this minimizer.

The main result involves the concept of R-minimizing solution, i.e. a solution $\bar{x}$ of the following problem

$$R(\bar{x}) = \min\{R(x)|Ax = y - b\}. \tag{9.32}$$

The authors assume that this solution exists and is unique and they prove that [7], theorem 4.8, for a suitable choice of the regularization parameter $\bar{\beta} = \bar{\beta}(\tau, \tilde{\mathcal{G}}_\tau)$, satisfying the conditions

$$\lim_{\tau\to\infty} \bar{\beta}(\tau, \tilde{\mathcal{G}}_\tau) = 0, \lim_{\tau\to\infty} \frac{\ln(\tau)}{\sqrt{\tau}\bar{\beta}(\tau, \tilde{\mathcal{G}}_\tau)} = 0, \tag{9.33}$$

the family $x^*_{\bar{\beta},\tau}$ generated by any minimizer of equation (9.31), defines a statistical regularization scheme in the following sense:

$$\lim_{\tau\to\infty} \mathbf{P}[\mathcal{D}_R^{\tilde{x}}(x^*_{\bar{\beta},\tau}, \bar{x}) > \epsilon] = 0, \forall \epsilon > 0. \tag{9.34}$$

Here $\mathbf{P}$ denotes probability, $\mathcal{D}_R^{\tilde{x}}$ is the Bregman distance associated to the regularizer R

$$\mathcal{D}_R^{\tilde{x}}(x, \bar{x}) = R(x) - R(\bar{x}) - <\tilde{x}, x - \bar{x}>, \tag{9.35}$$

and $\tilde{x}\in\partial R(\bar{x})$ is a subgradient of R in $\bar{x}$.

The authors propose a rule for the choice of the regularization parameter, which is based on an idea of Lepskii [10] and allows convergence rates to be obtained.

9.2.2 An alternative approach

In a very recent paper [20] the authors propose a statistical approach which is less general, but has some connections with the previous one. They propose a modified discrepancy principle, borrowed from that discussed in section 4.4 and allowing a proof of convergence of regularized solutions. In both the approaches the observed data are interpreted as a photon density with intensity $y\in L^1(S)$ and observation time τ so that they are described by a Poisson process $\mathcal{G}_\tau$ with intensity function $\tau\, y$.

The basic assumption is that, at each observation time τ, the datum consists of a good approximation y_τ of y that tends to y when τ approaches infinity.

Thanks to this assumption the following functional is considered

$$\mathcal{J}_{\beta,y_\tau}(x) = \Phi_0(x; y_\tau) + \beta R(x), \tag{9.36}$$

where Φ_0 is defined in equation (9.5) while β and $R(x)$ are a regularization parameter and a suitable regularization functional, respectively. Again, it is assumed $y = Ax$ where A is a linear bounded operator. The minimizers of this functional are denoted $x^*_{\tau,\beta}$.

In view of applying the discrepancy principle, discretized versions of the previous functional come into play with a discretization level depending on τ and becoming finer and finer as τ tends to infinity. First, one considers S as a union of boxes $\{S_i\}_{i=1}^m$ and the operator

$$(\Pi_m y_\tau)_i = \int_{S_i} y_\tau(\vec{r})d\vec{r}, \; i = 1, \dots, m, \tag{9.37}$$

mapping y_τ into an m-vector, is introduced. Second, a sequence of finite-dimensional subspaces Y_n is considered such that

$$Y_0 \subset Y_1 \subset \cdots \subset \overline{\bigcup Y_n} = L^2(S) \tag{9.38}$$

and the projections Π'_n onto these subspaces are introduced for mapping a function of L^2 into a function of Y_n. Third, one introduces the functionals

$$\mathcal{J}_{\beta,y_\tau;m,n}(x) = \Phi_0(\Pi_m y_\tau;\ \Pi_m A \Pi'_n x) + \beta R(\Pi'_n x), \tag{9.39}$$

and denotes $x^*_{\tau,\beta;m,n}$ minimizers of them. The existence of minimizers $x^*_{\tau,\beta}$ of $\mathcal{J}_{\beta,y_\tau}(x)$ and $x^*_{\tau,\beta;m,n}$ of $\mathcal{J}_{\beta,y_\tau;m,n}(x)$ can be proved from the results in [15].

The discrepancy principle is applied as follows: for any given observation time, select a discretization level $m = m(\tau)$ such that

$$\lim_{\tau\to\infty} \frac{m(\tau)}{\tau} = 0; \tag{9.40}$$

next, given constants $c_3 > c_2 > 1$, choose a value of the regularization parameter β such that

$$\frac{c_2 m(\tau)}{2\tau} \leqslant \mathrm{KL}(\Pi_m y_\tau,\ \Pi_m A \Pi'_n x^*_{\tau,\beta;m,n}) \leqslant \frac{c_3 m(\tau)}{2\tau}. \tag{9.41}$$

The authors first prove that the criterion is well-defined, in the sense that there exists a value $\beta = \beta(\tau)$ of the regularization parameter satisfying the criterion. The final result [20], proposition 3.16, is the following: given both a sequence τ_k of observation times (tending to infinity) and a sequence of parameters β_k (chosen according to the previous criterion with $\tau = \tau_k$), then the sequence $\Pi'_{n_k} x^*_{\tau_k,\beta_k;m_k,n_k}$ (formed by the minimizers of the functionals $\mathcal{J}_{\beta_k,y_{\tau_k};m_k,n_k}$) includes a subsequence that converges in the weak topology of $L^1(S)$. A limit of each convergent subsequence is an R-minimizing solution.

We point out that the authors give convergence results in terms of R-minimizing solutions, defined in equation (9.32), without assuming any relevant uniqueness.

9.3 Comments and concluding remarks

As follows from the previous outline, no well-developed regularization method seems to be available in infinite-dimensional spaces. From our point of view the statistical approaches are the most promising ones because they account appropriately for statistical properties of the data. They are not covered in a complete way in this book because they require the reader enjoys a mathematical background beyond what we assumed. In particular, familiarity with in-depth results in probability theory, functional analysis and calculus of variations. The most complete reference for the interested reader is [7], containing discussions and results of both deterministic and statistical methods as well as a very extensive list of references.

We think that linear inverse problems should be given more attention and different possible criteria of selection of the regularization parameter should be investigated and compared. We believe that the topic deserves attention by mathematicians working in the area of inverse problems and we hope that it will be the object of research in the future.

References

[1] Bardsley J M 2010 A theoretical framework for the regularization of Poisson likelihood estimation problems *Inverse Probl. Imaging* **4** 11–7

[2] Bardsley J M and Laobeul N 2008 An analysis of regularization by diffusion for ill-posed Poisson likelihood estimation *Inverse Probl. Sci. Eng.* **17** 537–50

[3] Bardsley J M and Laobeul N 2008 Tikhonov regularized Poisson likelihood estimation: theoretical justification and a computational method *Inverse Probl. Sci. Eng.* **16** 199–215

[4] Bardsley J M and Luttman A 2008 Total Variation-penalized Poisson likelihood estimation for ill-posed problems *Adv. Comput. Math.* **31** 35

[5] Engl H W, Hanke M and Neubauer A 1996 *Regularization of Inverse Problems* (Dordrecht: Kluwer)

[6] Hohage T and Werner F 2013 Iteratively regularized Newton-type method for general data misfit functionals and applications to Poisson data *Numer. Math.* **123** 745–79

[7] Hohage T and Werner F 2016 Inverse problems with Poisson data: statistical regularization, theory, applications and algorithms *Inverse Probl.* **32** 093001

[8] Kaipio J P, Kolehmainen V, Vauhkonen M and Somersalo E 1999 Inverse problems with structural prior information *Inverse Probl.* **15** 713–29

[9] Kingman J F C 1993 *Poisson Processes* (Oxford: Clarendon)

[10] Lepskii O V 1990 On a problem of adaptive estimation in Gaussian white noise *Theory Probab. Appl.* **35** 454–66

[11] Luttman A 2010 A theoretical analysis of L^1 regularized Poisson likelihood estimation *Inverse Probl. Sci. Eng.* **18** 251–64

[12] Mülthei H N 1993 Iterative continuous maximum likelihood reconstruction methods *Math. Methods Appl. Sci.* **15** 275–86

[13] Mülthei H N and Schorr B 1987 On an iterative method for a class of integral equations of the first kind *Math. Methods Appl. Sci.* **9** 137–68

[14] Mülthei H N and Schorr B 1989 On properties of the iterative maximum likelihood reconstruction method *Math. Methods Appl. Sci.* **11** 331–42

[15] Resmerita E 2005 Regularization of ill-posed problems in Banach spaces: convergence rates *Inverse Probl.* **21** 1303–14

[16] Resmerita E and Anderssen R S 2007 Joint additive Kullback-Leibler residual minimization and regularization for linear inverse problems *Math. Methods Appl. Sci.* **30** 1527–44

[17] Resmerita E, Engl H W and Iusem A N 2007 The expectation-maximization algorithm for ill-posed integral equations: a convergenge analysis *Inverse Probl.* **23** 2575–88

[18] Resmerita E, Engl H W and Iusem A N 2008 Corrigendum—the expectation-maximization algorithm for ill-posed integral equations: a convergence analysis *Inverse Probl.* **24** 059801

[19] Sawatzky A, Brune C, Kösters T, Wübbeling F and Burger M 2013 EM-TV methods for inverse problems with Poisson noise *Level Set and PDE Based Reconstruction Methods in Imaging* vol LNM 2090 ed M Burger and S Osher (Cham: Springer) pp 71–142

[20] Sixou B, Hohweiller T and Ducros N 2018 Morozov principle for Kullback-Leibler residual term and Poisson noise *Inverse Probl. Imaging* **12** 607–34

[21] Snyder D L and Miller M I 1991 *Random Point Processes in Time and Space* (New York: Springer)

[22] Werner F and Hohage T 2012 Convergence rates in expectation for Tikhonov-type regularization of inverse problems with Poisson data *Inverse Probl.* **28** 104004

IOP Publishing

Inverse Imaging with Poisson Data

From cells to galaxies

Mario Bertero, Patrizia Boccacci and Valeria Ruggiero

Appendix A

A.1 Linear algebra and analysis

Definition A.1. *Let A be an $M \times N$ matrix and B be a $P \times Q$ matrix. The Kronecker product of the two matrices A and B is the $MP \times NQ$ block matrix $A \otimes B$, given by:*

$$A \otimes B = \begin{pmatrix} a_{11}B & \cdots & a_{1N}B \\ \cdots & \cdots & \cdots \\ a_{M1}B & \cdots & a_{MN}B \end{pmatrix}. \tag{A.1}$$

Proposition A.1. *(Sherman–Morrison formula) Let A be a non-singular matrix of order N and u and v be vectors of $\mathbf{R}^N$. If $v^T A^{-1} u \neq -1$, then the matrix $A + uv^T$ is non-singular and its inverse is given by*

$$(A + uv^T)^{-1} = A^{-1} - \frac{A^{-1}uv^T A^{-1}}{1 + v^T A^{-1} u}. \tag{A.2}$$

Proof. It is immediate to verify that

$$(A + uv^T)\left(A^{-1} - \frac{A^{-1}uv^T A^{-1}}{1 + v^T A^{-1} u}\right) = \left(A^{-1} - \frac{A^{-1}uv^T A^{-1}}{1 + v^T A^{-1} u}\right)(A + uv^T) = I_N.$$

□

The next proposition on summable non-negative sequences is proved in [15].

Proposition A.2. *Let $\{a_k\}$, $\{\xi_k\}$ and $\{\eta_k\}$ be non-negative sequences of real numbers such that $a_{k+1} \leqslant (1 + \xi_k)a_k + \eta_k$ and $\sum_{k=0}^{\infty} \xi_k < \infty$, $\sum_{k=0}^{\infty} \eta_k < \infty$. Then, $\{a_k\}$ converges.*

Definition A.2. *Let* $F: \mathbf{R}^N \to \mathbf{R}$ *be a function.* F *is coercive if* $\lim_{\|x\| \to \infty} F(x) = \infty$.

The following property of the continuously differentiable functions with Lipschitz-continuous gradient is known also as descent lemma.

Proposition A.3. *[4, p 665] Let* $F: \mathbf{R}^N \to \mathbf{R}$ *be a continuously differentiable function, with gradient satisfying the Lipschitz condition with parameter* M_F

$$\|\nabla F(x) - \nabla F(y)\| \leqslant M_F \|x - y\|, \quad \forall x, y \in \Omega$$

where Ω *is a closed subset of* $\mathbf{R}^N$*. Then for all* $x, y \in \Omega$*, we have*

$$F(y) \leqslant F(x) + \nabla F(x)^T (y - x) + \frac{M_F}{2} \|y - x\|^2.$$

Proof. Let t be a scalar parameter and let $G(t) = F(x + t(y - x))$. The chain rule yields $\frac{dG}{dt}(t) = \nabla F(x + t(y - x))^T (y - x)$. Thus, we have

$$\begin{aligned}
F(y) - F(x) = G(1) - G(0) &= \int_0^1 \frac{dG}{dt}(t) dt = \int_0^1 \nabla F(x + t(y - x))^T (y - x) dt \\
&\leqslant \int_0^1 \nabla F(x)^T (y - x) dt \\
&\quad + \left| \int_0^1 (\nabla F(x + t(y - x)) - \nabla F(x))^T (y - x) dt \right| \\
&\leqslant \nabla F(x)^T (y - x) \\
&\quad + \int_0^1 \|\nabla F(x + t(y - x)) - \nabla F(x)\| \|(y - x)\| dt \\
&\leqslant \nabla F(x)^T (y - x) + \|(y - x)\| \int_0^1 M_F t \|y - x\| dt \\
&= \nabla F(x)^T (y - x) + \frac{M_F}{2} \|(y - x)\|^2,
\end{aligned}$$

and the assertion is proved. □

The next proposition is known as Zangwill's global convergence theorem.

Proposition A.4. *[12, p 187] Let* T *be a point to set mapping defined on some subset* $\Omega \in \mathbf{R}^N$*. Suppose that, given* $x^{(0)} \in \Omega$*, the sequence* $\{x^{(k)}\}$ *is generated in such a way to satisfy the condition* $x^{(k+1)} \in T(x^{(k)})$*. Let a set* $X^* \subset \Omega$ *be given and suppose that*

(i) *all points* $x^{(k)}$ *are contained in a compact set* $\Upsilon \subset \Omega$;
(ii) *there is a continuous function* $F\colon \Omega \to \mathbf{R}$ *such that*

$$\begin{aligned} F(y) &< F(x) \quad x \notin X^*, \\ F(y) &\leqslant F(x) \quad x \in X^*, \end{aligned}$$

for all $y \in T(x)$;
(iii) $T(x)$ *is closed at points outside* X^*, *i.e., when* $\lim_{k\to\infty} x^{(k)} = x$ *for* $x^{(k)} \in \Omega - \Upsilon$ *and* $\lim_{k\to\infty} y^{(k)} = y$, *with* $y^{(k)} \in Tx^{(k)}$, *then* $y \in T(x)$.

Then the limit point of any convergent subsequence of $\{x^{(k)}\}$ *is in the set* X^*.

Proposition A.5. *[14] Let* $\{x^{(k)}\}$ *be a sequence in a compact set* Ω *of a normed space. If* $\|x^{(k+1)} - x^{(k)}\| \to 0$ *as* $k \to \infty$, *then the set of limit points of the sequence is non-empty, compact and connected.*

A.2 Some notations on convex sets

Definition A.3. *Given a convex set* Ω, *let aff* Ω *be the smallest affine set containing* Ω. *We define the interior of* Ω *as*

$$int\ \Omega = \{x \in \Omega\colon \exists \epsilon > 0,\ x + \epsilon B(0, 1) \subseteq \Omega\}, \tag{A.3}$$

where $B(0, 1)$ *is the closed unit ball with respect to a prefixed vector norm, while the relative interior of* Ω *is*

$$ri\ \Omega = \{x \in aff\ \Omega\colon \exists \epsilon > 0,\ ((x + \epsilon B(0, 1)) \bigcap aff\ \Omega) \subseteq \Omega\}. \tag{A.4}$$

The concept of relative interior of a convex set is motivated by the fact that a line segment or a triangle embedded in $\mathbf{R}^3$ does not have a natural interior in the metric of $\mathbf{R}^3$. On the contrary, there exists a relative interior. For an open line segment Ω, ri $\Omega = \Omega$; for a closed line segment Ω, ri Ω is the open line segment.

A.3 Convex functions

Let $F\colon \mathbf{R}^N \to \mathbf{R}$ be a function. The effective domain of F is the set of vectors where F assumes a finite value. It is denoted by dom $F = \{x \in \mathbf{R}^N\colon F(x) < \infty\}$.

It is usual to consider extended functions, i.e. $F\colon \mathbf{R}^N \to \bar{\mathbf{R}}$, where $\bar{\mathbf{R}} := \mathbf{R} \cup \{-\infty, +\infty\}$.

Definition A.4. *[17] An extended function F is convex when the set defined by*

$$epi\ F = \{(x, y) \in \mathbf{R}^{N+1}\colon F(x) \leqslant y\} \tag{A.5}$$

is a convex subset of $\mathbf{R}^{N+1}$.

The above definition is consistent with the classic definition of convex function.

Definition A.5. *A function* $F(x)\colon \mathbf{R}^N \to \bar{\mathbf{R}}$ *is said to be convex if, for any* x_1, x_2 *we have*

$$F(\alpha x_1 + (1 - \alpha)x_2) \leqslant \alpha F(x_1) + (1 - \alpha)F(x_2),$$

with $\alpha \in [0, 1]$. *It is said to be strictly convex if strict inequality holds true for* $x_1 \neq x_2$

$$F(\alpha x_1 + (1 - \alpha)x_2) < \alpha F(x_1) + (1 - \alpha)F(x_2).$$

Finally, F is strongly convex with parameter $M_F > 0$ *when for any* x_1, x_2

$$F(\alpha x_1 + (1 - \alpha)x_2) \leqslant \alpha F(x_1) + (1 - \alpha)F(x_2) - \frac{1}{2}M_F\alpha(1 - \alpha)\|x_1 - x_2\|^2.$$

Equivalently, F is strongly convex with parameter $M_F > 0$ *if* $F(x) - \frac{M_F}{2}\|x\|^2$ *is convex.*

If the function F is twice continuously differentiable, F is strongly convex with parameter M_F if its Hessian is positive definite and M_F is a lower bound for the eigenvalues of $\nabla^2 F(x)$, for any $x \in \text{dom } F$.

Definition A.6. *A convex function F is said proper when dom* $F \neq \emptyset$.

Proposition A.6. *(Jensen's inequality) If* $F(x)$ *is a convex function, for any set of points* $\{x_1, x_2, \ldots, x_n\} \subseteq$ *dom F and positive weights* α_i, $i = 1, \ldots, n$, *such that* $\sum_{i=1}^n \alpha_i = 1$, *the following inequality holds true*

$$F\left(\sum_{i=1}^{n} \alpha_i x_i\right) \leqslant \sum_{i=1}^{n} \alpha_i F(x_i). \tag{A.6}$$

Proof. The proof easily follows by induction from the definition of a convex function.

□

A.4 Directional derivatives, stationary point and descent directions

We introduce the definition of directional derivative of a function.

Definition A.7. *The one-sided directional derivative of F at x with respect to a vector d is given by*

$$F'(x; d) = \lim_{\lambda \downarrow 0} \frac{F(x + \lambda d) - F(x)}{\lambda} \tag{A.7}$$

if the limit exists (∞ or $-\infty$ being allowed).

Note that

$$-F'(x; -d) = \lim_{\lambda \downarrow 0} \frac{F(x + \lambda d) - F(x)}{\lambda},$$

so that one-sided directional derivative of F is two-sided if and only if $F'(x; -d)$ exists and

$$F'(x; d) = -F'(x; -d).$$

If F is differentiable at x, the directional derivatives are all finite and two-sided and

$$F'(x; d) = \nabla F(x)^T d, \tag{A.8}$$

where $\nabla F(x)$ is the gradient of F at x.

Consider the following problem

$$\min_{x \in \Omega} F(x), \tag{A.9}$$

where $\Omega \subseteq \mathbf{R}^N$ is a non-empty convex set (possibly $\Omega = \mathbf{R}^N$). A vector $d \in \mathbf{R}^N$ is a feasible direction at x for problem (A.9) if $x + d \in \Omega$.

In view of equality (A.8), we can introduce the definition of stationary point for problem (A.9) when F is a differentiable function (smooth minimization problem).

Definition A.8. *Let F in problem* (A.9) *be a differentiable function. A point $x^* \in \Omega$ is a stationary point for problem* (A.9) *when for any $y \in \Omega$,*

$$\nabla F(x^*)^T (y - x^*) \geqslant 0. \tag{A.10}$$

As a consequence of this definition, we can introduce the definition of descent direction for smooth optimization problems.

Definition A.9. *Let F in* (A.9) *be a differentiable function. A vector $d \in \mathbf{R}^N$ is a descent direction at x for problem* (A.9) *if it is feasible and $\nabla F(x)^T d < 0$.*

For a convex function (possibly non-smooth), the directional derivative is well-defined for all vectors in the effective domain of F.

Proposition A.7. *[17, theorem 23.1] If F is a convex function and $x \in \mathrm{dom}\, F$, then for any $d \in \mathbf{R}^N$ the limit of the right-hand side of equation* (A.7) *exists and $F'(x; d) = \inf_{\lambda>0} \frac{F(x+\lambda d) - F(x)}{\lambda}$. Consequently, for $x \in \mathrm{dom}\, F$ and any $d \in \mathbf{R}^N$*

$$F'(x; d) \leqslant F(x + d) - F(x). \tag{A.11}$$

For a non-smooth objective function, the notion of stationary point can be given on the basis of the concept of directional derivative.

Definition A.10. *A point x^* is stationary for problem* (A.9) *if $x^* \in \mathrm{dom}\, F \cap \Omega$ and*

$$F'(x^*; d) \geqslant 0, \tag{A.12}$$

for all feasible directions $d \in \mathbf{R}^N$.

In view of proposition A.7, for a non-smooth convex function, a stationary point x^* is a minimum point. Furthermore, we can give the definition of descent direction at $x \in \mathrm{dom}\, F$.

Definition A.11. *A vector $d \in \mathbf{R}^N$ is a descent direction for a convex function F at x if it is feasible and $F'(x; d) < 0$.*

A.5 Semicontinuity and conjugate function

Definition A.12. *A function F is said lower semicontinuous (l.s.c.) at x if*

$$F(x) = \liminf_{y \to x} F(y) = \lim_{\epsilon \downarrow 0}(\inf\{F(y) |\, |y - x| \leqslant \epsilon\}).$$

A function F is said upper semicontinuous (u.s.c.) at x if

$$F(x) = \limsup_{y \to x} F(y) = \lim_{\epsilon \downarrow 0}(\sup\{F(y) |\, |y - x| \leqslant \epsilon\}).$$

Definition A.13. *A convex function F is said closed when epi F is a closed subset of $\mathbf{R}^{N+1}$.*

Proposition A.8. *[17, theorem 7.1] A convex function F is l.s.c. if and only if it is closed.*

We introduce the definition of conjugate function, which is an important tool in convex optimization.

Definition A.14. *Let* $F\colon \mathbf{R}^N \to \bar{\mathbf{R}}$ *be a convex function. The conjugate function of F is the function* $F^*\colon\mathbf{R}^N \to \bar{R}$ *defined by*

$$F^*(z) = \sup_{x\in\mathbf{R}^N}(x^Tz - F(x)) = -\inf_{x\in\mathbf{R}^N}(F(x) - x^Tz).$$

From the definition of conjugate function of F, it follows the Fenchel's inequality:

$$F^*(z) + F(x) \geqslant x^Tz \quad \forall\, x, z \in \mathbf{R}^N. \tag{A.13}$$

Proposition A.9. *[17, corollary 12.2.1] If F is l.s.c. and proper,* F^* *is l.s.c. and proper and* $(F^*)^* = F$.

A.6 Subdifferential and ϵ-subdifferential

Definition A.15. *Let* $F\colon \mathbf{R}^N \to \bar{\mathbf{R}}$ *be a proper, convex function and* $x \in \mathrm{dom}\, F$. *The subdifferential* ∂F *at x is the set*

$$\partial F(x) = \{p \in \mathbf{R}^N\colon F(y) \geqslant F(x) + p^T(y - x),\, \forall\, y \in \mathbf{R}^N\}. \tag{A.14}$$

An element $p \in \partial F(x)$ *is said a subgradient of F at x.*

When $\partial F(x) \neq \varnothing$, F is said subdifferentiable at x. Obviously, $\partial F(x) = \{p\}$ if and only if F is finite in a neighborhood of x, differentiable at x and $\nabla F(x) = p$.

In practice, the subdifferential of F is a multi-valued mapping from $\mathbf{R}^N$ to $\mathbf{R}^N$. As shown in figure A.1, a subgradient $p \in \partial F(x)$ defines a supporting hyperplane for epi F at $(x, F(x))$. For example, for $F(x) = \iota_\Omega(x)$, where Ω is a closed convex set, we have that for $x \in \mathrm{int}\,\Omega$, $\partial F(x) = \{0\}$ while for $x \in \Omega - \mathrm{int}\,\Omega$, $\partial\iota_\Omega(x) = \{p \in \mathbf{R}^N\colon p^T(y - x) \leqslant 0,\, \forall\, y \in \Omega\} = N_\Omega(x)$, that is the normal cone of Ω at x. For $F(x) = \|x\|$, we have

$$\partial F(x) = \begin{cases} \dfrac{1}{\|x\|}x & \text{for } x \neq 0, \\ B(0, 1) & \text{for } x = 0, \end{cases}$$

where $B(0, 1)$ is the unit closed ball with respect to the ℓ_2 norm. Analogously, for a general vector norm $\|\cdot\|_+$, we have that for $x \neq 0$, $\partial\|x\|_+ = \{p \in \mathbf{R}^N\colon p^Tx =$

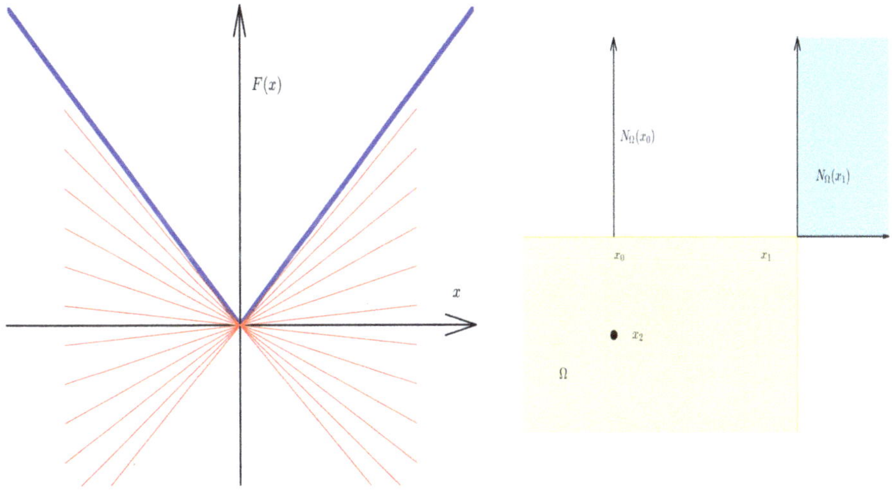

Figure A.1. Examples of subdifferential of F at x. In the left panel, we show the set of lines with slope equal to the values of $\partial|x|$ at $x = 0$, given by $[-1, 1]$; for $x > 0$, $\partial|x| = \{1\}$ and, for $x < 0$, $\partial|x| = \{-1\}$. In the right panel, for $F(x) = \iota_\Omega(x)$, $\partial F(x) = N_\Omega(x) = \{p \in \mathbf{R}^2\colon p^T(y - x) \leqslant 0, \forall y \in \Omega\}$: we consider three different cases, a point x_0 on an edge, a vertex x_1 and an inner point x_2, showing the related normal cone; in the last case the normal cone is the null vector.

$\|x\|_+$, $\|p\|_* = 1\}$, where $\|\cdot\|_*$ is the dual norm; in particular, $\partial\|0\|_+ = B^*(0, 1)$, where $B^*(0, 1)$ is the unit closed ball with respect to the dual norm $\|\cdot\|_*$.

Proposition A.10. *[17, theorem 23.4] Let F be a proper, convex function.*

For $x \notin$ dom F, $\partial F(x) = \emptyset$.

For $x \in$ ri dom F, $\partial f(x) \neq \emptyset$, $F'(x; d)$ is closed and proper as a function of d and

$$F'(x; d) = \sup_{z \in \partial F(x)} z^T d = \iota^*_{\partial F(x)}(d),$$

where $\iota^*_{\partial F(x)}(d)$ is the conjugate of the indicator function of the set $\partial F(x)$.

Furthermore $\partial F(x)$ is a non-empty, bounded set if and only if $x \in$ int dom F and, in this case, $F'(x; d)$ is finite for any d.

As a consequence, the following optimality condition holds for a convex optimality problem.

Proposition A.11. *Let F be a proper convex function. The vector x^* is a minimum point for F if and only if $0 \in \partial F(x^*)$.*

A vector x^* such that $0 \in \partial F(x^*)$ is said to be a zero of the operator ∂F. Consequently, the previous proposition can be reformulated by saying that, for any convex function F, the set of zeroes of the operator ∂F is equal to the set of the minimum points of F.

We report some results, useful in the computation of subgradients.

Proposition A.12. *The following statements hold.*

- *Let F be a proper, convex function. Then, $\partial F(\lambda x) = \lambda \partial F(x)$, for all $x \in \mathbf{R}^N$ and $\lambda > 0$. If $\partial F(x) \neq \emptyset$, the equality holds also for $\lambda = 0$.*
- *Let $F_1, \ldots, F_n$ be proper, convex functions and let $F(x) = \sum_{i=1}^{n} F_i(x)$. Then*

$$\partial F(x) \supset \sum_{i=1}^{n} \partial F_i(x) \quad \forall x \in \mathbf{R}^N.$$

If $\bigcap_{i=1}^{n}$ ri dom $F_i \neq \emptyset$, then

$$\partial F(x) = \sum_{i=1}^{n} \partial F_i(x).$$

- *Let $F\colon \mathbf{R}^M \to \bar{\mathbf{R}}$ be a proper, convex function and let A be a $N \times M$ matrix. Then*

$$\partial(F \circ A)(x) \supset A^T \partial F(Ax) \quad \forall x \in \mathbf{R}^N.$$

If ri dom $F \cap$ range$(A) \neq \emptyset$, *then*

$$\partial(F \circ A)(x) = A^T \partial F(Ax) \quad \forall x \in \mathbf{R}^N.$$

The following proposition relates the conjugate of a convex function with the theory of subgradients.

Proposition A.13. *For any proper, convex function F and any vector x, the following assertions on a vector x^* are equivalent:*

(a) $x^* \in \partial F(x)$;
(b) $z^T x^* - F(z)$ *achieves its supremum with respect to z at $z = x$;*
(c) $F(x) + F^*(x^*) \leqslant x^T x^*$;
(d) $F(x) + F^*(x^*) = x^T x^*$.

If, in addition, F is a l.s.c. function, the following conditions can be added:

(a) $x \in \partial F^*(x^*)$;
(b) $x^T z - F^*(z)$ *achieves its supremum with respect to z at $z = x^*$.*

The proof is based on the definitions of subgradient of F at x and of conjugate function of F. Indeed, the subgradient inequality defining (a) can be written as

$$(x^*)^T x - F(x) \geqslant (x^*)^T z - F(z) \quad \forall z,$$

that is equivalent to (b). Since $\sup_z (x^*)^T z - F(z) = F^*(x^*)$, (b) coincides with (c) or (d). Similarly, (e), (f) are equivalent to

$$(F^*)^*(x) + F^*(x^*) = (x^*)^T x$$

and this coincides with (d) when $F = (F^*)^*$, i.e. F is a l.s.c. function.

As a consequence, when F is a proper, convex and l.s.c. function, ∂F^* is the inverse of ∂F in the sense of multi-valued mappings, i.e. $x \in \partial F^*(x^*)$ if and only if $x^* \in \partial F(x)$. As a consequence of the last proposition and proposition A.11, for a proper, convex and l.s.c. function F, we have that $\arg\min_{x\in \mathbf{R}^N} F(x) = \partial F^*(0)$.

Definition A.16. *Let F be a proper convex function and $x \in \mathrm{dom}\, F$. For a given $\epsilon \geqslant 0$, the ϵ-subdifferential ∂F_ϵ at x is the set*

$$\partial_\epsilon F(x) = \{p \in \mathbf{R}^N\colon F(z) \geqslant F(x) + p^T(z - x) - \epsilon, \forall z \in \mathbf{R}^N\}. \tag{A.15}$$

An element $p \in \partial_\epsilon F(x)$ is said ϵ-subgradient of F at x.

For $c = 0$, the definition of subdifferential is recovered, while for $\epsilon > 0$, $\partial_\epsilon F(x) \supseteq \partial F(x)$; furthermore, for $\epsilon_1 > \epsilon_2 \geqslant 0$, we have $\partial_{\epsilon_1} F(x) \supseteq \partial_{\epsilon_2} F(x) \supseteq \partial F(x)$.

As shown in figure A.2, an ϵ-subgradient $p \in \partial_\epsilon F(x)$ defines a supporting hyperplane for epi $(F + \epsilon)$ at $(x, F(x))$.

We give some properties of the ϵ-subgradient.

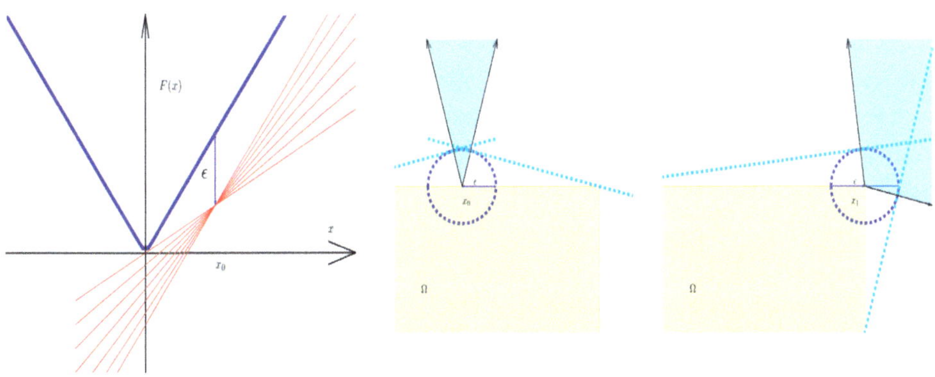

Figure A.2. Examples of ϵ-subdifferential of F at x. In the left panel we consider the function $F(x) = |x|$; at $x = 0.5$, $\partial F(0.5) = F'(0.5) = \{1\}$, while $\partial_\epsilon F(0.5) = [1 - 2\epsilon, 1]$. In the central and right panels, we show the ϵ-subdifferential of $F(x) = \iota_\Omega(x)$ at the point x_0 on an edge and at the vertex x_1 respectively. In both cases, we also show the two supporting hyperplanes, devised by the two ϵ-subgradients on the boundary of the ϵ-subdifferential.

Proposition A.14. *Let $F_1, \dots, F_n$ be proper, convex functions and let $F(x) = \sum_{i=1}^{n}\alpha_i F_i(x)$, with $\alpha_i \geqslant 0$, $i = 1, \dots, n$. If $x \in \bigcap_{i=1}^{n}$ dom F_i and $w_i \in \partial_{\epsilon_i}F_i(x)$, $\epsilon_i \geqslant 0$, then $\sum_{i=1}^{n}\alpha_i w_i \in \partial_\epsilon F(x)$, where $\epsilon = \sum_{i=1}^{n}\epsilon_i\alpha_i$.*

Proposition A.15. *[7] Let $F(x)$ be a proper, convex, l.s.c. function. If $x \in$ dom F and $x^* \in$ dom F^*, we have $x^* \in \partial_\epsilon F(x)$ with $\epsilon \geqslant F^*(x^*) + F(x) - x^T x^*$.*

Proof. Let $x \in \text{dom}\, F$, $x^* \in \text{dom}\, F^*$ and $z \in \mathbf{R}^N$. Then, we can write

$$\begin{aligned} F(x) + x^{*T}(z - x) &= F(x) - ((x^*)^T x - F^*(x^*)) + (x^*)^T z - F^*(x^*) \\ &\leqslant F(x) - ((x^*)^T x - F^*(x^*)) + \sup_{x^*}((x^*)^T z - F^*(x^*)) \\ &= \underbrace{F(x) + F^*(x^*) - (x^*)^T x}_{\leqslant \epsilon} + F(z). \end{aligned}$$

Then, in view of the definition (A.15), $x^* \in \partial_\epsilon F(x)$, with $\epsilon \geqslant F(x) + F^*(x^*) - (x^*)^T x$. □

With the same argument, the following proposition can be stated.

Proposition A.16. *Let F: $\mathbf{R}^M \to \bar{\mathbf{R}}$ be a proper l.s.c. convex function and A be an $N \times M$ matrix.*

Then, for any $x \in$ dom $(F \circ A)$ and $x^* \in$ dom F^*, we have $A^T x^* \in \partial_\epsilon F(Ax)$, with $\epsilon \geqslant F(Ax) + F^*(x^*) - (x^*)^T Ax$.

Proposition A.17. *[7] Assume that Ω is a compactly contained bounded subset of int dom(F). Then, the set $\bigcup_{x\in\Omega}\partial_\epsilon F(x)$ is non-empty, closed and bounded.*

Proof. Let $\lambda > 0$ be such that $\Omega + B(\Omega, \lambda) \subseteq \text{int dom}(F)$, where $B(\Omega, \lambda)$ is the ball of $\mathbf{R}^N$ with radius λ centered at a point of Ω and $\Omega + B(\Omega, \lambda) = \{u \in \mathbf{R}^N: \|u - x\| \leqslant \lambda, x \in \Omega\}$. By theorem 6.2 in [9] it follows that $\bigcup_{x\in\Omega}\partial_\epsilon F(x) \subseteq \bigcup_{x\in\Omega+B(\Omega,\lambda)}\partial F(x) + B(\partial F(x), \frac{\epsilon}{\lambda})$. The last term in the previous inclusion is non-empty, closed and bounded set (see [17, theorem 24.7]); thus the theorem follows. □

The following proposition states for the ϵ-subdifferential a result analogous to that for the subdifferential in proposition A.13, part (a) and (e).

Proposition A.18. *[19, theorem 2.4.4] Let F: $\mathbf{R}^N \to \bar{\mathbf{R}}$ be a proper, convex, l.s.c. function. Then, for any $\epsilon \geqslant 0$ and for any $x \in \mathbf{R}^N$, we have $x^* \in \partial_\epsilon F(x)$ if and only if $x \in \partial_\epsilon F^*(x^*)$.*

A.6.1 Scaled projected ϵ-subgradient method

We consider the problem

$$\min_{x \in \Omega} F(x), \tag{A.16}$$

where $F: \mathbf{R}^N \to \bar{\mathbf{R}}$ is a convex, proper, l.s.c. function and Ω is a non-empty, closed and convex subset of $\mathbf{R}^N$. Let $\bar{F} = \inf_{x \in \Omega} F(x)$.

A thoroughly investigated method for the numerical approximation of the solution of this problem is the scaled projected ϵ-subgradient method, whose basic iteration is

$$x^{(k+1)} = P_{\Omega, D_k}(x^{(k)} - \sigma_k D_k^{-1} u^{(k)}), \tag{A.17}$$

where $u^{(k)} \in \partial_{\epsilon_k} F(x^{(k)})$ for some $\epsilon_k \geqslant 0$, σ_k is a positive step-length and D_k is a s.p.d. matrix with bounded eigenvalues. For an unconstrained problem, i.e. $\Omega = \mathbf{R}^N$, we have the scaled ϵ-subgradient method, which does not require the projection on Ω. The choice $D_k = I_N$ and $\epsilon_k = 0$ for all k corresponds to the standard projected subgradient method, while $\epsilon_k > 0$ gives rise to the projected ϵ-subgradient method. For a survey about a convergence analysis of subgradient and ϵ-subgradient methods for different step-length choices see [5] and references therein, while in [1] analogous results are obtained with respect to non-Euclidean metrics ($D_k \neq I_N$).

In particular, we report the following convergence theorem for the ϵ-subgradient iteration, which resumes and generalizes a collection of results obtained under different assumptions in [2, 3, 6, 8, 10, 11, 13, 16] when $\{\sigma_k\}$ is chosen as a diminishing, divergent series, square summable sequence.

Proposition A.19. *Let $\{x^{(k)}\}$ be the sequence generated by iteration* (A.17), *where $u^{(k)} \in \partial_{\epsilon_k} F(x^{(k)})$, for a given sequence $\{\epsilon_k\} \subset \mathbf{R}$, $\epsilon_k \geqslant 0$. Assume that there exists a positive constant u_F such that $\|u^{(k)}\| \leqslant u_F$ for all $k \geqslant 0$. Furthermore, we assume that there exists a sequence of s.p.d. matrices $\{D_k\}$, $D_k \in \mathcal{D}_{L_k}$, such that $L_k^2 = 1 + \zeta_k$ with $\zeta_k \geqslant 0$ for all $k \geqslant 0$ and $\sum_{k=0}^{\infty} \zeta_k < \infty$. We also assume that the following conditions hold:*

$$\lim_{k \to \infty} \epsilon_k = 0, \tag{A.18}$$

$$\sum_{k=0}^{\infty} \sigma_k = \infty, \tag{A.19}$$

$$\sum_{k=0}^{\infty} \sigma_k^2 < \infty \quad \sum_{k=0}^{\infty} \epsilon_k \sigma_k < \infty. \tag{A.20}$$

Then, defining $\bar{X}$ *as the set of the solutions of problem* (A.16), *we have*

(a) $\liminf_{k\to\infty} F(x^{(k)}) = \bar{F}$;
(b) *if* $\{x^{(k)}\}$ *is bounded, there exists a limit point of it in* $\bar{X}$;
(c) *if* $\bar{X}$ *is non-empty, the sequence* $\{x^{(k)}\}$ *converges to a solution of problem* (A.16) *and* $\lim_{k\to\infty} F(x^{(k)}) = \bar{F}$;
(d) *if* $\bar{X}$ *is empty, the sequence* $\{x^{(k)}\}$ *is unbounded.*

The assumption on $\{u^{(k)}\}$ is satisfied for example when dom F^* is a bounded set. Indeed, it holds dom $F^* = \bigcup_{x\in\mathbf{R}^N} \partial_\epsilon F(x)$, for every $\epsilon > 0$ (see remark 2 in [18]).

A.7 Basic results of duality theory

We recall some basic results of the duality theory, with emphasis on the minimization problems addressed in chapter 6. For details, see [4, 17].

Given $F: \mathbf{R}^N \to \mathbf{R}$, $H_j: \mathbf{R}^N \to \mathbf{R}$, $j = 1, \dots, r$, and a subset Ω of $\mathbf{R}^N$, we consider the following minimization problem

$$\begin{aligned} &\min_{x\in\Omega} F(x), \\ &\text{subject to } H_j(x) \leqslant 0, j = 1, \dots, r; \end{aligned} \tag{A.21}$$

the form (A.21) of this problem is called *primal formulation*. We denote by $\bar{F}$ the optimal solution (possibly $\pm\infty$). The Lagrangian function related to the primal problem (A.21) is defined as

$$\mathcal{L}(x, p) = F(x) + \sum_{j=1}^{r} p_j H_j(x) = F(x) + p^T H(x), \tag{A.22}$$

A vector $p^* \in \mathbf{R}^r$ is said to be a Lagrange multiplier vector for the primal problem if $p^* \geqslant 0$ and $\bar{F} = \inf_{x\in\Omega} \mathcal{L}(x, p^*)$.

Proposition A.20. *[4, proposition 5.1.1] Let* p^* *be a Lagrange multiplier for equation* (A.21). *Then* x^* *is a global minimum of the primal problem if and only if* x^* *is feasible and*

$$x^* = \arg\min_{x\in\Omega} \mathcal{L}(x, p^*), \qquad p_j^* H_j(x^*) = 0 \quad j = 1, \dots, r.$$

We define the dual function G for any vector $p \in \mathbf{R}^r$ as

$$G(p) = \inf_{x\in\Omega} \mathcal{L}(x, p), \tag{A.23}$$

and we introduce the so-called *dual formulation* of problem (A.21)

$$\max_{p\geqslant 0} G(p); \tag{A.24}$$

we denote by $\bar{G}$ the optimal solution (possibly $\pm\infty$). Regardless of the structure of the objective function and constraints of the primal formulation, the dual formulation has nice properties, since G is a concave function and dom G is a convex set.

Proposition A.21. *(Weak duality theorem). We have*

$$\bar{G} \leqslant \bar{F}.$$

If $\bar{G} = \bar{F}$, we say that there is no duality gap and, in this case, the set of Lagrange multipliers is equal to the set of optimal dual solutions; on the contrary, the set of Lagrange multipliers is empty.

When there is no duality gap, there are two characterizations of primal and dual optimal solution pairs. Using the previous results and proposition A.20, we can obtain the following optimality conditions.

Proposition A.22. (x^*, p^*) *is an optimal primal solution-Lagrange multiplier pair if and only if*

$$x^* \in \Omega \quad H_j(x^*) \leqslant 0 \quad j = 1, \ldots, r \qquad \textit{primal feasibility,} \tag{A.25}$$

$$p^* \geqslant 0 \qquad \textit{dual feasibility,} \tag{A.26}$$

$$x^* = \arg\min_{x \in \Omega} \mathcal{L}(x, p^*) \qquad \textit{Lagrangian optimality,} \tag{A.27}$$

$$p_j^* H_j(x^*) = 0 \quad j = 1, \ldots, r \qquad \textit{complementarity slackness.} \tag{A.28}$$

Proposition A.23. *(Saddle point theorem).* (x^*, p^*) *is an optimal primal solution-Lagrange multiplier pair if and only if* $x^* \in \Omega$, $p^* \geqslant 0$ *and* (x^*, p^*) *is a saddle point of the Lagrangian, in the sense that*

$$\mathcal{L}(x^*, p) \leqslant \mathcal{L}(x^*, p^*) \leqslant \mathcal{L}(x, p^*) \quad \forall\, x \in \Omega, p \geqslant 0. \tag{A.29}$$

This proposition enables to obtain the *primal–dual formulation* of problem (A.21):

$$\min_{x \in \Omega} \max_{p \geqslant 0} \mathcal{L}(x, p). \tag{A.30}$$

From the weak duality theorem, if the primal problem is unbounded, i.e. $\bar{F} = -\infty$, the dual problem is infeasible. If the primal problem is infeasible, so that $\bar{F} = \infty$, the dual problem can be unbounded.

Furthermore, if for some j we have an equality constraint $H_j(x) = 0$, the signs of the corresponding dual variable and the Lagrange multiplier p_j and p_j^* are unrestricted.

When convexity conditions on the objective function and constraints of the primal problem are imposed, we have stronger results.

Consider the following problem

$$\begin{aligned} &\min_{x\in\Omega} \quad F(x) \\ &\text{subject to } H_j(x) \leqslant 0,\, j = 1, \dots, r, \\ &\qquad\qquad Ex = d \end{aligned} \tag{A.31}$$

where $E \in \mathbf{R}^{M\times N}$, $d \in \mathbf{R}^M$, Ω is a convex subset of $\mathbf{R}^N$ and F, H_j: $\mathbf{R}^N \to \mathbf{R}$ are convex on Ω.

Proposition A.24. *[4, proposition 5.3.2] (Strong duality theorem). Assume that the optimal value $\bar{F}$ of the problem* (A.31) *is finite and that there exists a vector $\tilde{x} \in ri\ \Omega$ such that $H_j(\tilde{x}) < 0$, $j = 1, \dots, r$, and $E\tilde{x} = d$. Then, there is no duality gap ($\bar{F} = \bar{G}$) and there exists at least one Lagrange multiplier.*

Under the assumptions of the previous proposition, since the primal problem has an optimal solution, the dual problem has at least one optimal solution and the two optimal values are equal. Furthermore, primal or dual problems are equivalent to determine a saddle point of the Lagrangian function.

The duality theorem of Fenchel is a special case of the duality theory which enables one to relate the minimization of the sum of two convex functions to the minimization of the sum of the related conjugate functions.

Proposition A.25. *[17] (Fenchel's duality theorem) Let F and G be two proper convex functions on $\mathbf{R}^N$. Then the following equality holds*

$$\inf_x (F(x) + G(x)) = \sup_{x^*}(-G^*(x^*) - F^*(x^*)),$$

if one of the following conditions is satisfied:

(a) *$ri\ dom\ F \cap ri\ dom\ G \neq \emptyset$;*

(b) *F and G are l.s.c. functions and $ri\ dom\ F^* \cap ri\ dom\ G^* \neq \emptyset$.*

If (a) *holds, the supremum is attained at some x^*, while, if* (b) *holds, the infimum is attained at some x; if both* (a) *and* (b) *hold, the infimum and the supremum are necessarily finite.*

The proof of the theorem is based on the Fenchel inequality (A.13):

$$F(x) + F^*(x^*) \geqslant x^T x^* \geqslant -G(x) - G^*(x^*) \quad \forall x, x^* \in \mathbf{R}^N,$$

so that we have

$$F(x) + G(x) \geqslant -G^*(x^*) - F^*(x^*).$$

Thus the assumption (a) or/and (b) enables one to prove that the equality holds.

In view of the previous theorem, we can address the following problem which is thoroughly investigated in the framework of inverse problems:

$$\min_{x \in \mathbf{R}^N} F(x) + G(Ax), \tag{A.32}$$

where $F: \mathbf{R}^N \to \bar{\mathbf{R}}$, $G: \mathbf{R}^M \to \bar{\mathbf{R}}$ are convex, proper, l.s.c. functions and A is an $M \times N$ matrix. Using the definition of conjugate function of $G(Ax)$, the primal–dual formulation of the problem (A.32) can be obtained

$$\min_{x \in \mathbf{R}^N} F(x) + \max_{p \in \mathbf{R}^M} p^T Ax - G^*(p) = \min_{x \in \mathbf{R}^N} \max_{p \in \mathbf{R}^M} F(x) - G^*(p) - p^T Ax, \tag{A.33}$$

and, by the definition of conjugate function of F, the corresponding dual formulation

$$\max_{p \in \mathbf{R}^M} - G^*(p) - F^*(-A^T p). \tag{A.34}$$

The vector x^* is a solution of problem (A.32) when $0 \in \partial(F + G \circ A)(x^*)$.

We observe that, since $\partial F + \partial(G \circ A) \subseteq \partial(F + G \circ A)$, any zero of $\partial F + \partial(G \circ A)$ is an optimal solution for the problem, but a zero of $\partial(F + G \circ A)$ may not be a zero of $\partial F + \partial(G \circ A)$. Thus, conditions assuring that $\partial F + \partial(G \circ A) = \partial(F + G \circ A)$ are needed.

In view of the Fenchel's duality theorem, we have that the conditions that guarantee that the primal and the dual problems have a finite solution and that the primal optimal $\bar{F}$ and the dual optimal $\bar{G}$ coincide, are that

- there exists $x \in \operatorname{ri} \operatorname{dom} F$ such that $Ax \in \operatorname{ri} \operatorname{dom} G$;
- there exists $p \in \operatorname{ri} \operatorname{dom} G^*$ such that $-A^T p \in \operatorname{ri} \operatorname{dom} F^*$.

Furthermore, under the above assumptions, we have that (x^*, p^*) is the pair of the optimal solutions of the primal and dual problems if and only if the following conditions hold:

- $F(x^*) + G(Ax^*) = -F^*(-A^T p^*) - G^*(p^*)$;
- x^* is optimal for equation (A.32) and $-A^T p^* \in \partial F(x^*)$, $p^* \in \partial G(Ax^*)$;
- p^* is optimal for equation (A.34) and $Ax^* \in \partial G^*(p^*)$, $x^* \in \partial F^*(-A^T p^*)$.

A.8 Poisson processes

A Poisson process is a particular case of a random point process, a mathematical model for a physical phenomenon characterized by highly localized events distributed randomly in a continuum. Each event is represented in the model by an idealized point to be conceived of as identifying the position of the event. For instance, in the case of imaging, a point corresponds to a place where a photon hits the detector plane.

A point process is a random process whose realizations consist of isolated points in some domain. In the case of imaging we can consider, for instance, a set S of the plane already considered in previous sections. More precisely, the realizations are defined as 'point patterns', namely countable sets of points with no limiting point in S. If S is bounded and closed (as a closed square in the plane) then the number of points is *finite*.

Every point process is a special case of a random measure representable as a sum of δ measures

$$\mathcal{G} = \sum_{k=1}^{n} \delta_{\vec{r}_k}, \tag{A.35}$$

where n is an integer valued r.v. and the $\vec{r}_k$ are random points of S. For each Borel subset B of S, $\mathcal{G}(B)$ is the r.v. whose realizations are the numbers of points in B. Then the expectation value $E(\mathcal{G}(B))$ is a measure, also known as the 'mean measure', that assigns to B the expected number of points in B. We set $\mu(B) = E[\mathcal{G}(B)]$.

Definition A.17. *A Poisson process $\mathcal{G}$ on a measurable set S is a point process satisfying the following conditions:*

- *if $S_1, \ldots, S_n$ are disjoint and measurable subsets of S, then the r.v.s $\mathcal{G}(S_1), \ldots \mathcal{G}(S_n)$ are statistically independent;*
- *given a measurable subset B of S, the r.v. $\mathcal{G}(B)$ is a Poisson r.v. P_μ, with $\mu = \mu(B)$, i.e.*

$$P[\mathcal{G}(B) = k] = e^{-\mu}\frac{\mu^k}{k!}. \tag{A.36}$$

It is possible to prove that $\mu(B)$ defines a measure on S which is non-atomic, i.e. $\mu(B) = 0$ if $B = \{\vec{r}_0\}$. A case which is interesting in imaging applications is that corresponding to a measure $\mu(B)$ which is given by an *intensity function* $y \in L^1(S)$, i.e.

$$\mu(B) = \int_B y(\vec{r})d\vec{r}. \tag{A.37}$$

An important quantity which can be computed in terms of the intensity function is the so-called *sample-function density* which leads to the likelihood of the following problem: given a realization of $\mathcal{G}$, estimate y.

To this end, consider a particular realization of $\mathcal{G}$ consisting of n points, i.e. $\vec{r}_1, \ldots, \vec{r}_n$. Since the points are isolated, consider around each point a ball centered on that point and containing only that point : $S_k = \{\vec{r} \mid \|\vec{r} - \vec{r}_k\| \leqslant \rho_k\}$. Then the probability of getting only one point inside S_k and 0 outside is given by:

$$\begin{aligned} P[\mathcal{G}(S_1) = 1, \mathcal{G}(S_2) = 1, \ldots, \mathcal{G}(S_n) = 1, \mathcal{G}(S) = n] \\ = \left[\prod_{k=1}^{n} \int_{S_k} y(\vec{r})d\vec{r} \exp\left(-\int_{S_k} y(\vec{r})d\vec{r}\right)\right] \\ \exp\left(-\int_{S-\bigcup_{k=1}^{n} S_k} y(\vec{r})d\vec{r}\right) \\ = \left[\prod_{k=1}^{n} \int_{S_k} y(\vec{r})d\vec{r}\right] \exp\left(-\int_{S} y(\vec{r})d\vec{r}\right). \end{aligned} \tag{A.38}$$

Dividing both sides by the product of the volumes of the n balls, letting the volumes tend to zero and assuming suitable smoothness properties of y, one obtains the following expression for the sample-function density of this particular realization:

$$p(\vec{r}_1, \ldots, \vec{r}_n) = \left[\prod_{k=1}^{n} y(\vec{r}_k)\right] e^{-\int_S y(\vec{r})d\vec{r}}. \tag{A.39}$$

A.9 The data-fidelity function of Hohage and Werner

Consider the situation described at the beginning of section 4.1 with the following additional assumptions: the ith detector, which detects y_i photons, has a finite surface corresponding to a subset S_i, with measure (area) $|S_i|$, of the image domain S; the expected values z_i are related to the underlying Poisson process with intensity function y by

$$z_i = \int_{S_i} y(\vec{r})d\vec{r}; \tag{A.40}$$

it is assumed that the positions $\vec{r}_k$, $k = 1, \ldots, n$, where the $n = \sum_i y_i$ detected photons hit the detection domain, are known. Then, consider the negative logarithm of the likelihood defined in equation (4.3), neglect the terms which depend only on y_i and add the term, also independent of y_i, $\sum_i y_i \ln|S_i|$; from equation (A.40) we obtain the function of y given by

$$y \rightarrow \sum_i \left[\int_{S_i} y(\vec{r})d\vec{r} - \ln\left(\frac{1}{|S_i|}\int_{S_i} y(\vec{r})d\vec{r}\right)\right]. \tag{A.41}$$

Letting the size of the detectors tend to zero (and their number tend to infinity) at the end within one sub-domain we obtain only 1 or 0 photons so that we get the function

$$\mathcal{S}(y; \vec{r}_1, \ldots, \vec{r}_n) = \int_S y(\vec{r})d\vec{r} - \sum_{k=1}^{n} \ln y(\vec{r}_k). \tag{A.42}$$

The maximum likelihood problem is defined as the problem of estimating y for a given realization of the photon positions $\vec{r}_1, \dots, \vec{r}_n$ by minimizing the previous functional. Because of the logarithm, this functional does not have minimizers without additional constraints or penalties. This problem is also a *semi-discrete* problem since the number of photons (the data) is finite while the unknown is a function.

The previous derivation is essentially heuristic even if it has the advantage of showing the relationship with the likelihood function used in the discrete case. A more direct derivation can be obtained from the sample-function density defined in equation (A.39). Indeed, if the intensity function y is given, then the equation gives the probability density of obtaining the points $\vec{r}_1, \dots, \vec{r}_n$ as a realization of the Poisson process; on the other hand, if the points are given, then the same expression is the likelihood for estimating y from the given realization. If we take the negative logarithm of the rhs of equation (A.39) we obtain precisely equation (A.42).

References

[1] Auslander A and Teboulle M 2009 Projected subgradient methods with non-Euclidean distances for non–differentiable convex minimization and variational equalities *Math. Program. Ser.* B **120** 27–48

[2] Alber Y I, Iusem A N and Solodov M V 1998 On the projected subgradient method for nonsmooth convex optimization in a Hilbert space *Math. Program.* **81** 23–35

[3] Bello Cruz J Y 2017 On proximal sub-gradient splitting method for minimizing the sum of two non-smooth convex functions *Set-Valued Var. Anal.* **25** 245–63

[4] Bertsekas D P 1999 *Nonlinear Programming* 2nd edn (Belmont, MA: Athena Scientific)

[5] Bertsekas D P 2015 *Convex Optimization Algorithms* (Belmont, MA: Athena Scientific)

[6] Bonettini S, Benfenati A and Ruggiero V 2016 Scaling techniques for ϵ-subgradient methods *SIAM J. Optim.* **26** 1741–72

[7] Bonettini S and Ruggiero V 2012 On the convergence of primal-dual hybrid gradient algorithms for total variation image restoration *J. Math. Imaging Vis.* **44** 236–53

[8] Correa R and Lemaréchal C 1993 Convergence of some algorithms for convex minimization *Math. Program.* **62** 261–75

[9] Ekeland I and Témam R 1999 *Convex Analysis and Variational Problems* (Philadelphia, PA: SIAM)

[10] Kiwiel K C 2004 Convergence of approximate and incremental subgradient methods for convex optimization *SIAM J. Optim.* **14** 807–40

[11] Larsson T, Patriksson M and Strömberg A-B 2003 On the convergence of conditional ϵ-subgradient methods for convex programs and convex-concave saddle-point problems *Eur. J. Oper. Res.* **151** 461–73

[12] Luenberger D G 1989 *Linear and Nonlinear Programming* 2nd edn (Reading, MA: Addison-Wesley)

[13] Nedić A and Bertsekas D P 2001 Incremental subgradient methods for nondifferentiable optimization *SIAM J. Optim.* **12** 109–38

[14] Ostrowski A M 1973 *Solution of Equations in Euclidean and Banach Spaces* (New York: Academic)

[15] Polyak B 1987 *Introduction to Optimization* (New York: Optimization Software—Inc.)

[16] Robinson S M 1999 Linear convergence of epsilon-subgradient descent methods for a class of convex functions *Math. Program., Ser.* A **86** 41–50

[17] Rockafellar R T 1970 *Convex Analysis* (Princeton, NJ: Princeton University Press)

[18] Villa S, Salzo S, Baldassarre L and Verri A 2013 Accelerated and inexact forward–backward algorithms *SIAM J. Optim.* **23** 1607–33

[19] Zalinescu A 2002 *Convex Analysis in General Vector Spaces* (River Edge, NJ: World Scientific)

www.ingramcontent.com/pod-product-compliance
Ingram Content Group UK Ltd.
Pitfield, Milton Keynes, MK11 3LW, UK
UKHW051904100926
8556IPUK00005B/754

9 780750 314381